EFFECTS OF IONIZING RADIATION

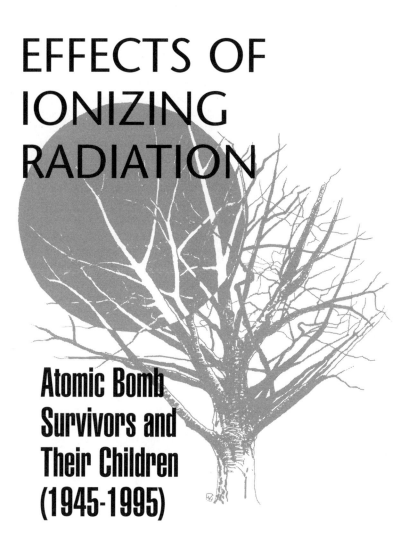

Atomic Bomb Survivors and Their Children (1945-1995)

Leif E. Peterson and Seymour Abrahamson, editors

JOSEPH HENRY PRESS
Washington, D.C. 1998

Joseph Henry Press • 2101 Constitution N.W. • Washington, D.C. 20418

The Joseph Henry Press, an imprint of the National Academy Press, was created with the goal of making books on science, technology, and health more widely available to professionals and the public. Joseph Henry was one of the founders of the National Academy of Sciences and a leader of early American science.

Library of Congress Cataloging-in-Publication Data
Effects of ionizing radiation: atomic bomb survivors and their
children (1945-1995) / Leif E. Peterson, Seymour Abrahamson.
        p.   cm.
Includes bibliographical references and index.
ISBN 0-309-06402-3 (alk. paper)
1. Radiation carcinogenesis. 2. Atomic bomb—Physiological
effect. 3. Atomic bomb victims—Diseases. 4. Children of atomic
bomb victims—Diseases.   I. Peterson, Leif E., 1957- .
II. Abrahamson, Seymour.
RC268.55.E36   1997
616.9'897—dc21   97-52238
                                                                        CIP

Cover art by Mari Omori.

Printed in the United States of America.

Contents

iii

PART III.   MUTAGENESIS AND CARCINOGENESIS

PART IV.   PSYCHOSOCIAL FACTORS

PART V.   FUTURE RESEARCH

Preface

When assessing epidemiology-based health effects of radiation, a critical question is, How good is the dosimetry? The problems in dosimetry are markedly different for Hiroshima and Nagasaki by virtue of two different types of bombs and their different radiation energy releases. Basically one needs to determine the transmission factors in air, the shielding characteristics of the exposed individual, and finally the individual's organ doses, and to make these determinations for over 100,000 people at varying distances and under a variety of conditions.

In the first of two chapters on dosimetry, Kerr traces the evolution of the dosimetry analysis, beginning with studies at the Nevada Test Site undertaken to empirically estimate radiation fields. Subsequent testing of fabricated Japanese houses provided information on transmission factors both inside and outside the houses. Incorporating survivor interview data on location and shielding led to a system of dose estimation known as T65D which was used to estimate and develop health risk analyses for approximately 15 years from its inception. Conflicts between these estimates and newer information led to an international review and the call for a more accurate dosimetry analysis. This reanalysis leading to the incorporation of the Dosimetry System 86 (DS86) is detailed both in the chapter by Kerr and the succeeding chapter by Kaul. In this later chapter an extensive discussion is provided on the new technologies introduced to assess dose; the new dosimetry, primarily by increasing gamma dose in Hiroshima and simultaneously reducing neutron dose contribution led to a total reduced biologically effective dose.

However, both Kerr and Kaul point out that there is still a discrepancy in Hiroshima neutron data, which remains to be resolved, and the extent of the increase in amount of neutrons' impact on dosimetry is unclear. There are additional concerns involving the shielding characteristics of factory workers in Nagasaki since their higher assigned doses do not correlate well with their lower cancer risks. Chromosome analysis of survivors, to be discussed below, also provides additional evidence of discordance between dosimetry and biological effects.

Beginning in the 1950s, it became apparent that the A-bomb survivors would be incurring continuing radiation risks from medical exposures, diagnostic as well as therapeutic procedures, particularly since routine examinations were not only part of the ABCC Adult Health Study (AHS) practice, but community medical examinations were also occurring. Kato's paper deals with radiological surveys conducted in both cities at the ABCC, and local hospitals and clinics beginning in 1961, which attempt to assess the cumulative doses received by the participants in the AHS study. When assessments of the risks of radiation from the atomic bombs are extrapolated into the low-dose range it clearly becomes important to know how much additional dose was received from routine examinations. Initially the number of radiographic examinations steadily increased, but there was a leveling off in the two cities because of the awareness of physicians of the necessity to avoid unnecessary radiation exposure. Nevertheless, the data accumulated both from survey of equipment and interviews have indicated that an appreciable proportion of the AHS participants did receive bone marrow doses in the range of 1–9 cGy by the end of 1952, doses equivalent to that received from the atomic bomb exposures. Just how these doses will be factored into the risk estimates has yet to be determined. The presentation is replete with figures and tables documenting the analyses. Additionally, new data on radiation therapy and second cancers are in the process of analysis.

The Nakamura paper deals with a number of different aspects of cytogenetic analysis of the A-bomb survivors. For the purpose of this preface we limit ourselves to those studies which relate directly to biodosimetry issues. Conventional cytogenetic analysis for stable reciprocal translocations has been completed on more than 2,500 survivors in both cities. There has been a consistent city difference: the Hiroshima dose-response curve is higher and more linear than the Nagasaki curve even with the new DS86 dosimetry. A discrepancy in dose response appears to exist in Nagasaki for those exposed in tenement houses or not in single family dwellings versus those in single family dwellings. No such serious difference exists in Hiroshima; moreover, there is good agreement for data from single family homes between the cities. In addition, chromosome analysis agrees with other biological endpoints in demonstrating that there are large random dose errors when comparisons of supposedly equivalent high-dose individuals with and without severe epilation are studied. In all cases the epilators show a two to three times greater response. These discrepancies are resolved with results from electron spin resonance (ESR) studies on tooth enamel. The ESR studies show high correlation with chromosome aberration data, and both have lower correlations with assigned doses. Therefore it appears that the differences are the result of distance-biased dose estimation, in this case presumably due to improper localization or shielding conditions, and not to differential radiosensitivities.

The next sequence of papers deals with RERF cancer studies, and we choose to start with Pierce's presentation on statistical aspects of this epidemiology program because he has provided an extremely clear historic exposition of the statistical program, its evolution to fit the needs of the emerging data, and its implications, topped off by examples from the then-unpublished Life Span Study Report 12 on cancer mortality. Statistical analysis has been fundamental to almost all of the ABCC/RERF research programs. Pierce describes the early use of the "contingency table method," wherein the specific exposure categories defined by city, sex, and age at exposure were used to determine both the expected and observed cancers seen, demonstrating that a dose effect was indeed observable. The limitations of this method then led to the next stage, regression models. Again the limitations are clearly presented, particularly the inability to study temporal patterns of excess risk as the follow-up period increased. This in turn led to the excess relative risk model (ERR), which incorporated stratification variables used in the earlier methods. The simplicity and elegance of the ERR model were important attributes for epidemiology. As Pierce points out, caution must be exercised both in asking the right questions and in preventing overinterpretation of the data. He next examines the limitations of this method and moves on to describing the usefulness of excess absolute risk (EAR) methodology, which is playing an increasingly important role in understanding the nature of radiation-induced cancer. He demonstrates the use of the ERR and EAR methods with respect to the recent mortality analysis, where age at exposure and sex suggest apparently greater difference by the relative risk method than the EAR method, in which the small number of background cases in both sexes and in the young may be distorting the results by the ERR method.

Leukemia was the first cancer recognized to be radiation-induced. Preston's paper on leukemia risks in A-bomb survivors traces the history of the leukemia studies, originating with the concerns of two Japanese physicians in Nagasaki, Takuso Yamawaki and Masanobu Tomonaga, in the late 1940s that an excess of leukemia was apparent in the survivors. This led to a number of surveys by the ABCC in 1951, 1953, and 1957. However, until the 1960s there was no consistent standard of ascertainment or review nor a defined population to study. Stuart Finch was responsible for the establishment of the ABCC Leukemia Registry and detailed shielding histories for all survivors within 2,000 meters of the bomb. By the mid-1980s, cases were reclassified by the French-American-British (FAB) system, a detailed classification of leukemia subtypes. The recent emergence of the two city tumor registries now serves as the primary source for new cases. Much of Preston's paper deals with the serial analyses of risks as data accumulated. Significantly, the first quantitative risk estimate was developed in 1957 by E.B. Lewis, who was named a Nobel Laureate during this symposium (for his studies on developmental genetics). The most recent analysis covering the period 1950–1987 is reviewed in the last section. It should also be noted that 50% or more of the cases of leukemia (ALL, AML, CML) are attributable to radiation exposure. Detailed models of all leukemias combined, and for ALL, AML, CML, ATL and other leukemias, are presented, as well as tabular data on changes of excess risk with time. The dose-response curve for combined leukemias is significantly non-linear, i.e., linear-quadratic in shape, with no convincing evidence for a threshold. The data indicate that while risks have fallen with time, they have continued to persist for those who were exposed as adults. Continued follow-up is required to see whether those who were young at the time of exposure and are now entering the "cancer prone" ages will show additional response.

The establishment of acceptable, high-standard tumor registries in the A-bomb cities was a hard-fought process. Although off to a promising start in the early 1950s, disagreements between important hospitals and members of the medical community and ABCC led to its decline. With the transition to RERF, old hostilities faded and the efforts of a dedicated group of staff reestablished the registries as modern population-based facilities. Mabuchi, a principal architect of this rejuvenation, describes the history, role, and usefulness of this program. The cancer incidence data collection, procedures, and staff were revitalized, and records from 1958 to 1987 were of such high quality that it was possible to carry out a comprehensive analysis. To develop quality data, hospital records must be collected, abstracted, linked to tissue registries, and RERF's Master File of Life Span Study participants, cross-checked with death certificate data and other clinical program data at RERF. Quality is measured by several indices, but high among these is the frequency of histological verification and the mortality-to-incidence rates. The registries now rank among the best in Japan.

While covering a shorter span of years than the mortality studies, the incidence studies yield more cases and more radiation-attributable cancer. Much of this comparative analysis is described in the companion paper by Ron. As should be emphasized, the solid cancer data is remarkably linear over the dose range studied, including very low doses, and the confidence limits of the data are very narrow indeed.

Followed over time, the earlier years were dominated by cancers appearing in the over-20 age-at-exposure group, but almost 50% of all total excess cancers were observed in the last 7 years of follow-up in the under-20 group, a cohort that has 85% of its members still alive and entering the cancer-prone period. The tables presented provide extensive details on sites, dose, excess cases, age, and sex.

There is a useful discussion of the limitations of registry data and the means to overcome such deficiencies.

Examples are given of how the registries are facilitating site-specific, case-control and nested case-control epidemiologic studies. For example, in liver cancer studies, what are the roles of hepatitis B and C viruses in relationship to radiation?

Until very recently, cancer mortality has served as the main indicator of radiation-induced somatic effects, based as it is on death certificate data readily obtainable from all of Japan because of the unique system known as the *koseki* family registry system. This assures virtually complete coverage. However, death certificate data is recognized to have serious inaccuracies which are circumvented in incidence data collection with its very high histological verification level. A major effort over the last decade has brought the two city tumor registries into a highly respected and accurate state. The paper by Ron et al. summarizes the first major analysis covering 1958–1987 and contrasts results with the mortality data from 1950 to 1987. It should be pointed out that incidence data is only collected from major hospitals and clinics within the two prefectures and is based on a smaller Life Span Study (LSS) population base. Nevertheless, the number of total cases and radiation-induced excess cases exceeds that in the mortality data by about 23% and 65%, respectively. Because of lower mortality, cancers of the breast, thyroid, skin, and salivary glands contribute to this difference. The two datasets are for the most part comparable with respect to risk of stomach, colon, liver, lung, breast, ovary, and urinary bladder cancer. By either risk analysis, the statistics ERR/Sv or 10^4PYSv for the incidence series have larger risks by 40% and nearly threefold higher, respectively. One of the most remarkable features of the data is the exquisite linearity with dose over the wide range studied. Using EAR comparisons, the female risk of cancer in both studies is about twice that of males, and there is a much higher EAR of 22.4 for incidence compared with mortality (3.8) for those exposed under the age of 20. The difference is not as great for those over 20 at the time-of-bombing (ATB). These datasets will continue to play a major role in the evaluation of cancer risks, particularly since over 80% of the under-20 age group are alive and entering the cancer-prone age period over the next two decades. In conjunction with the tumor registry, the tissue registries of the two cities provide important access to the material needed for site-specific studies, which in turn allows more in-depth analysis of cancer-specific subtypes and their radiosensitivity, and facilitates incipient studies on the molecular basis of cancer, and perhaps whether radiation-specific molecular lesions at the DNA level will be recognizable.

Ethel Gilbert's paper on nuclear workers exposed to low-level radiation has immediate relevance to the preceding papers on risk estimates of the A-bomb survivors because it provides a check on the validity of the extrapolations to low dose from the latter study, which have served as a standard for radiation protection regulation over several decades throughout the world. While the presentation examines the individual worker studies of the United States, Canada, and the UK, it is the combined international study carried out by the International Agency for Cancer Research (IARC) that provides the most powerful database on which to compare results for male workers monitored for dose, aged 20 to 60, against the corresponding subset of the LSS data. For example, the estimated ERR/Sv for all cancers excluding leukemia in the international worker study is close to zero, but the upper 95% confidence interval (CI) overlaps that of the A-bomb study; in fact individual national studies can exceed the Japanese study, but there is less precision in such data. For leukemias excluding chronic lymphocytic leukemia (CLL), the two risk estimates were remarkably close: 2.2 for workers and 3.7 for survivors with similar upper 95% CI, but a much lower but non-zero estimate for the lower 95% CI for workers. The remaining discussion deals with uncertainties associated with confounding factors and dosimetry issues which could affect the worker analysis. Given the even more complex issues surrounding the A-bomb studies, it is very reassuring that the single acute-dose data and much higher overall dose data agree so well with the low-dose and mostly low-dose rate data. The A-bomb data have greater precision at present than the worker data, which is important for a number of reasons, primarily for risk estimation. The combined worker analysis also indicates that the A-bomb data have not underestimated risk at low doses; in fact, the upper confidence limits of the worker study quite effectively rule out underestimates by one order of magnitude, as some investigators have charged. It is clear that continued follow-up of both datasets will make important contributions to the low dose risk assessment issue.

Land's presentation deals with the interaction of radiation dose and other risk factors. He uses smoking/lung cancer and reproductive factors/breast cancer as a two-model system in which to analyze the interactions in the A-bomb survivors. Both cancer endpoints are clearly demonstrated to be linearly dose related. Sex and age have readily been demonstrated to affect some cancer rates and are easily observable risk factors. As is well known, smoking is a principal contributor to lung cancer production in non-radiation-exposed populations; and reproductive performance, age at first birth, number of births, and cumulative lactation history are well-known moderating factors for "naturally" occurring breast cancer.

For lung cancer, Land points out that heavy cigarette smoking is a much greater risk factor than even the highest dose received by the A-bomb survivors. But the question of interaction deals with how the two agents influence the final response. Will the excess risk be the sum of the two risks, an additive response, or greater than additive, a multiplicative response (a × b), or somewhere in between, as developed by statistical analysis? While the published data at RERF are unable to distinguish between these models, the data of US uranium miners suggest that a mixture model intermediate between additive and multiplicative gives the best fit.

For breast cancer the analyses (based on case-control studies) were much more clear cut. That is, the interaction results consistently conform to the model predicted by the multiplication formula and are significantly different from the additive model. As stated above, age at first pregnancy, number of births, and cumulative lactation time are major interactive factors. Based on rodent experimental studies, there is the hypothesis that cells that may have been initiated toward the cancer states are reprogrammed to differentiated milk producing cells by the pregnancy process, thus removing them from the pool of cells still cancer prone.

In the area of modeling of radiation-induced cancer production, the following series of papers provide different approaches to this issue.

The Mendelsohn paper has developed a modified Armitage-Doll model derived from the RERF solid cancer data. The Armitage-Doll model describes the temporal pattern of many cancers as a power function of age, dependent upon a number of irreversible steps, in the range of five to six in their model. This would now be supported by molecular analysis which has implicated five to six mutational steps in the development of colon cancers. Mutations in oncogenes and tumor-suppressor genes and other types represent the pool from which a subset of mutational events may result in a cancer. Mendelsohn's model is simply that radiation has a certain dose-dependent probability of introducing a mutational step in the process, thereby reducing the number of mutational steps by one. The model predicts that the number of induced cancers will follow the pattern for the overall non-exposed cancer rate but be accelerated in time by one step. Mendelsohn credits the well-defined linear dose-response observations as critical to the evolution of his thinking. However, it seems to us that it is the induced event (and not the specific dose-response relationship, linear or linear quadratic, for example) that is critical to the model; the simplicity of this central induced event imparts a sense of biological elegance to the model. Caveats are well spelled out in the final section.

As we have seen in the Mendelsohn paper, the Armitage-Doll model provided a mechanistic approach to cancer risk. Little's presentation also utilizes the A-D model as a starting point to his analysis and describes various modifications to the model developed by himself and colleagues as well as other workers. Since the A-bomb study provides the largest source of cancer data, most of the models attempt to fit the mathematical modeling to the observed results in terms of excess relative risk. Little develop his "optimal solid cancer model," a three-stage model, restricted to the first stage being radiation affected (compare to Mendelsohn's unrestricted stage). The leukemia data are also fitted to a three-stage model with the first and second stages responsive to radiation. Extensive discussion of these models is provided, including their limitations. He then reviews the two-stage model of Knudson and the Moolgavkar, Venzon, Knudson (MVK) generalized model and compares the conditions to a generalized multistage model (modified A-D). In the MVK model, the conditions involve the number of stem cells, with varying mutation rates and mutation steps, cell division rates, and elimination rates, while somewhat different elements exist in the multistage model. By varying these parameters, Little attempts to show how they effectively predict cancer observations in the Life Span Study, and he concludes that three or more mutation steps reconcile with the solid cancer epidemiologic data.

We wonder if Occam's razor is not being blunted by so many ways of varying the parameters.

In the paper on the genetic effects of the atomic bombs presented by Neel, the founder of the genetics program at ABCC and first director of the entire program, we are provided with a summary of all the major studies since the inception of the genetics program. Remarkably, Neel and Schull have provided guidance to this, the largest human genetic prospective study of its kind, throughout the life of the program. The paper begins with the early history of data collection on the children of the survivors, some 70,000, of whom 31,150 were born to parents who were exposed. In recent years, reanalysis of the major studies was undertaken with respect to DS86 dosimetry, namely in terms of untoward pregnancy outcomes, mortality of liveborn for the first 26 years of life, tumor development through age 20, cytogenetic abnormalities, and mutations altering the electrophoretic behavior of some 30 different blood proteins. It should be recognized that to date there has been no significant increase in any of the genetic endpoints studied, consistent with the expectations when the program was established. The doubling dose (DD) concept was used as a measure of radiation damage, i.e., the estimated contribution of spontaneous mutations summed over these endpoints was divided by the sum of the dose-related regression coefficients. This value provides an estimate of risk from acute radiation. The DD ranged between 1.7–2.2 Sv and was assumed to be about twice as large if converted to the chronic exposure condition. Depending on which experimental mouse data are used for comparison, Neel estimates that both humans and mice are much less radiosensitive than previously thought, or that humans are less sensitive than the estimate of 1 Sv DD derived from specific-locus mouse studies. In the present work, which is attempting to assess genetic damage at the DNA level, several techniques are elaborated, including: scoring for nucleotide substitutions using denaturing gradient gel electrophoresis, detection of changes in unique minisatellite DNA sizes, and detection of increases or decreases in thousands of DNA nucleotides. This study may have significance to a new array of genetic diseases involving increases in trinucleotide repeats. Another approach involves two-dimensional DNA gel electrophoresis, which allows optical scanning of thousands of gene fragments and their computer analysis, for parents and children, and which will detect 50% intensity changes in the specific spots. These studies are still in the developmental stages, but await the final resolution of the question, Were mutations induced in the parents' germ cells? Neel also speculates on new DNA approaches to correlate genetic and somatic risk estimates. He ends with two important issues: the approach one would take if these studies were done again and, secondly, the independence of the ABCC studies from any government pressures, because of the buffer provided by the National Academy of Sciences management, an issue not appreciated by revisionist historians.

Trosko's provocative presentation to this symposium is a theoretical examination of the molecular factors involved in the evolution of a cancer. Three main stages are examined: initiation, promotion, and progression before the ultimate appearance of cancer. In order for the final event to occur, the breakdown of a myriad of defense mechanisms must occur at the molecular, cellular, tissue organ, and organ system levels in a creature such as ourselves. One issue that Trosko is primarily concerned with is at what stages low radiation levels (i.e., very low doses) can affect the enormous variety of defense mechanisms that have evolved in aerobic organisms to protect against oxidative stresses from both endogenous metabolic pathways and exogenous agents. While there is no question that radiation can induce mutations (i.e., primarily deletions) in DNA of recessive tumor-suppressor genes and rearrangements of dominant acting oncogenes, Trosko asks whether radiation is capable of affecting where promotion and progression stages occur. He presents an enormous array of cellular events that would indicate low-dose radiation would not act as a tumor promoter. The question is not completely resolved (to our minds) because, among the cascade of events that could result from even a single ionization track in non-genetic cellular regions, the possibility that some epigenetic events can affect the stem cell of interest and its cellular descendants in ways that can be considered promotional is not resolved, particularly since every cell of interest may be in different metabolic susceptibility states. Trosko raises these points of concern as well. A major discussion in this paper is the important role of intercellular communication in maintaining cellular homeostasis, thereby preventing metabolic disruptions. It appears that many stem cells, however, do not express the gap-junction genes that initiate intercellular communication. Trosko argues that sustained chronic exposure could only bring about promotional activity if a series of conditions regarding oxidative stress could be met that would exclude cell death and allow cell division. The essence of his analysis is that it is unlikely that low doses of radiation can affect all the steps necessary to the development of a malignant cancer. The reader can contrast this view with that presented in the paper by Mendelsohn reviewed earlier; then the question becomes: Is it necessary for radiation to affect all these steps, if it can advance the process by just one step? This brief review does not do justice to Trosko's elaborate presentation, for which we apologize.

It would be presumptuous of us to try to summarize the very difficult concepts presented by Bond, wherein he compares the factors involved in a classical medical pharmaceutical dose response in an individual, of a threshold sigmoid shape, with the population-based epidemiological response to cancer from radiation. The measure of dose on a population basis is rederived. From population to organ-tissue to cell—Bond concludes that hypotheses involving linear dose responses with no threshold are untenable. The reader will have to determine if he or she can agree with the model Bond provides.

The next two papers deal with the use of somatic mutation assays at two different genetic loci. Albertini provides a concentrated review of the X-linked *hprt* gene, hypoxanthine-guanine phosphoribosyltransferase gene, which has been under investigation for over two decades. Jensen reviews the more recently developed autosomal glycophorin A (MN) blood group locus, which has greater relevance to the A-bomb survivor study.

The *hprt* gene mutation or variant is scored in T lymphocytes by two methods, a short-term assay that defines variant frequencies, untestable at the DNA level, and mutant frequencies determined by clonal assays. The system has been used to monitor humans for environmental exposures, smoking, chemicals, and radiation. Unfortunately, the persistence of mutants induced in adults is only three to four years at most because the mutant T lymphocytes in this case are not derived from the stem cell compartment but from the peripheral lymphoid tissue. Nevertheless, extensive molecular analysis has revealed considerable differences in the spectra of spontaneous and radiation-induced events. It is hoped that different mutagens may give different and recognizable spectra, and some data suggests such characterizations will be possible. Albertini suggests in a very intriguing discussion that *hprt* mutations of specific non-random DNA break sequences serve as surrogates for certain lymphoid malignancies because of identical DNA sequences.

These events are tied to illegitimate V(D)J recombinase activity responsible for the enormous immune system repertoire. These rearrangements at *hprt* are found at high frequencies in newborn and young children's blood, paralleling the childhood lymphoid cancer changes. He expects that there will soon be a myeloid *hprt* screening system available that will allow assessing mechanisms underlying adult hematopoietic malignancy.

Unlike the *hprt* system, which can be measured in any individual, the glycophorin system is applicable only to MN heterozygotes, namely about 50% of the population. Electronic screening systems can detect loss of either the M or N antigen expression in red blood cells which normally carry both and are fluorescently stained. Jensen describes the evolution of the assay system and its application to the A-bomb population, wherein a linear dose response was demonstrated in a subset of the population. One clear advantage then is that this red blood cell screening system has memory similar to the cytogenetic analysis; variants induced in the erythroid stem cell of the marrow are recovered in the mature RBC progeny. But such cells lack a nucleus, and the molecular nature of the event is not determinable in the recovered cell variant. Based on recent studies in Chernobyl, dose rate effects have been distinguished, i.e., much lower rates among chronically exposed versus acutely exposed individuals. It is noteworthy that at RERF the GPA assay has larger individual variation per unit dose than the cytogenetic assay, and given the dose-rate observation just cited, leads to the suggestion that the observed linear response in A-bomb survivors may be masking a linear-quadratic response.

In the final paper of this volume, Zimbrick discusses the future of the Radiation Effects Research Foundation (RERF). Taking up the issue of RERF's history, he summarizes the recent events that led the RERF visiting Japanese-American Scientific Council to recommend in June 1995 the establishment of a blue ribbon panel of experts from the international community to evaluate the scientific program. That committee met in February 1996 and issued its final report in June 1996 to both governments. Zimbrick also discusses the establishment of the two-year US Department of Energy contract with the National Academy of Sciences to continue its management role of RERF for the US side and remain a buffer against direct government intrusion.

With respect to future research programs, Zimbrick cites continued surveillance of the in utero exposed population regarding their potentially greater radiosensitivity than those exposed as adults, the need to assess more accurately the evidence for increased non-cancer mortality risks, particularly through clinical studies of the Adult Health Study program, wherein confounding factors can be assessed. In addition to the first generation studies described in Neel's chapter, Zimbrick mentions the need to establish a well-designed clinical study on a subset of this group to determine the possible onset of late-acting detrimental mutations. He describes the enormous reservoir of stored biological samples at RERF and their importance to future studies using biomarkers to assess disease onset, mutational changes in tumor tissue, radiation repair capacity, dose-response relationships, and other mechanistic issues. Finally he describes RERF's role in international collaboration with countries of the former Soviet Union that have also experienced radiation catastrophes.

In the five decades that have passed since the bombings of Hiroshima and Nagasaki, a number of scientific and economic developments have taken place. On a scientific note, there has been an explosive growth of new technologies related to the molecular characterization of disease and exposure. This has resulted in a virtual avalanche of knowledge on the role of exogenous and endogenous factors in the development of radiation-induced health effects. Not surprisingly, new findings from epidemiologic research cannot keep up with advancements being made in molecular-based work. On another note, economic and political changes have had significant impact on the approach of the Radiation Effects Research Foundation, the US Department of Energy, and the National Academy of Sciences toward the joint US-Japan program.

The present volume assesses the five decades of research on the survivors and their children and focuses on proceedings of the first Schull Symposium. Five broad areas of interest and activity were addressed: physics, cancer statistics and cancer epidemiology, genetics, molecular biology, and psychosocial effects and social responsibility. To reflect the focus of the symposium, the present volume is entitled *Effects of Ionizing Radiation: Atomic Bomb Survivors and Their Children (1945–1995)*. Emphasis is placed on unifying the individual disciplines that serve to condition and shape the current outlook on human exposure to ionizing radiation. The current text devotes much more attention than earlier texts to the issues and problems faced by investigators in the early years of the Atomic Bomb Casualty Commission (ABCC), especially those relating to the National Academy activity abroad. Changes in international economic issues are also given broad coverage in the text.

The main themes of contemporary research are presented specifically in each of the chapters. The premise that exploration and analyses in radiation research are more meaningful when discussed in a historical context is reflected in the chapters as each author used a common thread to bridge existing gaps in knowledge. The successes and failures experienced by the contributors in their research over the past fifty years have had much bearing on reactions to challenges of the scientific and economic environments prevailing in the world today.

An international conference conceived and executed within twelve months owes a debt of gratitude to a number of individuals who made countless contributions to its success. The list is long but warrants inclusion.

Institutional sponsors include the Consulate General of Japan at Houston; the Atomic Energy Control Board of Canada; Offices of the President and Vice President of the University of Texas Health Science Center at Houston; the Executive Vice President for Administration and Finance of the University of Texas Health Science Center at Houston; the School of Public Health, Graduate School of Biomedical Sciences, and the Medical School of the University of Texas Health

Science Center at Houston; The Methodist Hospital; Baylor College of Medicine; the Department of Medicine at the University of Texas Branch at Galveston (Division of Oncology, General Medicine, and Office of Texas Department of Criminal Justice Affairs). The contribution from Dr. Joe Goldman and the International Center for Solutions of Environmental Problems provided not only office space and support but immeasurable moral support and guidance for the local organizing group of academic and community volunteers.

The academic committee consisted of Dr. Richard Wainerdi, Dr. Antonio Gotto, Dr. Jim Williamson, Dr. R. Palmer Beasley, Dr. Bill Butcher, and Dr. Don Powell.

Student scholarships for conference attendance were provided by Mr. and Mrs. Paul Bertin, Judith Booker, Dr. James Crow, Dr. Steve Daiger, Dr. Tommy Douglas, Tyrell Flawn, Dr. and Mrs. Arthur Garson, Dr. Lu-yu Hwang, Dr. David Hewett-Emmett, Mr. and Mrs. Joseph Meyer, Dr. Masotoshi Nei, David and Margaret Noble, Dr. Leif Peterson, Dr. Anthony Pisciotta, Melva Ramsay, Catherine Roberts, Mr. and Mrs. John Sellingsloh, Dr. Emoke Szathmary, Dr. K. Okamoto and Dr. T. Aoyama.

Founding contributions to the conference were Ms. Sara Barton, Mr. Terry Bertin, Dr. Patricia Buffler, Dr. Ranajit Chakraborty, Dr. Darwin Labarthe, Dr. Don Powell, and Dr. Kim Dunn.

Community volunteers from the Learning Center for Sustainable Living and Foundation for Global Community were coordinated by Mrs. Catherine Roberts. The volunteers included Vickie Ratello, Carole Breckbill, Jim and Kate Conlan, Christine Economides, Don and Sharon Hill, Francis Jones, Mia and Victor Lamanuzzi, Eileen McGovern, Charles Orelup, Katherine Prelat, Barbara Stein, Bill and Majorie Tracy, and Yumi Yonehara. Students volunteers included Lisa Amelse, Mike Badzioch, Molly Bray, Elizabeth Bruckhoimer, Elena Capsuto, Dawn Chandler, Charles Earley, Karen Earley, Margaret French, Rob Harson, Julia Krushkal, Chun Hsin Lin, Grier Page, Brinda Raua, Wen Shui, Allison Stock, and Paul Wong.

The steering committee provided invaluable direction. It was composed of Dr. Phillip McCarthy, T.J. Dunlap, Armin Weinberg, Geraldine Gill, Paula Knudson, Betsy Chadderdon, Carolyn Milton, and Melva Ramsay. Dr. Fred Tuthill provided legal counsel. Mr. Spencer Knapp, Amersham Corporation, is gratefully acknowledged for supporting the publication of this proceedings.

The planning group, without whose efforts this conference would not have occurred, consisted of Margaret Dybala, Deb Hall, Margaret Irwin, Catherine Roberts, Amelia Kurth, Terry Bertin, Sara Barton, and Dr. Kim Dunn.

LEIF E. PETERSON AND SEYMOUR ABRAHAMSON
April 1998

EFFECTS OF IONIZING RADIATION

PART 1
RADIATION PHYSICS AND DOSIMETRY

1

Development of A-Bomb Survivor Dosimetry

GEORGE D. KERR

SUMMARY

An all-important datum in risk assessment is the radiation dose to individual survivors of the bombings in Hiroshima and Nagasaki. The first set of dose estimates for survivors was based on a dosimetry system developed in 1957 by the Oak Ridge National Laboratory (ORNL). These Tentative 1957 Doses (T57D) were later replaced by a more extensive and refined set of Tentative 1965 Doses (T65D). The T65D system of dose estimation for survivors was also developed at ORNL and served as a basis for risk assessment throughout the 1970s.

In the late 1970s, it was suggested that there were serious inadequacies with the T65D system, and these inadequacies were the topic of discussion at two symposia held in 1981. In early 1983, joint US-Japan research programs were established to conduct a thorough review of all aspects of the radiation dosimetry for the Hiroshima and Nagasaki A-bomb survivors. A number of important contributions to this review were made by ORNL staff members. The review was completed in 1986 and a new Dosimetry System 1986 (DS86) was adopted for use.

This paper discusses the development of the various systems of A-bomb survivor dosimetry, and the status of the current DS86 system as it is being applied in the medical follow-up studies of the A-bomb survivors and their offspring. [1]

[1] Research sponsored by the US Department of Energy under contract DE-AC05- 84OR21400 managed by Lockheed Martin Energy Systems, Inc.

3

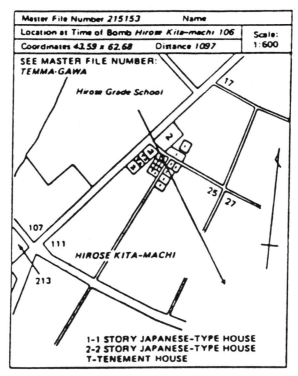

FIGURE 1.1(a)   Example of a shielding history for a survivor exposed inside a one-story Japanese-type house in Hiroshima.

INTRODUCTION

Some individuals survived close to the bombs because they were protected by buildings (Noble, 1967). The amount of protection provided by buildings and other nearby structures is commonly referred to as shielding (see Figure 1.1). Other individuals survived because they were located at very great distances from the hypocenters of the bombs. Thus, an individual's distance from the hypocenter or ground range was one of the important criteria used in the selection of the major study populations of the Radiation Effects Research Foundation (RERF) and its predecessor, the Atomic Bomb Casualty Commission (ABCC) (Beebe and Usagawa, 1968).

A summary of the major study populations is provided in Table 1.1. Originally the Life Span Study (LSS) consisted of 100,000 survivors and control subjects, but additional members were added to this mortality study in 1967 and 1985 so that the LSS extended sample (designated more appropriately as the LSS-E85 sample)

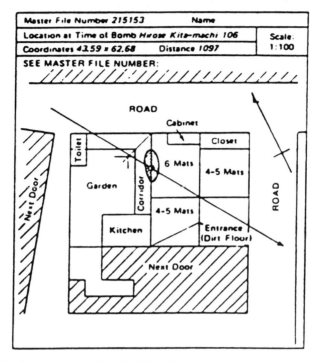

FIGURE 1.1(b) Example of a shielding history for a survivor exposed inside a one-story Japanese-type house in Hiroshima.

now consists of 120,000 survivors and control subjects (Preston et al., 1987). The Adult Health Study (AHS) originally consisted of 20,000 members, all of whom were part of the LSS sample. Attrition by death and migration reduced this clinical study sample to about 50% of its original size by 1984. Thus, the current AHS sample was also enlarged to include more of the highly exposed survivors in the LSS sample, the in utero clinical sample, which is part of the larger in utero sample, and appropriate control subjects.

The survivors (and offspring of survivors) who were located at ground ranges of less than 2,500 m are the core group of the major study samples and are often referred to as the proximal exposed group. Other groups used as controls are the distal exposed group of survivors who were located at ground ranges between 2,500 and 10,000 m, and the non-exposed group of survivors who were located at more than 10,000 m. The non-exposed group is referred to commonly as the not-in-city (NIC) group because it contains a large number of individuals who were not in

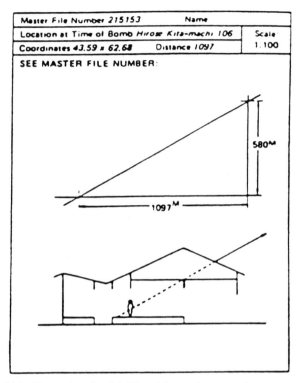

FIGURE 1.1(c) Example of a shielding history for a survivor exposed inside a one-story Japanese-type house in Hiroshima.

the cities at the time of bombing (ATB) but took up residency in Hiroshima or Nagasaki prior to 1950 (Ishida and Beebe, 1959).

Shielding histories were compiled for proximal exposed survivors starting in 1951 in Nagasaki and 1954 in Hiroshima (Noble, 1967; Beebe and Usagawa, 1968). The strategies used in the two cites were somewhat different because there were more proximal exposed survivors in Hiroshima than in Nagasaki (see Table 1.2). In Nagasaki, shielding histories were compiled on all survivors who were located at ground ranges of less than 2,000 m. However, the approach in Hiroshima was to take shielding histories out to 2,000 m for only those survivors included in smaller study samples (e.g., the AHS and in utero samples). It was decided initially that shielding histories would only be taken on the LSS subjects who were located at less than 1,200 m. After shielding histories were compiled on the LSS subjects under 1,200 m, however, the 100 percent criteria was extended

TABLE 1.1 Major study populations.

Study	Approximate number of subjects	Year base populations established	Year studies initiated
Survivors:			
Life Span Study (LSS)	120,000	1950[a]	1958
Adult Health Study (AHS)	23,500	1950[a]	1958
In utero sample	2,800	1945–46[b]	1960
Offspring of survivors:			
First generation (F_1) mortality	75,000	1946–85[c]	1960
Chromosome aberrations	33,000	1946–85[c]	1967
Biochemical Genetics Study (BGS)	45,000	1946–85[c]	1975

[a] Special supplement to the 1950 National Census in Japan.
[b] Birth records from August 1945 through May 1946.
[c] Birth records from June 1946 through December 1985.

to those LSS subjects under 1,300 m, and so on, ending at 1,600 m (Milton and Shohoji, 1968).

When the T65D system became available (Auxier et al., 1966), it appeared that the radiation doses were approximately equal at 1,600 m in Hiroshima and 2,000 m in Nagasaki. At that time, shielding histories were available for most of the Nagasaki survivors who were located at ground ranges of less than 2,000 m. In Hiroshima, shielding histories were available for most of the LSS subjects under 1,600 m, 30% of the LSS subjects between 1,600 and 2,000 m, and most other sample subjects under 2,000 m. In both cities, there were a number of cases in which shielding histories were either incomplete or unavailable because of the migration or death of survivors before 1965. For some cases, however, information such as "exposed inside a Japanese house" or "exposed outside unshielded" was available from earlier studies and surveys (Ishida and Beebe, 1959).

In the late 1970s, the T65D estimates for Nagasaki survivors were recalculated using a different hypocenter and burst height for the bomb (Kato and Schull, 1982), and the dose estimates for survivors in both cities were redesignated as T65D Revised (T65DR). There were no changes in the dose estimates for Hiroshima survivors, and the changes for the Nagasaki survivors were small (less than 10%). Hence, it needs to be noted that sometimes the T65DR estimates for survivors may

TABLE 1.2   Inventory of shielding histories for proximal exposed survivors by shielding category.

City	Shielding category	Number	Percent
Hiroshima	Outside–unshielded	2,490	12.2
	Outside–partially shielded	547	2.7
	Outside–shielded by terrain	46	0.2
	Outside–shielded by a house	2,463	12.1
	Inside–Japanese house	14,130	69.4
	Inside–concrete building	329	1.6
	Inside–factory building	33	0.2
	Inside–air raid shelter	46	0.2
	Miscellaneous shielding	275	1.4
	Total all categories	20,359	100.0
Nagasaki	Outside-unshielded	513	6.1
	Outside-partially shielded	625	7.5
	Outside-shielded by terrain	392	4.7
	Outside-shielded by a house	1,125	13.5
	Inside-Japanese house	3,660	43.8
	Inside-concrete building	616	7.4
	Inside-factory building	1,047	12.5
	Inside-air raid shelter	336	4.0
	Miscellaneous shielding	41	0.5
	Total all categories	8,355	100.0

be referred to simply as T65D estimates (Kerr, 1989). The hypocenters and burst heights used in the various dosimetry studies are discussed by Milton and Shohoji (1968), Kato and Schull (1982), and Kerr et al. (1987).

TENTATIVE 1957 DOSES (T57D)

During Operation Teapot at the Nevada Test Site (NTS) in 1955, ORNL, in co-operation with the Los Alamos National Laboratory (LANL), conducted a series of experiments which provided a much better understanding of weapon radiation fields. The results of these Operation Teapot experiments indicated the possibility of a definitive description of radiation fields from the Hiroshima and Nagasaki bombs (Kerr et al., 1992).

Consequently, in early 1956, a survey team visited ABCC in Hiroshima and Nagasaki to determine the feasibility of a dosimetry study. This survey team included Sam Hurst and Rufus Ritchie of ORNL, Payne Harris of LANL, Bill Ham of the Medical College of Virginia, and Bob Corsbie of the US Atomic Energy Commission (AEC). Several ABCC studies had already reported an elevated incidence of cataracts and leukemia in the surviving populations, especially in Hiroshima.

After reviewing records and examining shielding configurations for survivors, the survey team recommended that a dosimetry program be initiated. Emphasis was to be placed on the shielding provided by Japanese-type houses because of the high structural uniformity of the houses and the large number of survivors who were inside such structures when exposed. As a result of the survey team's findings, an AEC-funded program was established at ORNL and designated as Ichiban—a Japanese word meaning first or number one—because it was considered to be one of the top-priority programs at ORNL during the late fifties and early sixties.

After completion of the analysis of data from Plumbbob, a summary of all dosimetry information applicable to the survivors was prepared and transmitted to the shielding group at ABCC. Designated as T57D, this tentative dosimetry served as a guide for determining dose values from the shielding histories of the exposed individuals. With the assignment of Ed Arakawa to Hiroshima from 1958 to 1960, the shielding results from the nuclear weapons test at NTS were applied in medical follow-up studies of the survivors by ABCC (Arakawa, 1960). These studies, together with the equations for the weapon radiation fields by E.N. York of the Air Force Special Weapons Center (York, 1957), led to the assignment of gamma-ray and neutron doses to individual medical records of survivors instead of the previously used broad dose-value categories based on distance from the hypocenter of the bombs. The T57D system of dose estimation was first used in 1959 to derive dose response curves for leukemia among lightly shielded survivors of the two cities (Heyssel et al., 1959; Tomonaga et al., 1959).

The procedure for estimating radiation doses to survivors exposed in houses that were either unshielded or lightly shielded by neighboring houses was as follows. From Figure 1.1(a), the distance from the hypocenter or ground range in Hiroshima was computed (i.e., 1,097 m) and the radiation levels in the open at the given ground range in Hiroshima were then determined from the equations defining the dose curves shown in Figure 1.4. With the help of Figure 1.1(c), the house penetration distance was measured (i.e., the dashed line from the roof to the survivor), and the house transmission factors were then determined from Figure 1.3. The computations of in-air tissue kerma within the house were performed separately for each radiation component (i.e., neutrons and gamma rays) and for each exposed individual in Hiroshima or Nagasaki.

York's dose curves in Figure 1.4 were in general agreement with the results of an earlier 1951 study by R.R. Wilson (1956). In this earlier study, Wilson concluded that the ratio of neutrons to gamma rays was quite different at Hiroshima and Nagasaki, and one might hope to separate the radiological effects due to gamma

10

FIGURE 1.2 One of the two Japanese-type houses constructed at the Nevada Test Site during Operation Plumbbob in 1957.

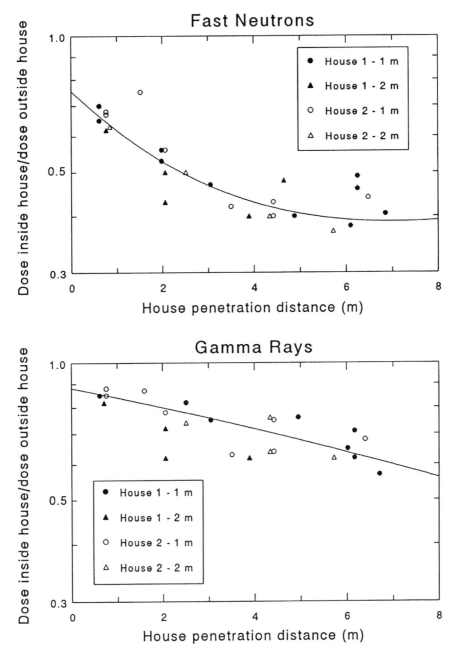

FIGURE 1.3 Transmission factors for single-story Japanese-type houses as measured during Operation Plumbbob in 1957.

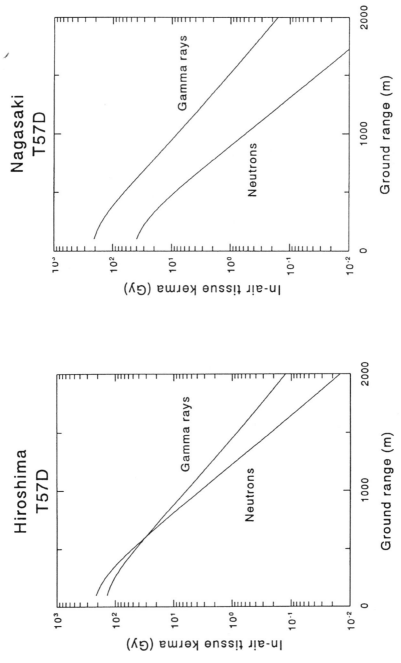

FIGURE 1.4 T57D values for the radiation fields in the open at Hiroshima and Nagasaki. The bomb yield and burst height were assumed to be 18.5 kiloton and 580 m (1,900 ft.) at Hiroshima and 23 kiloton and 490 m (1,610 ft.) at Nagasaki.

rays and neutrons by a comparative study of the effects in the two cities. At Hiroshima, neutron effects might predominate, whereas at Nagasaki, the situation was reversed and nearly all the radiation dose was due to gamma rays. To improve the radiation dose estimates for atomic bomb survivors, Ritchie and Hurst (1959) suggested that it would be necessary to (1) establish more accurate source terms for neutrons and gamma rays from the Hiroshima and Nagasaki bombs and (2) obtain information on radiation shielding by more general house configurations.

TENTATIVE 1965 DOSES (T65D)

Following Operation Plumbbob, laboratory studies of the shielding coefficients of Japanese and American building materials were conducted by John Auxier, Fred Sanders, and Wendell Ogg. Cement asbestos board, commercially available in large sheets, was found to be suitable as a substitute for the mixture of clay, oyster shells, and seaweed wall plaster and for the embedding clay and tile roofs of Japanese houses for both neutrons and gamma rays. The wood framing used in Japan fitted well with the substitution of cement-asbestos board, and domestic materials were used to construct Japanese house replicas for shielding studies during later weapon tests at NTS (see Figure 1.5).

During Operation Hardtack II in 1958, a large number of collimators were used for measuring the angular distributions of the neutrons and gamma rays from a nuclear weapon, and seven replicas of Japanese houses were constructed for the shielding studies. Three different floor plans which represented about 90% of all single family dwellings in Hiroshima and Nagasaki were used to construct the seven house replicas (i.e., three small one-story tenement houses, two medium-size one-story houses, and two large two-story houses). Emphasis was placed on evaluating the shielding as functions of house size, orientation, and position relative to other nearby houses. Because of the durability of the wall board, six of the seven houses were repaired and used three times and the seventh was used twice. The measurements at weapon test sites ended with the Limited Test Ban Treaty of 1962.

Consequently, it was decided to do a definitive study of the radiation fields at large distances from a small unmoderated and unshielded reactor. Designated as Operation BREN (Bare Reactor Experiment Nevada), the experiments were conducted under the leadership of John Auxier and Fred Sanders during the spring and early summer of 1962. The reactor was mounted on a hoist car, which was in turn mounted on a 465-m tower at NTS and operated at various heights above ground to simulate the prompt neutron and gamma-ray fields from a nuclear weapon. At 465 m (1,527 ft.), the BREN tower was taller than the Washington Monument at 169 m (555 ft.), the Eiffel Tower at 300 m (984 ft.), and the Empire State Building at 448 m (1,472 ft.). A ^{60}Co source of about 1,200 curies was to simulate the delayed gamma-ray field from the fireball of a nuclear weapon following the completion of the reactor studies.

During Operation BREN in 1962, extensive measurements were made of the radiation fields in the open, in Japanese houses, and in clusters of Japanese houses. By the use of data from Operation BREN and earlier weapon tests, Cheka et al. (1965) developed a set of nine-parameter linear regression equations that could be used to calculate the transmission factors for survivors exposed inside houses in either city. The nine-parameter equations allowed these factors to be calculated as functions of such things as house size and orientation, location of a survivor inside the house, house penetration distance, location of the house with respect to nearby structures, and proximity of a survivor to an unshielded window facing toward the bomb.

The primary techniques for obtaining transmission factors in the T65D system of dose estimation for A-bomb survivors were as follows:

- The nine-parameter formulas for survivors exposed inside either one- or two-story Japanese houses or smaller tenement houses (Milton and Shohoji, 1968, pp. 42–43).

- The globe technique of determining transmission factors by a combination experimental and calculational approach, using measured angular distributions as input data (Noble, 1967, pp. 28–29 and 79–80).

- The ad hoc assignment of transmission factors based on a review of shielding histories or groups of similar shielding histories (Milton and Shohoji, 1968, pp. 8–9).

The globe technique was used for survivors who were outside but shielded by houses or terrain and for some survivors who were inside concrete buildings. Several important examples of ad hoc assignments within the T65D system were as follows:

- The use of averaged transmission factors for survivors who were known to be inside Japanese houses but for whom shielding histories were either incomplete or unavailable.

- The assignment of transmission factors of 0.9 or 1.0 for survivors inside factory buildings at Nagasaki and either shielded or unshielded by heavy equipment and machine tools, respectively.

- The assignment of transmission factors of 1.0 for all survivors who lacked shielding histories and were located at ground ranges of more than 1,600 m in Hiroshima and 2,000 m in Nagasaki.

Transmission factors and radiation doses were neither calculated nor assigned for 3,017 proximal exposed survivors because their shielding conditions were either extremely complex or unknown.

Following Operation BREN in 1962, Auxier et al. (1966) also developed a new set of dose curves for the weapon radiation fields in the open that were designated

15

FIGURE 1.5 Photograph made in 1958 during a weapons test at the Nevada Test Site. The Japanese house replicas are in the foreground, and the collimators used to measure the angular distributions of the neutrons and gamma-ray fields are in the background.

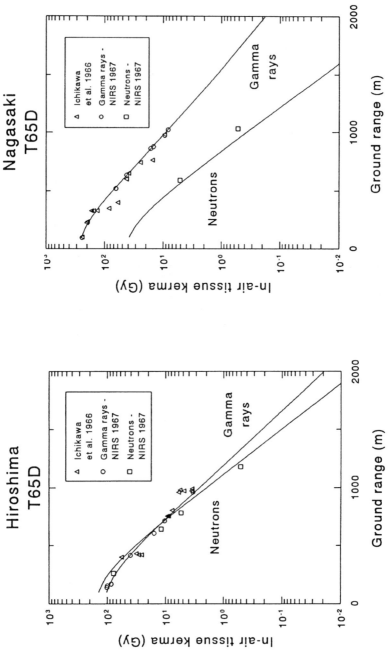

FIGURE 1.6 T65D values for the weapon radiation fields in the open at Hiroshima and Nagasaki. The bomb yield and burst height were assumed to be 12.5 kiloton and 580 m (1,900 ft.) at Hiroshima and 22 kiloton and 490 m (1,610 ft.) at Nagasaki.

as T65D (see Figure 1.6). Ideally, these dose curves would have been established from test firings of exact duplicates of the Japanese weapons. Some information was available from early tests of several Nagasaki-type weapons, but the Hiroshima bomb was the only one of its type that was ever fired, and the weapon radiation fields in Hiroshima had to be constructed using indirect evidence from calculations and experiments with nuclear reactors. However, the dose curves generated by the T65D equations were found to agree closely with results of independent studies by Hashizume et al. (1967) at the Japanese National Institute of Radiological Sciences (NIRS) and by Ichikawa, Higashimura, and Sidei (1966) at the University of Kyoto (see Figure 1.6).

The gamma-ray doses of Ichikawa and co-workers were derived using thermoluminescence of quartz crystals from roof tiles. Some rather large uncertainties were involved in the estimated ground ranges. Since roof tiles were used only on Japanese houses and all houses close to the hypocenter were destroyed, the exact location of each roof tile ATB was in doubt. The gamma-ray and neutron doses in the NIRS study were derived using the gamma-ray-induced thermoluminescence of quartz crystals in decorative tiles and bricks and the neutron-induced ^{60}Co radioactivity in steel reinforcing bars (rebars) taken from commercial buildings that had been repaired and used for a number of years after the bombings. The exact location of each sample ATB was well known, and the uncertainties in the estimated ground ranges were minimized. The NIRS study seemed to confirm the T65D results, and the T65D dose curves for the weapon radiation fields in both cities were used with a great deal of confidence in risk assessment throughout the 1970s.

Finally, the transport of radiation in the body of the survivors was calculated by Troyce Jones and co-workers in the 1970s (Jones et al., 1975), and the results were provided to the RERF as sets of organ dose factors which allowed one to account for the self-shielding of internal organs by overlying tissues of the body (Kerr, 1979). For leukemia, the organ of interest was the active bone marrow, and for other cancers, the specific organs of interest were the female breast tissue, lungs, stomach, etc. For studies of survivors exposed in utero, the radiation dose to the fetus was needed, and for studies of first generation (F_1) offspring of survivors, the radiation doses to the testes and ovaries of the F_1 parents were important. The absorbed doses to the deeply seated internal organs and fetus were significantly less than the T65D estimates of in-air tissue kerma for survivors, which served only as an approximation to the maximum absorbed dose at the surface (skin) of the body (see, for example, Committee on the Biological Effects of Ionizing Radiations, 1972, p. 101).

In the late 1970s, it was suggested that there were serious inadequacies with the T65D system, and these inadequacies were discussed at two symposia held in 1981 (Sinclair and Failla, 1981; Bond and Thiessen, 1982). The starting point for these discussions was the source term calculations for the Hiroshima and Nagasaki bombs by W.E. Preeg of LANL (Bond and Thiessen, 1982, pp. 125–130). In early

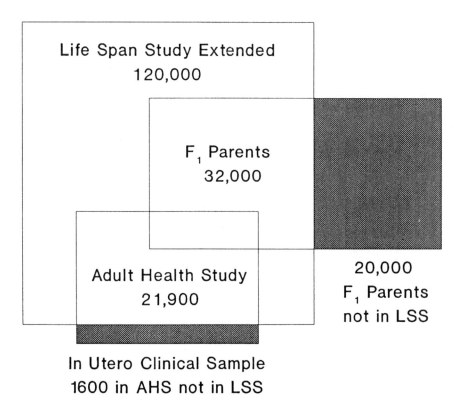

FIGURE 1.7 Illustration of the overlap among members and parents of offspring in the major study samples.

1983, joint US-Japan research programs were established to conduct a thorough review of all aspects of the radiation dosimetry for the Hiroshima and Nagasaki A-bomb survivors (RERF, 1983a; RERF, 1983b). The review was completed in 1986 and the new DS86 system of dose estimation was adopted for use (RERF, 1987; Shimizu et al., 1989).

Two conditions were set on the cohort selected for dose estimation using the DS86 methods: (1) each individual must have a T65D (or T65DR) estimate, and (2) each individual must be a member or parent of an offspring in a major study sample (see Figure 1.7). The total cohort due to overlap among the populations of the various samples is approximately 141,600 individuals (i.e., the 120,000 survivors of the LSS sample plus the 1,600 survivors of the in utero clinical sample and the 20,000 F_1 parents who are not part of the LSS sample).

To facilitate the application of the DS86 methods of dose estimation for individual survivors, a modular computer code system was developed (RERF, 1987, Vol. 1, pp. 405–431). The DS86 methods were embodied in this code system as follows:

- A database for the weapon radiation fields in the open, which specifies the differential energy and angular fluences of neutron and gamma rays at four different heights above ground and at 25-m intervals from 100 to 2,500 m of ground range in both cities (Kerr et al., 1987).

- A database for house-shielding cases, which describes how the differential neutron and gamma-ray fluences were modified at over 50 locations inside a Japanese house (or house cluster) and at a similar number of locations in which a survivor was outside and either partially shielded or totally shielded by a Japanese house (Woolson et al., 1987).

- A database for organ dosimetry, which describes how the differential neutron and gamma-ray fluences were further modified at 15 internal organ sites as functions of a survivor's orientation and posture ATB (RERF, 1987, Vol. 1, pp. 306–404). Age-dependent organ tissue doses can be made for infants (less than 3 years old ATB), children (3 to 12 years old ATB), and adults (more than 12 years old ATB).

Since the DS86 Final Report was published, two additional shielding databases have been added to the modular computer code. One of these databases was developed for application to terrain-shielded survivors at Nagasaki (the database for terrain shielding in the DS86 Final Report was never used and it was later replaced by a more refined database for terrain shielding), and the other database was developed for application to factory-shielded survivors at several sites in Nagasaki (the Ordnance Plant at Oshashi, Steel Works at Mori-machi, and Dockyards at Saiwai-cho).

Suppose a survivor was exposed inside a small one-story Japanese-type house as illustrated in Figure 1.8. First, the house was positioned about the survivor to simulate his or her actual shielding configuration ATB, and both the house and individual were positioned at the correct ground range from either the Hiroshima or Nagasaki bomb. Next, the differential particle fluences from the database for the radiation fields in the open for the appropriate city were coupled with the adjoint particle fluences from the database for house shielding to obtain the radiation fields inside the house. Finally, the adjoint particle fluences from the database for organ doses were coupled with the radiation fields inside the house to provide organ tissue doses (as functions of age, posture, and orientation of the individual ATB) and in-air tissue kermas (inside and outside the house). This same procedure was used with the databases for other shielding situations (i.e., outside shielded by a house, inside a factory building, etc.).

Because of the expense of a re-examination of the shielding histories, the DS86 database for house shielding was constructed to use computerized shielding data

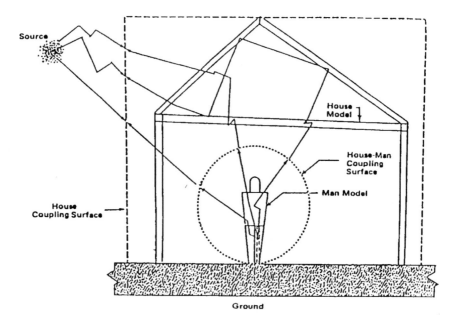

FIGURE 1.8 Illustration of the overall DS86 coupling procedure for dose estimation for individual A-bomb survivors with shielding histories.

that had been coded for T65D. Unlike T65D, however, DS86 does not make use of
transmission factors or organ-dose factors per se, and organ tissue doses and in-air
tissue kermas were calculated directly for survivors with shielding histories if the
survivors' location ATB fit one of the following categories:

- Inside a Japanese-type house or tenement for which nine-parameter data
  were coded (18,315 individuals).

- In the open but shielded by a Japanese-type house or tenement for which
  globe data were coded (3,806 individuals).

- In the open but shielded by terrain features for which globe data were coded
  (361 individuals).

- In the open and unshielded with thermal flash burns reported on exposed
  portions of the face, neck, or arms (1,297 individuals).

- Inside a steel frame factory building of light construction in Nagasaki without any additional shielding by heavy equipment or machine tools (815
  individuals).

For survivors without shielding histories, it was necessary to develop supplemental techniques for indirect computation of the in-air tissue kerma and organ tissue doses. The various supplemental techniques for dose estimation are as follows:

- The use of an in-air tissue kerma of zero at ground distances of more than 2,560 m in Hiroshima and 2,760 m in Nagasaki, where the in-air kermas to shielded survivors were less than 0.005 Gy (45,405 individuals).

- The use of averaged transmission factors and organ doses for survivors with limited shielding data which identified them as being exposed inside Japanese-type houses or tenements ATB (25,962 individuals).

- The use of averaged transmission factors and organ doses for survivors with limited shielding data which identified them as being exposed outside with or without shielding by houses or terrain ATB (10,034 individuals).

For this latter group, the in-air tissue kermas and organ tissue doses are less than 0.2 Gy in Hiroshima and less than 0.1 Gy in Nagasaki. Most of these survivors have been used traditionally as control subjects, and their addition to the LSS sample offers little in the assessment of somatic risks. The primary reason for this extension of DS86 was to meet the special needs of the genetic studies. An offspring of two exposed parents may not be part of the control group for the assessment of genetic risks, and the offspring may be classified as a DS86 unknown-dose case if the radiation doses are not available for both parents. Thus, the use of this extension of DS86 is considered optional in the analyses of data from the different study populations (see Table 1.1).

Currently, DS86 estimates using either direct or indirect methods are available for 105,995 individuals or 92% of the total of 115,019 survivors who are members of the LSS and in utero samples, or parents of offspring in the F_1 mortality sample (see Table 1.3). However, there are now 9,024 survivors without DS86 estimates (DS86 unknown) compared to only 3,017 survivors without T65D estimates (T65D unknown). Most of the DS86 unknown-dose cases are proximal exposed survivors who reported being in the open without reporting flash burns, inside very heavily shielding structures (e.g., concrete buildings or air-raid shelters), or in very complex shielding situations (e.g., inside street cars, etc.).

Figure 1.9 presents a comparison of the DS86 and T65D curves for the radiation fields in the open at Hiroshima and Nagasaki, and Table 1.4 presents a comparison of DS86 and T65D values of the radiation doses for an adult A-bomb survivor exposed inside a Japanese-type house at 1,200 m of ground range in both cities. The latter DS86 and T65D values can be compared directly because they were derived in a consistent manner (i.e., the DS86 values are based on house transmission factors and organ dose factors derived as averages from direct DS86 estimates for all survivors who were exposed inside Japanese-type houses or tenements). The principal differences between these two dosimetry systems can be summarized as follows:

TABLE 1.3 Inventory of DS86 estimates for individual survivors.

Estimation method	Number of individuals
Direct DS86 estimates	24,594
Indirect DS86 estimates	81,401
Total DS86 estimates	105,995
Unknown DS86	9,024
Total DS86 cohort	115,019
Not-in-city (NIC)	26,616
Sample totals	141,635

- In Hiroshima, the gamma-ray kerma is larger than before, due in part to a change in estimate of bomb yield from the T65D value of 12.5 kiloton to the DS86 value of 15 kiloton, whereas in Nagasaki, values of bomb yield and gamma-ray kerma were similar to previous values (see Figure 1.9).

- In both Hiroshima and Nagasaki, the neutron kermas are significantly less than before for a couple of reasons (see Figure 1.9). One reason is that the newer source terms suggest that the prompt neutrons coming from the Hiroshima bomb are more degraded in energy than thought previously (RERF, 1983a, pp. 13–39), and the other reason is that the high humidity in Hiroshima and Nagasaki was not adequately taken into account before (i.e., modern weapon calculations indicate a reduction by a factor of two in neutron kerma because of the higher humidities, all other things being equal). The DS86 values suggest that neutrons are no longer considered a significant contributor to the radiation doses for Hiroshima survivors, and survivors in both cities were exposed mainly to gamma rays (see Table 1.4).

- The newer DS86 values for the transmission of gamma rays by houses are about half of the T65D values [i.e., the average transmission factors are 0.90 (T65D) versus 0.46 (DS86) in Hiroshima and 0.81 (T65D) versus 0.48 (DS86) in Nagasaki]. Thus, the T65D system seriously overestimated the transmission of gamma rays by houses (see Table 1.4). The newer DS86 values for the transmission of neutrons by houses have remained about the same as the T65D values [i.e., the average factors for the two systems are 0.36 (T65D) versus 0.46 (DS86) in Hiroshima and 0.35 (T65D) versus 0.41 (DS86) in Nagasaki].

- Organ dosimetry was not included as part of the original T65D system, but techniques for the estimation of organ tissue doses were added later. These techniques have now been found to seriously underestimate the tissue transmission of gamma rays by a factor of as much as two in the case of the deeply seated internal organs (Kerr, 1989). However, the increased tissue transmission for gamma rays in the DS86 system is largely or wholly offset by the changes in the house transmission factors, except in the case of a few superficial organs such as the female breast (see Table 1.4).

DISCUSSION

One major task of the recent dose reassessment studies was to test the DS86 calculations to the maximum extent feasible. For example, considerable effort was made to extend the thermoluminescent (TL) measurements of the 1960s to longer ground ranges in both cities (RERF, 1987, Vol. 1, pp. 143–184). These TL measurements as well as the newer TL measurements of the DS86 study, are shown in Figure 1.10 as functions of the slant range, which is the distance from the weapon's burst point in air to a point of interest at or near the ground. The slant range is used here to account for differences in the heights of the various TL samples above the ground. It should be noted that the DS86 calculations for the gamma-ray fields in the open are in close agreement with the TL measurements at the longer ranges where there are considerable differences between the DS86 and T65D values (see Figure 1.9). The TL measurements are discussed in reports by Ichikawa and colleagues (Ichikawa et al., 1966; Ichikawa et al., 1987; Hoshi et al., 1987) and the staff of the NIRS in Japan (Hashizume et al., 1967; Maruyama et al., 1987).

A major disappointment of the DS86 study was the poor agreement between calculations and measurements of cobalt activation by slow neutrons at both Hiroshima and Nagasaki (Loewe, 1985; Kerr et al., 1990; Kimura and Hamada, 1993). Recently, it has been found that the DS86 neutron calculations were in close agreement with a newer and more extensive set of measurements at Nagasaki for chlorine activation by slow neutrons (Straume et al., 1994). However, the new chlorine-activation data have also confirmed the previously noted discrepancy in the cobalt activation at Hiroshima (see Figure 1.11). At Hiroshima, the calculated-to-measured ratios for cobalt activation are always greater than one at shorter ranges (i.e., less than 600 m of ground range or 840 m of slant range) and always less than one at longer ranges. Thus, the DS86 neutron calculations for Hiroshima appear to be too low by factors ranging from 2 to 10 at the ranges of most interest in the A-bomb survivor studies (i.e., 1,000 to 2,000 m of ground range or 1,160 to 2,080 m of slant range).

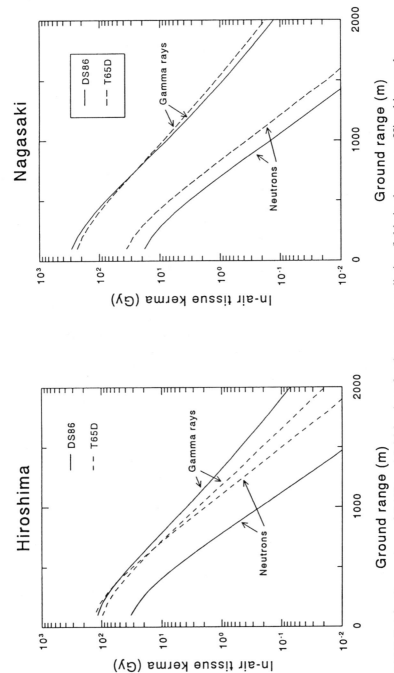

FIGURE 1.9 Comparison of T65D and DS86 values for the weapon radiation fields in the open at Hiroshima and Nagasaki. The DS86 values were calculated for a bomb yield and burst height of 15 kiloton and 580 m (1,900 ft.) at Hiroshima and 21 kiloton and 503 m (1,650 ft.) at Nagasaki.

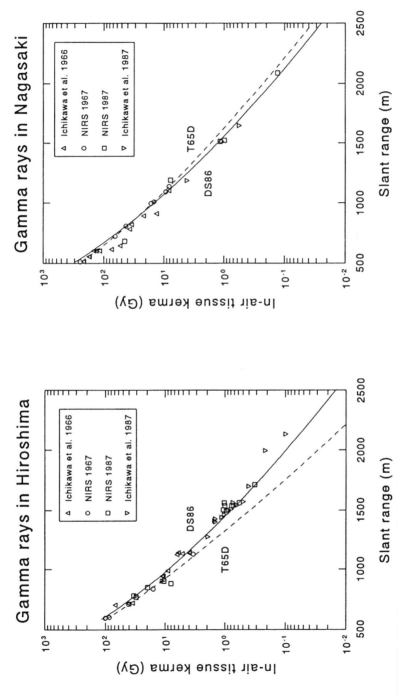

FIGURE 1.10 Comparison of calculated and measured values for the gamma-radiation fields in the open at Hiroshima and Nagasaki.

TABLE 1.4 Comparison of radiation doses for an adult A-bomb survivor exposed inside a Japanese-type house at 1,200 m of ground range.

City	Dosimetric quantity	Dose system	Average radiation dose (Gy)		
			Total	Gamma rays	Neutrons
Hiroshima	In-air tissue kerma	DS86	0.783	0.762	0.021
		T65D	1.041	0.855	0.186
	Female breast tissue	DS86	0.663	0.650	0.013
		T65D	0.794	0.692	0.102
	Active bone marrow	DS86	0.634	0.626	0.008
		T65D	0.543	0.491	0.052
	Lung	DS86	0.619	0.612	0.007
		T65D	0.482	0.441	0.041
	Stomach	DS86	0.584	0.578	0.006
		T65D	0.450	0.415	0.035
	Large intestine	DS86	0.573	0.569	0.004
		T65D	0.382	0.356	0.026
Nagasaki	In-air tissue kerma	DS86	1.556	1.544	0.016
		T65D	3.219	3.186	0.037
	Female breast tissue	DS86	1.311	1.301	0.010
		T65D	2.570	2.550	0.020
	Active bone marrow	DS86	1.274	1.267	0.007
		T65D	1.797	1.787	0.010
	Lung	DS86	1.236	1.230	0.006
		T65D	1.604	1.596	0.008
	Stomach	DS86	1.175	1.170	0.005
		T65D	1.507	1.500	0.007
	Large intestine	DS86	1.160	1.156	0.004
		T65D	1.282	1.277	0.005

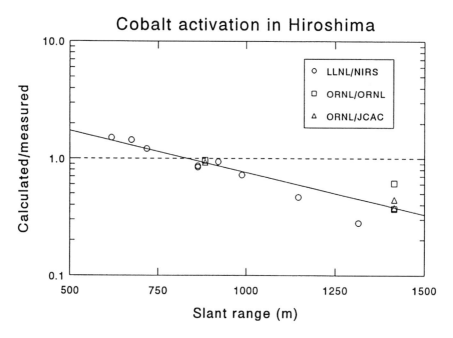

FIGURE 1.11 Calculated-to-measured ratios for cobalt activation in Hiroshima. The ratios shown in the figure are derived from studies at the Lawrence Livermore National Laboratory (LLNL), Japanese National Institute for Radiological Sciences (NIRS), Japan Chemical Analysis Center (JCAC), and Oak Ridge National Laboratory (ORNL).

A new set of ORNL calculations has been made recently by Rhoades et al. (1994) to obtain neutron kermas that were consistent with the various neutron activation data at Hiroshima (see Figure 1.12). Once a neutron source term was found that reproduced the neutron activation by fast neutrons in sulfur and by slow neutrons in cobalt, chlorine, and europium to within approximately 20% at all ranges, it was used to calculate the in-air tissue kerma from neutrons and from neutron-produced gamma rays in air and ground, which were then added to the other gamma-ray components of the weapon radiation fields in the open. Good agreement is maintained with the DS86 gamma-ray values (and the TL measurements shown in Figure 1.10), even though the new ORNL calculations are a factor of five greater than the DS86 neutron values and only a factor of two less than the T65D neutron values over the ranges of most interest in the A-bomb survivors studies (i.e., 1,000 to 2,000 m of ground range or 1,160 to 2,080 m of slant range). Based on these new ORNL calculations, it seems particularly important to give a top priority to a more complete resolution of the causes for

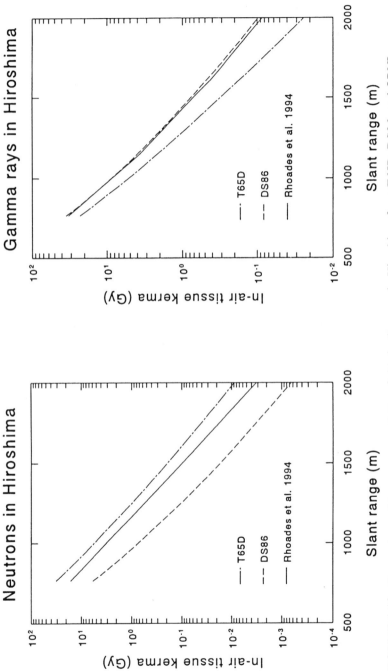

FIGURE 1.12 Comparison of the weapon radiation fields in the open in Hiroshima from T65D, DS86, and ORNL calculations by Rhoades et al. (1994). A source term was selected in these 1994 ORNL calculations which reproduced the correct neutron activation by fast neutrons in sulfur and by slow neutrons in cobalt, chlorine, and europium.

the DS86 neutron discrepancy and to incorporate the revised neutron doses at Hiroshima into an updated DS86 system (see Figure 1.12). This should enable neutron quality factors to be derived directly from human data rather than using the more exaggerated values sometimes seen in animal or cell studies (Grimwood and Charles, 1994).

ACKNOWLEDGMENTS

The author has served as a consultant to the RERF since 1975, and many individuals at the RERF and elsewhere have contributed to the work summarized in this report. Those who deserve special acknowledgements for their contributions are Dale Preston (RERF), Shoichiro Fujita (RERF), Jill Ohara (RERF), Dean Kaul (SAIC), Bill Woolson (SAIC), and the many ORNL staff members who have contributed to the development of the radiation dosimetry for the A-bomb survivors.

2

The Origin of DS86

DEAN C. KAUL

SUMMARY

After 15 years of existence, the first integrated system of dose estimates for Japanese A-bomb survivors, called T65D, was challenged. The challenge was based on the potential impact of epidemiological findings based on T65D, an almost wholly empirical system, and the advent of new capabilities to calculate dose values. Ultimately, this led to the creation of Dosimetry System 1986 (DS86). The primary effect of DS86 was an increase in total dose at Hiroshima, relative to that at Nagasaki, and a decrease in Hiroshima neutron dose to a level similar to that at Nagasaki. Comparisons with measurements have confirmed DS86 gamma-ray dose values but have suggested neutron dose values at Hiroshima should be increased, possibly by as much as a factor of five. In addition, testing of DS86 shielding treatment suggests the existence of large random scatter and bias, possibly due to limitations in the means of shielding categorization inherited by DS86 from T65D. Studies of survivors asked to characterize their location at time of burst twice over an eight year period show a very low dose uncertainty due to variations in survivor recall. This suggests that dosimetry improvements would substantially reduce random scatter in the dose data with a potential for more precise epidemiological assessments. Further, the availability of biological observables, such as chromosome aberrations, in the 51% of survivors still living provides a means of assessing dose accuracy for individual survivors and a possible alternative to dose as a basis for high-precision epidemiological study.

31

REASSESSING A-BOMB SURVIVOR DOSIMETRY

The Dosimetry System 1986 (DS86) (Roesch, 1987) is a computer program for assigning radiation doses to individual survivors of the atomic bombing of Hiroshima and Nagasaki, Japan. It succeeds the previous system, T65D (Auxier et al., 1966), which has been used since 1965 for the same purpose. The need for a successor to T65D grew out of a combination of circumstance and technological advancement, the circumstance being that conclusions based on epidemiological data caused T65D to be questioned, and the technological advance being a new means of assessing survivor dose.

T65D was almost entirely empirical in nature, based on measurements made at atmospheric tests of atomic weapons and at tests involving the deployment of a fast reactor and cobalt sources on a tower at the Nevada Test Site (NTS) to simulate weapons radiation. Therefore, it was a rather straightforward system of data and multiplicative factors, given as

$$Organ\,Dose_{man} \;=\; Dose_{free\,field} \times Shielding\,factor_{structure,terrain} \times$$
$$Organ\,Dose\,Factor_{man} \qquad (2.1)$$

where $Dose_{free\,field}$ refers to the free-field unshielded tissue kerma above a flat plain, $Shielding\,factor_{structure,terrain}$ is the shielding factor due to structure and terrain, and $Organ\,Dose\,Factor_{man}$ is the transmission factor relating air kerma and organ dose. While the form of T65D was straightforward, its means of obtaining values for its shielding factors were not, but involved sophisticated correlations between reported survivor locations and those for which experimental data had been obtained at NTS. The organ dose factors were not part of the original T65D, but were added in the mid-1970s (Kerr, 1979). Calculations of weapons radiation propagation played little role in T65D development because of limitations in radiation transport code development, nuclear data availability, and computer power.

In 1976, Los Alamos National Laboratory (LANL) reported a new calculation of the neutron and gamma-ray leakage from the Hiroshima and Nagasaki weapons (Preeg, 1976). Such calculations were made possible by significant advancements in areas of technology that had been so limited at the time T65D was developed. Subsequently, Rossi and Mays (1978) cited evidence for a substantial increase in the supposed effectiveness of neutron radiation in causing leukemia, based on epidemiological data from the A-bomb survivors. This circumstance caused great consternation at Lawrence Livermore National Laboratory (LLNL), which saw this as an unbearable constraint on its development of nuclear weaponry. Therefore, LLNL proceeded to investigate the validity of that evidence, particularly the dosimetry from which it was derived. In 1980, using the LANL weapon leakage data and modern codes and cross sections, LLNL publicly challenged the accuracy of the free-field component of T65D (Loewe and Mendelsohn, 1980).

In the meantime, the Defense Nuclear Agency (DNA) had sponsored a comprehensive program of transport code development and the measurement and evaluation of a complete set of cross-section data for air, ground and shielding which was completed in time for release in the ENDF/B-4 evaluated cross-section library in 1974 (Garber et al., 1975). At the same time, the US Army sponsored the development of a sophisticated shielding calculation system called the Vehicle Code System (VCS) (Rhoades, 1974). With DNA support, Science Applications International Corporation (SAIC) used the Army code and DNA cross-sections to calculate radiation transmission into houses and human organs which differed markedly from that predicted by T65D (Kaul and Jarka, 1977).

In 1981, in the midst of the controversy which ensued from the LLNL and SAIC findings, the US Department of Energy (DoE) sponsored the conference "Reevaluations of Dosimetric Factors, Hiroshima and Nagasaki" to air all differences (Bond and Thiessen, 1982). This conference gave rise to an effort, sponsored by the DoE, to replace T65D.

THE CREATION OF DS86

The effort to create a revised dosimetry system began in earnest with a conference of all participating organizations in Nagasaki, Japan, in February 1983 (Thompson, 1983a), which was immediately followed by another Hiroshima conference in November of that same year (Thompson, 1983b). Participating organizations and their contributions are summarized in Table 2.1. It should be clear from this table that the re-creation of the A-bomb survivor dosimetry had a very broad technical base.

The creation of the new dosimetry system was characterized from its outset by a calculational approach rather than the empirical approach taken by T65D. The intent was to take advantage of the wealth of detail made possible by modern radiation transport technology, while firmly anchoring the results by comparison with measurements, including the very measurements used as the basis of T65D.

Along with its technical duties, SAIC was responsible for creating the framework of the new dosimetry system independently of other agencies. After considerable discussion it was decided not to create a system of scalar multipliers in the vein of T65D. Instead, SAIC proposed a more ambitious system in which each component—free-field, shielding, and organ dosimetry—would be determined independently and would not be affected by a change in any other component. This level of sophistication was deemed necessary because calculations had shown that the energy and angular characteristics of the radiation fluence changed with distance from the hypocenter. This could not be taken into account on a continuous basis with a multiplicative system. Further, the suggested approach allowed a single set of shielding and organ dosimetry data to serve both cities, since they were determined independently of the free-field data.

34

TABLE 2.1 Participants in the development of DS86.

Subject	Responsible organization
Weapon radiation leakage and yield	Los Alamos National Laboratory
Radiation transport in the atmosphere	Oak Ridge National Laboratory Science Applications International Corporation
Structure and terrain shielding	Science Applications International Corporation
Organ dosimetry	Science Applications International Corporation Oak Ridge National Laboratory
Overall dosimetry system design	Science Applications International Corporation
Survivor data (location, shielding)	Radiation Effects Research Foundation
Shielding material characteristics	National Institute of Radiological Sciences (Japan)
Gamma-ray TLD measurements	National Institute of Radiological Sciences (Japan) Nara University, Hiroshima University (Japan) Oxford University (UK), University of Utah
Neutron activation measurements	National Institute of Radiological Sciences (Japan) Kanazawa University, Hiroshima University (Japan) Oak Ridge National Laboratory
Hiroshima weapon replica measurements	Los Alamos National Laboratory Oak Ridge National Laboratory Science Applications International Corporation Defence Research Establishment Ottawa (Canada)

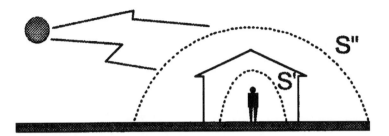

FIGURE 2.1 DS86 dosimetry component schematic.

In principle, the new approach was simply an extension of VCS shielding technology. However, it took that technology a step further by calculating a shield within a shield. Thus, the calculation of structure and terrain shielding was required to retain the full energy and angle differential detail of the fluence in order that the organ dosimetry be calculated in a succeeding step. This process is described by the following integrals describing use of the three independent components of the system, given by

$$\Phi_{shield}(S', E', \Omega') =$$

$$\int_{S'', E'', \Omega''} \Phi_{freefield}(S'', E'', \Omega'') \Phi^*_{shield}(S'', E'', \Omega'', S', E', \Omega') \, n'' \bullet S'' \, dS'' \, dE'' \, d\Omega'' \quad (2.2)$$

where $\Phi_{free\,field}$ describes the energy (E'') and angle (Ω'') differential fluence calculated over the surface, S'', as depicted in Figure 2.1. The adjoint fluence Φ^*_{shield} defines the importance of the fluence over S'' for producing a fluence Φ_{shield} distributed over the surface S' when multiplied by $\Phi_{free\,field}$ and integrated over all energy, angle and locations on the surface S''. This process is repeated to obtain Φ_{man}, integrating the product of Φ_{shield} and Φ^*_{man} in the form

$$\Phi_{man}(E) = \int_{S', E', \Omega'} \Phi_{shield}(S', E', \Omega') \Phi^*_{man}(S', E', \Omega', E) \, n' \bullet S' \, dS' \, dE' \, d\Omega' \quad (2.3)$$

where Φ_{man} is the mean fluence in a specific human organ. Finally, dose is obtained by use of the form

$$Dose = \int_E K(E) \Phi_{man}(E) \, dE \quad (2.4)$$

which integrates the product of the appropriate fluence-to-dose conversion factor, K, and the fluence Φ_{man} over all energy E. In actual operation these integrals

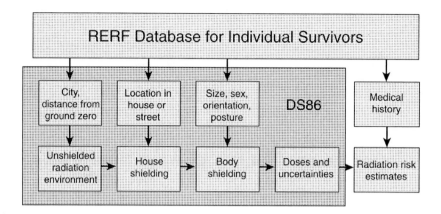

FIGURE 2.2   The DS86 dosimetry system.

are accomplished numerically, using matrix multiplication to combine fluence values with adjoint fluences sampled using three-dimensional Monte Carlo radiation transport.

Figure 2.2 shows the sequence in which DS86 retrieves information from the RERF survivor database, performs the calculation to obtain dose values, and estimates uncertainties for risk estimates. Within the box marked DS86, the small boxes in the upper row represent the three types of information obtained from the survivor data, including city and location, shielding, and personal characteristics. Each type of information determines which database within DS86 will be accessed to perform the dosimetry calculation. Table 2.2 shows values from both the T65D and DS86 dosimetry systems for a hypothetical survivor at approximately 1,500 m from the Hiroshima hypocenter (the location on the ground, directly below the explosion). The changes from T65D to DS86 for such a survivor include an increase by nearly a factor of two in total dose but a decrease by a factor of ten in neutron dose. These changes are almost entirely due to changes in the radiation free-field, since changes in the house and body shielding nearly offset each other.

In 1986, at the conclusion of the DS86 dose reconstruction effort, participants could point to success in creating a system that identified significant changes in survivor dosimetry and provided a superior level of detail to T65D as well. The new system had also been thoroughly compared to measured data, including data taken after 1965. For the most part these comparisons were favorable to DS86. For example, as shown in Figure 2.3, the DS86 gamma-ray component, which dominates the free-field radiation, matches TLD measurements in quartz crystals from locations at Hiroshima and Nagasaki better than does T65D, particularly at Hiroshima. It should be noted that nearly all the measured data beyond 1 km have been obtained since 1965.

TABLE 2.2 T65D and DS86 dose values (cGy) at 1,485 m from the Hiroshima hypocenter.

	Total neutron		Total gamma-ray		Total	
Location	T65D	DS86	T65D	DS86	T65D	DS86
Free-field	12.0	0.93	24.0	51.7	36.0	52.6
House-shielded	3.4	0.37	21.0	30.8	24.4	31.2
Organ (marrow)	0.95	0.11	13.0	24.4	14.0	24.5

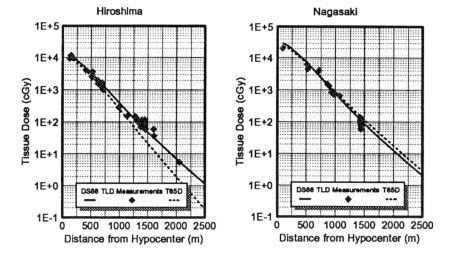

FIGURE 2.3 T65D and DS86 gamma-ray dose calculations and TLD measurements.

The relatively good agreement between T65D, DS86 and the measurements should be expected, given the empirical origins of T65D. There were several tests of weapons similar to the Nagasaki weapon, at both the Nevada and Pacific test sites. However, in the case of the Hiroshima weapon, no basis for such empiricisms existed, and the perturbation to known gamma fields used by T65D to allow for

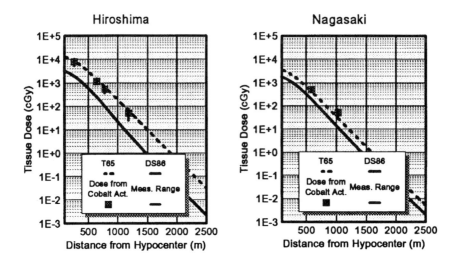

FIGURE 2.4 T65D and DS86 neutron dose calculations and doses inferred from thermal neutron activation measurements at Hiroshima and Nagasaki.

expected differences caused by the unusual design of that weapon clearly caused a loss of too many high energy gamma rays.

No measurements of neutron kerma akin to the TLD gamma-ray dose measurements exist at the two Japanese cities. However, measurements of thermal neutron activation of cobalt do exist for both cities. These measurements were also available at the time of inception of T65D. Figure 2.4 shows T65D, DS86, and neutron dose deduced from thermal neutron activation measurements, assuming a degraded fission spectrum similar to that emitted from a bare fast reactor, like the one used to support T65D development. Given the fast reactor spectrum assumption, the basis for the T65D neutron dose estimates is clear. However, DS86 is considerably at odds with the T65D values and those inferred from the measurements. This situation is worse at Hiroshima than it is at Nagasaki, where there are only two measurements. However, there is an additional set of measurement data available at Hiroshima. These are sulfur activation measurements, which measure neutrons having energies above approximately 3 MeV. The DS86 calculation-to-measurement ratio for sulfur activation at Hiroshima is shown in Figure 2.5. The solid stars in Figure 2.5 show the ratio as it exists with the current DS86 weapon yield of 15 kT. The best agreement is obtained with a yield of 12.7 kT, which implies a decrease in fast neutrons required to produce neutron dose, rather than the increase inferred from the thermal neutrons.

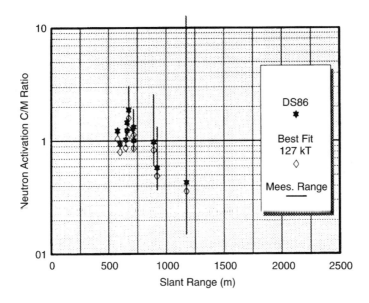

FIGURE 2.5 Neutron sulfur activation at Hiroshima, 1986 comparison with calculations.

Given the contradicting indicators, the thermal activation discrepancy did not cause approval of DS86 to be withheld. Thus, at the time of approval and installation of DS86 at RERF in the spring of 1987, the new dosimetry system could claim improved gamma-ray dose, shielding treatment, and organ dosimetry over its predecessor. However, some problems remained. These included the unresolved neutron activation and lack of coverage of approximately 15% of the Life Span Study members, due to special shielding circumstances, such as occupancy of small factory buildings not treated by DS86, or persons recorded to have been unshielded but without flash burns.

FROM 1986 TO 1995

Since the creation of DS86, small programs in the United States and Japan have continued to assess the adequacy of information embodied in the code system. In the US, ORNL and SAIC have used the new ENDF/B-6 cross section and nuclear data library (Rose and Dunford, 1991) and more powerful computers to recalculate the prompt and delayed radiation free fields at Hiroshima and Nagasaki. In addition, new measurements of thermal neutron activation have been performed and made available by scientists in both the US and Japan (Straume et al., 1992; Straume et al., 1994). The result of this new work has led to reconciliation of calculations and measurements at Nagasaki, as shown in Figure 2.6. The 1986 portion of this figure shows the large discrepancy between the actual calculation/measurement ratio, a linear fit which is represented by the solid line, and the ideal ratio, represented by the dotted line at unity. As shown in the 1995 portion of Figure 2.6, calculations are now within approximately 20% of the measured values, out to over 1,250 m slant range. Improvements in the calculations removed the discrepancy near the burst point, where DS86 calculates too high a level of activation, and new measured data were obtained which were consistent with calculations and large distances.

To show that such an improvement at Nagasaki was no fluke, equivalent calculations were performed to compare with measurements made on Nagasaki-like weapons deployed at NTS. Figure 2.7 shows the good agreement obtained for both the cadmium-difference gold measurements of thermal neutrons and fast (E>3 MeV) neutron measurements made with sulfur. Fortunately, while the calculation of observables has been improved, this has done very little to the dose at Nagasaki, as shown in Figure 2.8, which depicts the ratio of 1995 free-field dose to DS86 free-field dose as a function of distance from the hypocenter.

In the case of Hiroshima, new nuclear data and computer power had little effect on the agreement between calculation and measurement of either neutron thermal activation or gamma-ray TLD measurements. However, these factors significantly changed the agreement with the sulfur data from suggesting a yield of 12.7 kT to suggesting a yield of nearly 17 kT as shown in Figure 2.9. This evidence, taken with that from thermal radiation and blast observations, indicates that it is highly likely that the Hiroshima weapon yield was higher than the 15 kT used in DS86. The effects on free-field dose of such a yield change, together with the change provided by new cross sections and computer power are shown in Figure 2.10. The yield change causes an increase of approximately 20% in the gamma-ray dose, while the cross section changes cause an increase of as much as 50% in the neutron dose. Thus, as we near the end of 1995, it seems clear that any future changes to DS86 are likely to leave Nagasaki free-field doses relatively unchanged, whereas it is very likely that free-field doses at Hiroshima will be increased, perhaps substantially.

A GLIMPSE OF THE FUTURE

Investigations have been carried out to determine whether a change in the Hiroshima weapon neutron leakage could provide a solution to the outstanding disagreement between calculations and measurements of thermal neutron activation. Figure 2.11 shows that it is indeed possible to improve the agreement by changing the source, though there remain some measurements which cannot be reconciled with the calculations using source neutrons of any energy. The change in source results in a considerable increase in the neutron free-field dose, as shown in Figure 2.12. As shown previously in Figure 2.10, the yield increase results in an

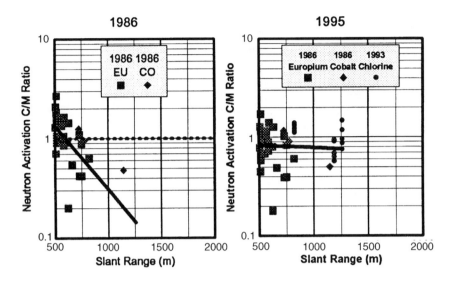

FIGURE 2.6 Comparison of thermal neutron activation calculation and measurement in 1986 and 1995.

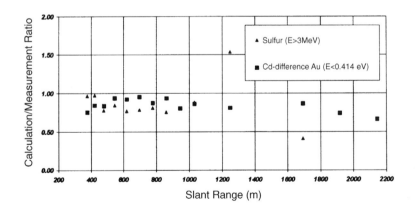

FIGURE 2.7 Ratio of calculated-to-measured sulfur (fast) and Cd-difference gold (thermal) neutron fluence measurements for a Nagasaki-like weapon at NTS.

increase of approximately 20% over DS86. However, the neutron free-field dose increases by nearly a factor of 5.

Unfortunately, at present the only way to obtain simultaneous agreement with all three sets of observations at Hiroshima, gamma-TLD, sulfur neutron and thermal neutron activation measurements, is to invent a source which has no possibility of existing under any circumstances. As shown in Figure 2.13, compared to the existing Hiroshima neutron leakage, the revised source must have ten times as many neutrons in a narrow band from 2.3 MeV to approximately 3 MeV. A substantial reduction in leaking neutrons is required in the range from 0.5-2 MeV. While this source is capable of reconciling all existing observations at Hiroshima to an acceptable degree, it is not considered to be a reliable predictor of dose. Hence, investigations along these lines merely indicate the expected direction of dose changes, i.e., upward, but cannot definitively determine their amount.

The fact that only a physically impossible source can reconcile calculations with measurements at Hiroshima suggests that it is the transport cross sections rather than the source which is in error. Of course, any changes to the cross sections would have to leave calculations of fluence from Nagasaki-like weapons unchanged.

TABLE 2.3 Potential change in Hiroshima dose resulting from reconciliation with measurements.

Component	Absorbed Dose (cGy)	Dose equivalent (cSv)
DS86		
Neutron	0.85	8.50
Gamma-ray	48.80	48.80
Total	49.65	57.30
% Neutron	1.74	17.42
Revision 1995?		
Neutron	3.83	38.25
Gamma-ray	58.56	58.56
Total	62.39	96.81
% Neutron	6.53	65.32
% Change in total:	25.65	68.95

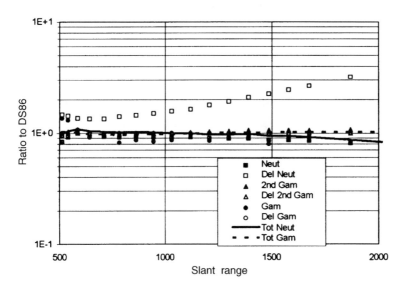

FIGURE 2.8 Ratios of 1995 to DS86 free-field dose components for Nagasaki.

The solution to the neutron discrepancy at Hiroshima will not be easy or inexpensive to obtain. This raises an inevitable question: Is it worthwhile to pursue a solution to this and any other outstanding problems with the dosimetry? The answer to this question can only be guessed at based on the evidence at hand. If the dose values resulting from the optimized source may be trusted sufficiently to suggest the order of a likely increase in neutron dose resulting from the reconciliation of calculations and measurements at Hiroshima, then the overall dosimetry changes would likely be similar to those shown in Table 2.3. At 1,500 m the absorbed dose would increase by approximately 25%, and the dose equivalent (assuming a quality factor of 10) would increase by nearly 70%. While it is claimed that with the advent of DS86 there is no statistically significant difference between risks for various cancers at the two cities, nevertheless it is clear that an increase in dose at Hiroshima would improve intercity agreement as represented for DS86 leukemia mortality (Shimizu et al., 1989), as shown in Figure 2.14.

Random error in assigning doses to survivors is a very important factor in differentiating between the two cities or establishing the shape of the risk curves, particularly at low dose. Of course no dosimetry system, no matter how precise, is any better than the data taken from the survivors themselves. An attempt was made to quantify the error associated with survivor recall of location and shielding by having 88 survivors give a second survivor history after an eight-year interval (Kaul and Egbert, 1989).

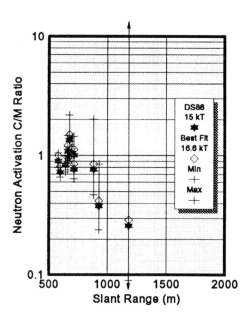

FIGURE 2.9   Sulfur activation measurements at Hiroshima, 1995 comparison with calculations.

When the data from the first and second histories for each survivor are compared in the context of DS86, it is found that over 60% of this group showed no change in dose due to variations in either location relative to the hypocenter or with respect to shielding. Overall, it is found that the coefficient of variation in dose associated with location with respect to the hypocenter is approximately 14% for both cities and that associated with shielding is approximately 11%. Thus, the minimum random uncertainty that may be expected to be associated with survivor total doses is approximately 20%.

It has already been shown that changes due to any reconciliation with neutron activation measurements at Hiroshima will likely be greater than the minimum uncertainty expected based on the quality of data taken from survivors themselves, but what of other aspects of DS86, like shielding? There is reason to believe that some important sources of error exist in DS86, which affect its use in epidemiological studies. One important potential error is that associated with Nagasaki. In Figure 2.14 it can be seen that the risk of leukemia death goes to zero at approximately 40 cGy. The cohort in this dose regime at Nagasaki is dominated by a large group of survivors in the Mitsubishi Heavy Industries Shipyard factory, approximately 1,700 m from the hypocenter. Shielding analysis performed as part

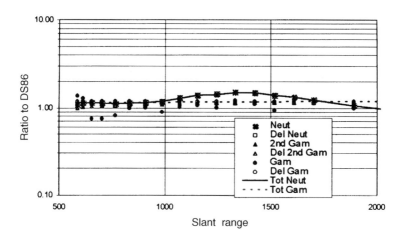

FIGURE 2.10   Ratios of 1995 to DS86 free-field dose components for Hiroshima.

of DS86 determined that these persons were very lightly shielded. A possible explanation for the discrepancy between the risks at the two cities is that the actual shielding of the factory workers may be much greater than currently estimated.

Benchmark studies have been performed, comparing DS86 shielding assessment to rigorous calculations for survivors based on their individual shielding histories. These studies suggest that the random error in terms of coefficient of variation caused by limitations in DS86 is as great as 31%. This is considerably greater than the baseline error estimate of 20% due to survivor recall variations. Most of DS86 random shielding error has been identified as arising from the use of shielding descriptions inherited from T65D. One of the ground rules for the creation of DS86 was that it rely on data coded for use in T65D, because to revisit the survivor shielding histories was assumed to be impractical with existing RERF staff and probably unnecessary. The results of the benchmark studies posed a significant challenge to the reliance on T65D data. For example, there is the matter of how T65D determined whether a survivor was or was not shielded by an adjacent house. According to T65D, if the house was more than two house heights away from the survivor, that house was not counted as providing shielding. This criterion was based on experiments performed at NTS with a fast reactor located on a high tower to simulate the bomb. Fast neutrons from that reactor created gamma-rays in a large volume, limiting the value of shielding in the shadow of test structures.

In fact, at Hiroshima and Nagasaki, there were few fast neutrons and the gamma-rays could be characterized as coming from a point. Thus, any structure in the line

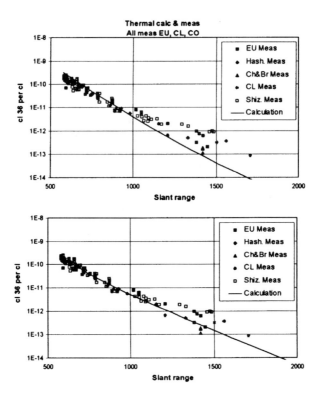

FIGURE 2.11 Measured thermal neutron activation at Hiroshima and that calculated in 1995 with new nuclear data and yield (left) and with an optimized source (right).

of sight between the survivor and the bomb, no matter how far away, would provide some shielding. Thus, the T65D category with no adjacent house shielding should be broken up into two categories, one in which there is actually a house in the line of sight, providing approximately 50% more shielding than currently accounted for, and another in which there is truly no house in the line of sight, providing half as much shielding.

The presumed impracticality of returning to the shielding histories to obtain more accurate data for use in DS86 is debatable. In fact, RERF staff returned to those histories to obtain some data, such as survivor orientation and terrain shielding information, for use in DS86. Thus, given the proper instructions and the use of templates for assessing physical information from shielding drawings, it does not

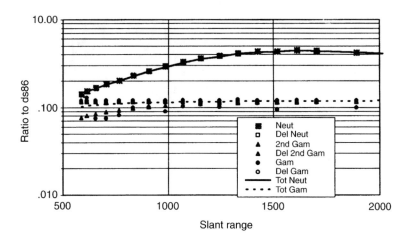

FIGURE 2.12　Ratios of DS86 values to free-field dose components calculated using an optimized source, at Hiroshima.

seem unreasonable to believe that a more precise job of shielding classification could be performed.

Finally, given that uncertainties in the assigned dose will always exist to some extent, is there any alternative to current dosimetry methods or any better way of assessing their accuracy? The answer to both these questions is yes. Straume (1995) has demonstrated that very high quality epidemiologic regressions may be obtained for excess relative risk of total solid tumors based on the incidence of some observed quantal response, such as chromosome aberrations, as illustrated in Figure 2.15. Because such incidence can also be produced in vitro, the relationship to radiation intensity and quality may be obtained independently.

Of course, this approach has significant limitations. Assessment of chromosome aberrations or of any other observable would be a very expensive undertaking and would be limited to living survivors or those for whom blood or tissue samples have been retained. Some of the benefits of this approach can be realized, such as the independence from radiation quality considerations, by using DS86 fluence spectra to calculate the biological quantal response, based on relationships established in

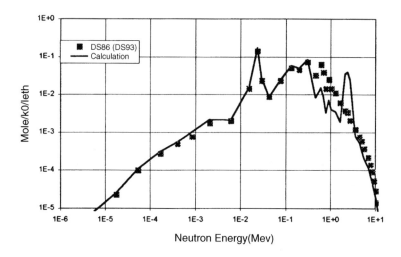

FIGURE 2.13 Calculated Hiroshima weapon leakage spectrum (symbols) and a spectrum optimized to match observation (line).

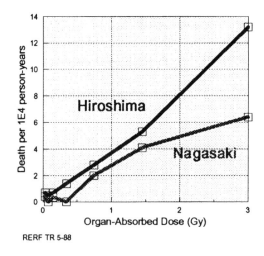

RERF TR 5-88

FIGURE 2.14 Leukemia mortality based on DS86.

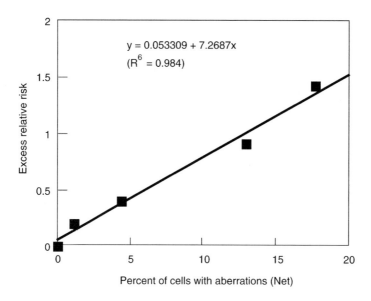

$$y = 0.053309 + 7.2687x$$
$$(R^6 = 0.984)$$

Regression analysis courtesy of Dr, Tore Straume, LLNL
x: Preston et al. 1989
y: Thompsol et al. 1994

FIGURE 2.15 Relative risk of total solid tumors versus chromosome aberration incidence.

vitro between a biological observable and hit size as defined by microdosimetric theory. This approach has the further advantage that it can be compared directly to observations in individual survivors, thereby providing a basis for direct assessment of the accuracy of DS86 and newer systems to come.

ACKNOWLEDGMENTS

The author would like to acknowledge the technical contributions made to work reported in this paper by S.D. Egbert and W.A. Woolson, SAIC. Portions of this work were supported under Defense Nuclear Agency contract DNA001-94-C-0135.

3

Medical Radiation Exposures of Atomic Bomb Survivors

KAZUO KATO, WALTER J. RUSSELL, AND
KAZUNORI KODAMA

SUMMARY

The medical dosimetry program of the Radiation Effects Research Foundation (RERF), formerly the Atomic Bomb Casualty Commission (ABCC), in Hiroshima and Nagasaki has been conducted since 1961. Information concerning the medical radiation exposures of the Adult Health Study (AHS) participants was obtained continually through interviews by specially trained personnel at ABCC/RERF. A series of hospital and clinic surveys in the two cities have revealed the activity of the medical radiation exposures among the atomic bomb survivors. Repeated phantom dosimetry has been conducted to determine the radiation doses to the subjects. Preliminary calculations revealed that 39% of the members of the AHS population received cumulative active bone marrow doses exceeding 1 cGy by the end of 1982. This means that the doses from diagnostic x-ray examinations had already become significantly great contaminates of the radiation doses from the atomic bombs. The radiation therapy received by the atomic bomb survivors is also studied in this dosimetry program. Radiation treatments of 1,670 members of the Life Span Study population were detected in surveys through 1984, and the studies are focused on the primary cancers that have developed after radiation therapy.

INTRODUCTION

Risk evaluation in the low-dose region using the atomic bomb data is indispensable for estimating the risk of ionizing radiation exposures in catastrophes such

51

as nuclear reactor accidents and nuclear warfare. However, the atomic bomb data have relatively large uncertainties in the low-dose region, as introduced by Kato et al. (1987). The atomic bomb survivors have not only received ionizing radiation from the A-bombs but also from medical procedures. Such medical radiation exposures have probably contributed to some increases in cancer incidence among the A-bomb survivors, and this is one of the reasons why the atomic bomb data have large uncertainties in the low-dose region. Quantitative assessments of such contributions have been continuing at the Atomic Bomb Casualty Commission (ABCC)/Radiation Effects Research Foundation (RERF) since 1961. The assessments included several series of surveys of medical radiation exposures, which revealed the trends of medical radiation exposures at the institutions in Hiroshima and Nagasaki. These surveys were performed mainly in order to estimate the individual cumulative medical doses received by the members of the Adult Health Study (AHS) who routinely visited the ABCC/RERF. The number of medical radiation exposures by body site and by type of radiation source were documented from the interviews of the AHS participants during such routine visits. Every radiological examination conducted on the AHS participants at RERF has been documented in detail. The doses incurred by medical radiation exposures were estimated by phantom dosimetry experiments and measurements of outputs of radiological units at the hospitals and clinics in both cities. The results of the interview data and phantom dosimetry enable us to evaluate the individual doses to the AHS participants from the diagnostic x-ray examinations.

During radiation therapy, patients receive relatively large doses which are important for evaluating the radiation risks using epidemiological data of the atomic bomb survivors. In our studies, a large-scale survey was performed and devoted to investigating radiation therapy among the members of the Life Span Study (LSS) at RERF. During this survey, records of members who received radiaiton treatments were transcribed and later reproduced using human phantoms to estimate the organ doses. This study is now in progress.

HOSPITAL AND CLINIC SURVEYS

Practices of Hospitals and Clinics

Ihno et al. (1963) conducted a hospital-clinic survey in 1962 and 1963 in Hiroshima and Nagasaki. They obtained information concerning the number of exposures by site, type of radiological equipment, and exposure conditions of examinations taken at the hospitals and clinics. The total number of diagnostic x-ray apparatuses surveyed was 50 in hospitals and 54 in clinics. In order to obtain more precise information, Sawada et al. (1971a) carried out a large-scale survey in 1964. The facilities for this survey were all large institutions and 40% of small institutions in both cities. Here, a large institution was a hospital or clinic using more than 10,000 films per year. The number of large institutions was 14 in Hiroshima and 11 in Nagasaki, in 1963. The number of small institutions was 487 in Hiroshima and 334 in Nagasaki. The survey produced records of yearly totals for films, radiographic, fluoroscopic and photofluorographic examinations, and number of treatments by x-ray and telecobalt from 1945 to 1963. Subsequently, trends in radiological practice in Hiroshima and Nagasaki from 1964 to 1970 were surveyed by Sawada et al. (1979). The survey covered 17 large and 153 randomly sampled small institutions in Hiroshima, and 13 large and 108 randomly sampled small Nagasaki institutions.

A large-scale survey of hospitals and clinics was conducted in 1974 (Sawada et al., 1979; Antoku et al., 1986). During two-week periods in that year, all records of medical x-ray examinations conducted in Hiroshima and Nagasaki large institutions and a 40% random sample of small institutions were collected to confirm exposures and estimate doses by type, frequency, and exposure conditions. The members of the RERF population involved were identified among the examinees. The technical factors used in Hiroshima and Nagasaki medical institutions were also analyzed.

Use of radiation therapy for A-bomb survivors has also been assessed during these surveys mentioned. However, the information lacked sufficient detail, and three additional surveys were conducted only on radiation therapy among the atomic bomb survivors. The first survey was conducted in 1970 by Russell and Antoku (1976). The second survey was performed in 1980 by Pinkston et al. (1981). These two surveys confirmed that 187 members of the AHS population received radiation therapy exposures. After these two surveys of radiation therapy and two pilot studies, a third survey was conducted in 1985, which provided the information necessary concerning radiation therapy received by the LSS population. During the third survey, a survey team visited the responsible hospitals and all radiation therapy records in each hospital were examined. The radiation therapy data were transcribed from the medical records. The number of hospitals surveyed were 8 large hospitals in Hiroshima (one of which was actually located in the city of Kure) and 3 large hospitals in Nagasaki. Other smaller hospitals had ceased administering radiation therapy by the time this survey was conducted. During the three radiation therapy surveys, the radiation therapy data were obtained from the medical records of 22 institutions in Hiroshima and 8 institutions in Nagasaki, and the data on the radiation therapy provided to a total 1,670 members were obtained.

No survey has been conducted since 1985. During this period, modern diagnostic x-ray equipment such as computed tomography and computed radiography have been installed in many institutions. The quality of films has also been improved, and the technical exposure factors seem to have changed. It is not certain whether the medical records of the radiation therapy have been kept for more than 10 years. Only hospital and clinic surveys enable us to obtain the information about the technical exposure factors used in the diagnostic x-ray examinations, and about how radiation treatments were conducted for the atomic bomb survivors.

Trends in radiological practice in the early days were assessed in the two surveys that were performed by Sawada et al. (1971a, 1975). The result of these surveys is shown in Figure 3.1. The numbers of radiographic examinations in the early days may be uncertain due to the long time lapse before the first survey was undertaken. The figure shows that the number of radiographic examinations steadily increased until 1950 in Hiroshima and until 1953 in Nagasaki. This increase seems to have been caused by more diseases among the atomic bomb survivors in both cities. Takeshita et al. (1978) reported that the number of radiographs taken in Japan rapidly increased from 1960 through 1970. The increases in Hiroshima and Nagasaki in those days were more moderate than those in all Japan as shown in Figure 3.1. This implies that the physicians and patients minimized the use of medical x-ray in both cities. The trends of fluoroscopic examinations in the early days were different from those of radiography. Fluoroscopy was frequently used in Hiroshima. No steady increase in the numbers of fluoroscopy was seen in Nagasaki and in the whole of Japan. The difference between Hiroshima and Nagasaki probably can be explained by the difference in the number of large institutions in the two cities.

The number of examinations by population, sex, and city in 1974 was examined by Sawada et al. (1979). Figure 3.2 shows that the numbers of radiographs per capita in Nagasaki became greater than in Hiroshima. The frequency of fluoroscopy in Nagasaki was still less than in Hiroshima, and both were nearly the same as those in 1970. The number of radiographic and fluoroscopic examinations for the Nagasaki non-AHS male population was significantly larger than those for the other non-AHS populations and for the general population. The reason for this is not clear. The relatively high frequency of x-ray examinations in the AHS population is due mainly to the RERF AHS "routine" x-ray examinations they received. If the examinations at RERF were excluded from analysis, the mean annual x-ray examination frequency for the AHS population would be nearly the same as for those of non-AHS participants.

The radiological examinations detected in the surveys and interviews of the AHS participants mentioned below were tallied by body sites and by city (Ihno et al., 1963; Sawada et al., 1979; Russell et al., 1963). The number of chest examinations far exceeded all others. The number of exposures per examination was examined by Sawada et al. (1979) and was 12–16 in gastrointestinal series and less than 5 in most other examinations. The fewest exposures were made during chest examinations. Thus, more than half the radiographic exposures were from upper GI series due to the many spot films made during fluoroscopy.

Trends in Radiation Therapy

The number of radiation treatments in Hiroshima and Nagasaki found in the hospital and clinic surveys (Sawada et al., 1971a; Sawada et al., 1975) is shown in Figure 3.3. All relatively large hospitals and clinics were selected for these surveys. The number of radiation treatments steadily increased until 1962, but the increase suddenly ceased around 1964 in both cities. The reason for the declines in radiation therapy in those days is now in question.

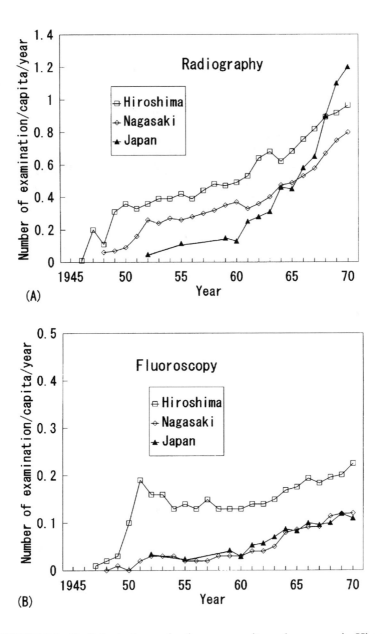

FIGURE 3.1 Radiological examinations per capita and per year, in Hiroshima and Nagasaki. Radiographic (A) and fluoroscopic (B) examinations (Sawada et al., 1971; Sawada et al., 1975; Takeshita et al., 1978).

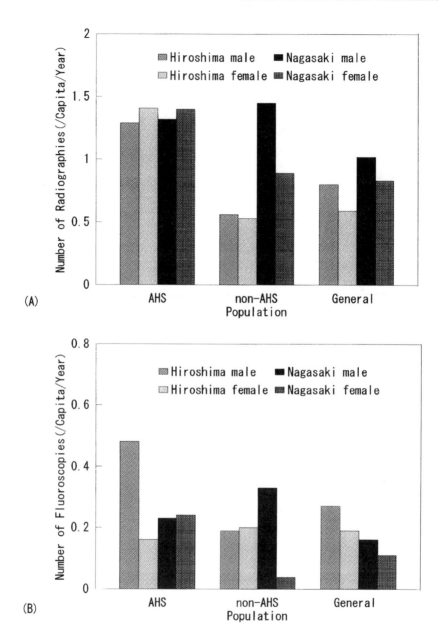

FIGURE 3.2 Average number of radiographic (A) and fluoroscopic (B), exposures by population and by sex in Hiroshima and Nagasaki (Sawada et al., 1979). The non-AHS population means the Life Span Study population excluding the Adult Health Study (AHS) population.

In the 1964-65 interviews, a total of 150 treatments by radiation were reported by Hiroshima AHS participants, where 102 (68%) were for ailments thought by the subjects to have been benign (Sawada et al., 1967). Of the remaining 48 (32%) of the exposures, only 15 (10%) were thought by the participants to have been for malignant disease. Only two exposures to radiation therapy were reported by Nagasaki AHS participants, both for unknown diseases.

The number of the AHS participants who reported receiving radiation therapy before 1979 was 525 in Hiroshima and 135 in Nagasaki (Pinkston et al., 1981). The first and second radiation therapy surveys confirmed treatment for 143 patients in Hiroshima and 47 in Nagasaki. The confirmation rate was 38% in both cities. The major reasons for this low confirmation rate were unavailability of hospital records, confusion of therapeutic exposures with diagnostic ones, and no hospital or clinic reported. This low confirmation rate was one of the reasons why the third hospital clinic survey for radiation therapy was performed between 1981 and 1984.

By means of the three radiation therapy surveys (Russell et al., 1976; Pinkston et al., 1981; Kato et al., 1996), a total of 1,670 members of the LSS population were confirmed to have received radiation therapy. Among them, 364 patients were neither in Hiroshima nor in Nagasaki at the times of the atomic bombs. The treatments were conducted not only for malignant neoplasms, but also for benign diseases. Among the malignant lesions, carcinoma of the cervix was most frequent, followed by carcinoma of the breast and lung carcinoma. Among the benign lesions treated by radiation were many skin diseases and tuberculosis.

Interviews of the Atomic Bomb Survivors

Ishimaru and Russell (1962) conducted a three-month survey, from April to June, 1961. This was the first survey of medical x-ray exposures of the atomic bomb survivors. It involved a medical questionnaire designed to determine the frequency, location, date, kind of medical x-ray, and part of the body exposed during the past year. The results of the questionnaires aided in determining the numbers of radiographic and photofluorographic units, their location by type of institution and by city in 1961. The total number of subjects who responded to this survey was 1,862: 1,303 in Hiroshima and 559 in Nagasaki. For the 1,303 Hiroshima participants included in this study, their ABCC/RERF medical records were compared with answers shown on the questionnaires. The reporting rate was 40%, where the reporting rate was defined as the percentage of reported and confirmed exposures to those documented in ABCC/RERF charts.

Sawada et al. (1967) analyzed the responses to interviews with 5,293 AHS participants in Hiroshima and with 2,221 in Nagasaki in 1964 and 1965. From their responses, the frequency of exposures was 0.34 among the Hiroshima males, 0.25 among Hiroshima females, 0.13 among the Nagasaki males, and 0.09 among Nagasaki females.

Sawada et al. (1971b) conducted two pilot studies on the responses of the AHS participants regarding radiological examinations. The first study was performed

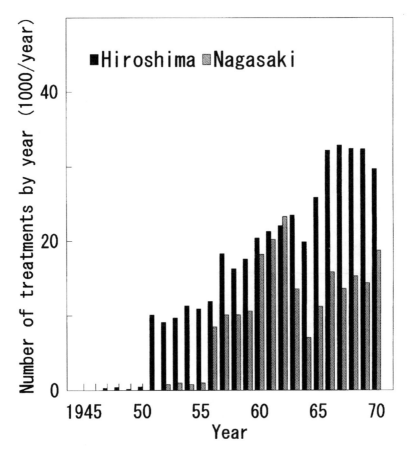

FIGURE 3.3 Radiation treatments by year in Hiroshima and Nagasaki (Sawada et al., 1971; Sawada et al., 1975).

in 1966, where a self-administered questionnaire was completed by 100 AHS participants. Eighteen participants either reported fluoroscopic examination at ABCC/RERF, or were recorded as having had at least one. Among them, 7 participants did not report any fluoroscopic examinations at ABCC/RERF. The reports by the remaining 11 participants were not confirmed by the records, or had some errors as to number, date, body site, or type of examination. This showed that an appropriate interview is necessary to obtain correct responses from AHS participants. The second pilot study was performed in 1966 and 1967. In this study, one primary specially trained member of the clerical staff and one alternate inter-viewed 454 AHS participants. Agreement of responses and records was markedly improved over the first pilot study. For example, 101 participants had records of fluoroscopy at ABCC/RERF, and 86 participants correctly reported that they had received fluoroscopy examinations at ABCC/RERF.

Yamamoto and Fujita (1977) analyzed the data from the medical history ques-tionnaires completed by the AHS participants from 1 April 1965 to 31 December 1967. For diagnostic x-ray, 50% of the total from both cities reportedly experienced exposure since their last ABCC/RERF examination, 20% within the three months prior to interview. The affirmative response rates for radiation therapy exposure were 2.6% in Hiroshima and 1.6% in Nagasaki; and for occupational exposure, 0.5% and 0.2%, respectively. Neither radiation therapy nor occupational exposure rates differed significantly by A-bomb dose or age.

The dates of GI examinations reported by 240 AHS participants in 1982 were compared with those of the medical records at the hospitals involved (Yamamoto et al., 1988). Positive responses were regarded as correct when they agreed completely with hospital records, and within three months before or after of the examination date. In this study, there was 81% agreement of responses.

From December 1967 in Hiroshima and January 1968 in Nagasaki, data on medical radiation exposures of AHS participants were continuously obtained by interviews. The routine interviews were conducted mainly by clerks who had special knowledge of radiology.

DOSIMETRY

Dosimetry of Diagnostic X-Ray Exposures

In 1963, Ihno and Russell (1963) estimated the doses to gonads and active bone marrow by diagnostic x-ray examinations. This work consisted of the application of experimental data by Epp et al. (1961, 1963) to the records of radiographic examinations at the ABCC/RERF. For example, the mean bone marrow dose for a chest projection using a tube voltage of 100 kVp, current multiplied by exposure time of 7.5 mAs, and source-to-film distance of 72 in was estimated to be 28 μGy,

In order to determine the active bone marrow doses from measurements of absorbed doses, Russell et al. (1966) measured the active bone marrow distribution in the human body. A phantom comprised of a human skeleton was used whose size and body weight were approximately 156 cm and 53.5 kg, which were those of the average Japanese adult in 1964. The results were used in the dosimetry work at RERF.

Bone marrow and gonadal doses from diagnostic x-ray exposure were deter-mined in a previous series of dosimetry studies (Antoku et al., 1965, 1971, 1972, 1973, 1980; Takeshita et al., 1972). The phantom used was a Mix-D phantom containing a human skeleton and beeswax impregnated cellulose to represent lung tissue. The absorbed doses were measured using Memorial polystyrene condenser ionization chambers made in the Department of Medical Physics of the Sloan-Kettering Institute, New York. Surface doses and x-ray outputs were measured using an Electronic Instruments Limited dose meter, model 37A, equipped with a 35-cc polystyrene ionization chamber. Several examples of the doses incurred from the diagnostic x-ray examinations at ABCC/RERF are shown in Table 3.1. The outputs of the x-ray equipment at ABCC/RERF were measured and the doses were determined for the actual examination conditions. The outputs of the diagnostic x-ray equipment at the institutions in Hiroshima are shown in Figure 3.4. As shown in this figure, the outputs varied. This means large errors might have resulted in estimating doses without correction for output. To minimize the discrepancies in radiation output and quality between the units in the hospitals and clinics and the experimental units, 218 radiography units and 49 photofluorography units in both cities were assessed for radiation output and quality. Together with the exposure factors obtained in the hospital and clinic surveys, the doses to patients from x-ray examinations conducted in Hiroshima and Nagasaki were determined. A part of the results is shown in Table 3.2.

Kato et al. (1991a, 1991b, 1991c) continued phantom-based dosimetry research to estimate doses to the salivary glands, thyroid gland, lung, breast, stomach, and colon incurred by radiological examinations. In these studies, an adult female human phantom made by Alderson Research Laboratories was used. Radiological examinations were reproduced and the organ doses were measured by Mg_2SiO_4(Tb) thermoluminescent detectors (Kasei Optonix, MSO-S).

Each detector was calibrated using an ionization chamber (Exradin A2 Shonka-Wycoff chamber). Experiments were performed using the radiological equipment at RERF and also those at several institutions in Hiroshima. Organ doses from typical radiography and fluoroscopy at RERF are shown in Table 3.3. The doses from upper gastrointestinal series were adjusted so that the maximum skin dose was the same as that determined in the prior study of dosimetry. The critical organ doses were determined for the other types of examinations. For example, doses from computed tomography (CT) were determined as shown in Table 3.4. In this text, active bone marrow dose is absorbed dose averaged over the whole active bone marrow in the human body. Active bone marrow exposed to the CT direct beam receives relatively high doses. For example, a rib dose of 5 cGy/examination was observed during a chest CT.

The first record of a CT examination of an AHS participant appeared in 1977, and the numbers of records rapidly increased. The AHS participants reported approximately 200 examinations in 1983, as shown in Figure 3.5. They resulted in large increases of doses to the AHS participants. Recently, computerized radiography such as digital subtraction angiography, which reduces dose to a patient, has become more popular. The dosimetry for these radiographs was also conducted in our study, and the results are useful for evaluating individual cumulative doses.

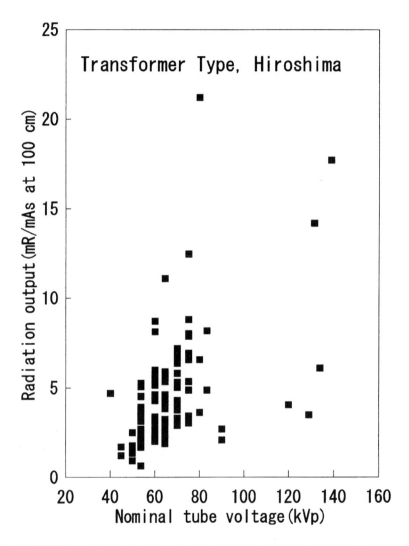

FIGURE 3.4 Correlation of tube voltage and output from transformer type of diagnostic x-ray equipment in Hiroshima (Antoku et al., 1972).

TABLE 3.1 Radiation doses from radiography and fluoroscopy at ABCC/RERF (Antoku et al., 1971).[a]

Projection / site	Film size (inch)	Tube voltage (kVp)	Current × exposure time (mAs)	Mean bone marrow dose (μGy/exp)	Gonadal dose (μGy/exp) Male	Female	Skin dose (μGy/exp)
PA chest	14 × 17	100	5	16.5	0.2	0.4	92.1
Lat. chest	14 × 17	110	15	31.5	0.3	0.8	370.0
AP abdomen	14 × 17	100	20	169.2	116.0	511.0	1,590.0
AP C-spine	8 × 10	90	20	27.4	0.2	0.2	981.0
AP T-spine	11 × 14	100	30	174.0	1.9	2.6	2,490.0
AP L-spine	11 × 14	100	30	245.7	147.0	704.0	2,210.0
Flx-A upper GI[b]	–	90		325.0/min	124.0/min	811.0/min	12,200.0/min
Flx-B upper GI[c]	–	90		68.9/min	3.5/min	128.0/min	3,280.0/min
Upper GI spot	8 × 10	90	PHT	12.9	0.7	26.8	1010.0

[a] All projections have added filtration of 2.5 mm Al, except for Flx-B upper GI and upper GI Spot, which are equal to 3.0 mm Al.
[b] Flx-A: Conventional fluoroscopy.
[c] Flx-B: Image-intensifier fluoroscopy.
Source-to-film distance for chest films is 72 in; abdomen, C-spine, T-spine and L-spine are 40 in.

TABLE 3.2 Mean dose from radiography and fluoroscopy at institutions in Hiroshima and Nagasaki (Antoku et al., 1972, 1980).

Site of examination	Mean bone marrow dose (μGy/exam)	Gonadal dose (μGy/exam) Male	Female	Surface dose (μGy/exam)
Chest	48	1.6	6.1	382
Abdomen	375	119.0	1,500.0	7,850
C-spine	289	<0.1	<0.1	18,900
T-spine	510	4.6	21.3	17,000
L-spine	1,281	259.0	2,340.0	53,500
Upper GI	1,000	66.0	2,000.0	55,000
Mass chest	390	2.9	16.2	6,460
Mass GI	2,310	210.0	1,400.0	57,300

Dosimetry of Radiation Therapy Exposures

Kato et al. (1994) proceeded to simulate radiation therapy, in which the Alderson Rando adult human female phantom mentioned above was exposed to various kinds of therapeutic radiation. This facilitated measurement of the doses to the salivary and thyroid glands, the lungs, breasts, stomach, colon, ovary, and active bone marrow. Each radiation therapy exposure was studied using its own characteristics and technical conditions. It was therefore necessary to perform various simulated treatments. The radiation sources used were ^{60}Co needles, 3000 Ci ^{60}Co gamma-ray therapy apparatus, and x-ray equipment for radiation therapy at the Research Institute for Nuclear Medicine and Biology, Hiroshima University, and the 4 MeV and 10 MeV linear electron accelerators at Hiroshima University Hospital. Doses were measured using thermoluminescence dosimeters which were used in the study of diagnostic x-ray dosimetry. Ionization chambers (JARP HU5 and Capintec PM-30) were also used for determining the sensitivities of thermoluminescence dosimeters. Radiation therapy for cervical cancer was the most frequent among the treatments found in the surveys. Most of the treatment doses to the cervix were larger than 50 Gy; hence, organs outside the exposure field were also exposed to intense scattered radiation. The method for teletherapy for cervical cancer is relatively uniform, and our results can be compared with those obtained by other studies. Preliminary results are nearly identical to those reported by Stovall (1983) except for the lower and middle energy range of x-ray exposure.

TABLE 3.3 Radiation doses from radiography and fluoroscopy at RERF, Hiroshima, in units of μGy/exposure (Kato et al., 1991b).

Projection/site	Surface	Salivary	Thyroid	Breast	Lung	Stomach	Colon
PA chest	140.0	9.4	12.0	20.0	66.0	23.0	2.5
Lat. chest	330.0	61.0	80.0	110.0	120.0	21.0	3.5
AP abdomen	2,940.0	8.8	13.0	55.0	56.0	800.0	960.0
AP C-spine	1,940.0	890.0	1,710.0	31.0	220.0	2.5	1.3
AP T-spine	3,160.0	91.0	170.0	2,030.0	710.0	630.0	87.0
AP L-spine	3,000.0	9.9	13.0	54.0	53.0	1,710.0	1,350.0
Flx upper GI*	3,280.0	8.3	12.0	340.0	180.0	2,340.0	770.0
Spot upper GI	1,010.0	2.5	3.6	110.0	56.0	720.0	240.0

*Flx denotes fluoroscopy, dose in units of μGy/min.

TABLE 3.4 Radiation doses during CT (unit: mGy/examination) (Kato et al., 1991c).

Tissue	CT examination				
	Head	Chest	Upper abdomen	Lower abdomen	Whole abdomen
Surface	42.00	48.00	45.00	37.00	47.00
Salivary	3.16	4.60	0.34	0.07	0.31
Thyroid	1.65	6.67	0.47	0.07	0.40
Breast	0.20	48.10	3.00	0.60	3.00
Lung	0.50	43.00	7.00	0.40	6.70
Stomach	ND	5.30	35.30	2.70	42.10
Colon	ND	0.80	5.40	17.00	38.10
Female gonads	ND	0.20	0.40	24.20	25.40
Active bone marrow	4.10	7.30	5.60	4.60	10.30

ND: Not detected.

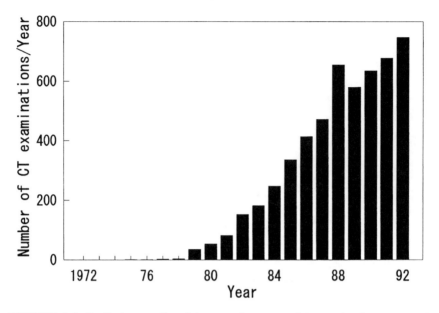

FIGURE 3.5 Preliminary tally of computed tomography examinations reported by the AHS participants. The more precise data by irradiation site, by city, and by year will be reported with organ doses in the future.

MEDICAL RADIATION DOSE AND ITS SIGNIFICANCE

Cumulative Dose from Diagnostic X-Ray Exposures

Kato et al. (1991b) calculated the cumulative doses by organ and by age for a female AHS participant, as shown in Figure 3.6. The integration was done only for the examinations at RERF until 1982. She received many upper gastrointestinal series, and her stomach doses were relatively high. Doses received by each AHS participant were incurred during examinations at RERF and at the other institutions. Figure 3.7 shows cumulative bone marrow doses received by 25 AHS participants. Cumulative doses indicated in this figure were calculated for examinations reported before the end of 1986. The participant who received the highest dose among these 25 participants received 18 fluoroscopic examinations. Her cumulative skin dose was 264.5 cGy. Figure 3.8 shows the distribution of cumulative bone marrow doses from diagnostic x-ray examinations that were conducted for the total 15,111 AHS participants before the end of 1982 (Yamamoto et al., 1988). The average bone marrow dose was 1.2 cGy for the A-bomb exposed group, and 0.89 cGy for the control group (those not in the cities at the time of bombing). At the same evaluations, the mean cumulative gonadal dose was 1.7 cGy for the female A-bomb exposed group, and 1.3 cGy for the female control group. The percentage of members whose bone marrow doses were larger than 1 cGy was 40.4% in Hiroshima and 32.8% in Nagasaki. The bone marrow doses from the atomic bombs calculated by Shimizu et al. (1989) are also shown in Figure 3.8. This figure suggests that numerous atomic bomb survivors have received medical radiation doses which are comparable with their atomic bomb radiation doses.

Radiation Therapy Doses in the Double Primary Cancer Cases

An analysis using the radiation therapy survey data and the data for cancer incidence before the end of 1983 showed that 134 patients developed malignant neoplasms after their radiation treatments. A precise analysis is in progress. In several cases, the mean therapy dose and the mean atomic bomb dose received in each target organ are compared. Average target organ doses from therapy were much larger than those from the atomic bombs.

Radiation therapy for cervical cancer was typically conducted using high radiation doses on the order 5,000 cGy and typical time-integrated activity of intracavitary therapy by Ra-needle of 3200 mghr. As suggested by Boice et al. (1985), such a large treatment dose might destroy the marrow in the pelvis. However, the doses outside the pelvis were also large. Previous investigations by Russell et al. (1976) and Pinkston et al. (1981) and a preliminary study of the second cancers show that 134 cases of second primary cancers included 9 leukemia cases. Among them, two acute myelocytic leukemias were diagnosed after radiation therapy for cancer of the cervix. A case of sarcoma 11 years after external beam radiation therapy for cervical cancer was also reported by Pinkston and Sekine (1982).

Dose Response to Chromosome Aberrations

It is well known that an increased frequency of radiation-induced chromosomal aberrations has been observed among atomic bomb survivors (Preston et al., 1988). It is also known that diagnostic x-ray exposures may be associated with chromosomal aberrations. The results of a study by Awa and Sawada (1990) seem to confirm this, as shown in Figure 3.9.

However, the frequency of stable cells with aberrations caused by diagnostic x-ray examinations seems to have been higher than that estimated from the linear dose-response model for the atomic bomb doses. Possible reasons for this are as follows:

- Partial active bone marrow is exposed to x-rays in radiography. For example, the maximum bone marrow dose from a CT examination is approximately ten times the bone marrow dose averaged over total bone marrow. This is in contrast to the bone marrow dose from atomic bomb radiation exposure, and the dose responses might be different from one another.

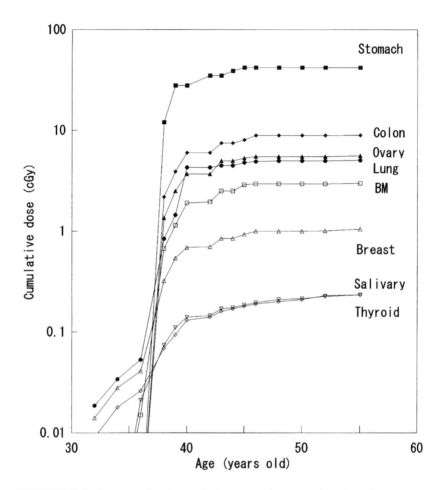

FIGURE 3.6 An example of cumulative organ doses received by a female Adult Health Study participant from diagnostic x-ray examinations at ABCC/RERF (Kato et al., 1991b).

69

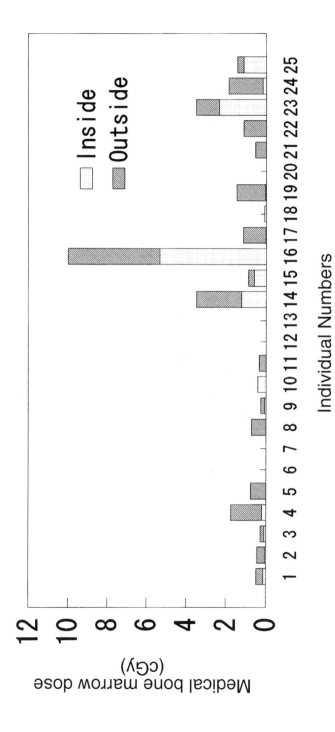

FIGURE 3.7 Samples of cumulative medical bone marrow doses received by the Adult Health Study participants.

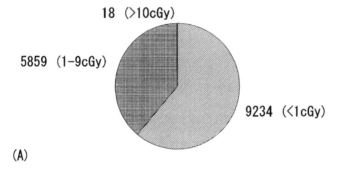

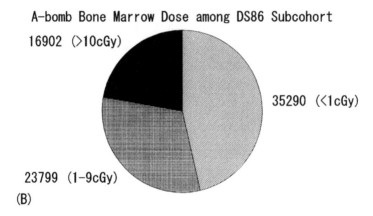

FIGURE 3.8 Distribution of cumulative individual bone marrow doses incurred by diagnostic x-ray examinations (A) and by the atomic bombs (B). The active bone marrow doses from diagnostic x-ray examinations were evaluated for 15,111 Adult Health Study participants until the end of 1982 (Yamamoto et al., 1988). The doses from the atomic bombs were evaluated for 76,000 members in the LSS population (Shimizu et al., 1989).

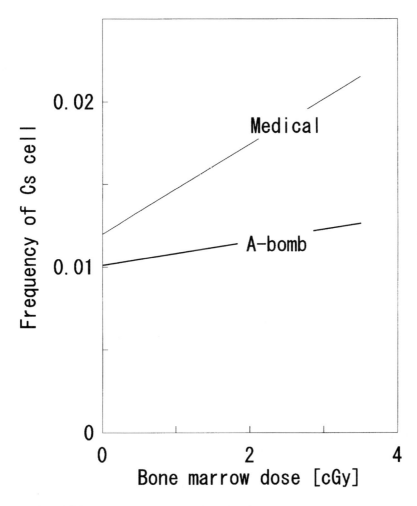

FIGURE 3.9 Dose-response relationships for stable chromosome aberrations. The line indicated by "Medical" was for bone marrow doses from diagnostic x-ray examinations in Hiroshima (Awa and Sawada, 1990). The line labelled "A-bomb" was for the bone marrow doses from the Hiroshima A-bomb (Preston et al., 1988).

- In the early stages of the present studies, the reporting rate by the AHS participants was 40%. The individual bone marrow doses might have been underestimated due to lack of information about examinations conducted in the early stages. The data from the hospital and clinic surveys during the early years will be useful in supplementing them.

- As shown in cell survival data by Spadinger and Palcic (1992), the biological effects of low-energy photons used in radiological examinations may be greater than those of high-energy photons such as the A-bomb gamma rays.

The relationship between radiation therapy and chromosome aberrations is also under study. The results will be reported in the future.

The present study of medical dosimetry has shown that the doses from the radiological examinations have increased the radiation dose levels of the atomic bomb survivors. The study of radiation therapy showed that many patients treated by radiation received large doses to critical organs near the treatment sites. These medical radiation exposures have been performed since 1945. Relatively long follow-up studies (50 years at maximum) have already become possible. Without the atomic bomb radiations, it may be sufficient to determine the effects of medical radiation exposure, as suggested by the study of the fluoroscopy to the thorax (Boice et al., 1978). Since the quality of medical radiation exposure is different from that of the atomic bomb exposures, it is difficult to assume that they both have the same effects in carcinogenesis. We therefore suspect that the A-bomb zero-dose members in the AHS population will play an important role in solving this equation. The size of the AHS population is relatively small; hence, we must carefully evaluate the subjects' medical radiation doses.

ACKNOWLEDGMENTS

The authors are grateful to Drs. Shigetoshi Antoku, Shozo Sawada, Osamu Yamamoto, John S. Laughlin, and Lowell Anderson for their helpful comments and suggestions. The assistance of Mrs. Grace Masumoto in preparing the manuscript is deeply appreciated.

4

Biodosimetry of Atomic Bomb Survivors by Karyotyping, Chromosome Painting, and Electron Spin Resonance

NORI NAKAMURA, DAVID J. PAWEL, YOSHIAKI KODAMA,
MIMAKO NAKANO, KAZUO OHTAKI,
CHYUZO MIYAZAWA, AND AKIO A. AWA

SUMMARY

Radiation dose estimation is one of the important tasks for radiation epidemiology. Frequency of chromosome aberrations in lymphocytes has long been recognized as a useful biomarker. Whereas unstable-type aberrations are clearly detectable, they disappear with time, and were not useful in dose estimation for atomic bomb survivors because more than 20 years had passed when the cytogenetic study was initiated. Consequently, stable-type aberrations, known to persist for decades but requiring greater skill to detect, were the only choice.

Recent results for over 2,300 survivors show a dose-related increase of chromosome aberration frequency with DS86 estimated dose, whereas variations among those with similar DS86 doses were common, as have been recognized. In addition, Hiroshima survivors showed a dose response about twice as steep as in Nagasaki survivors. The intercity difference greatly diminished when survivors exposed in Japanese houses were compared. The results suggest a systematic bias in dose estimation according to the shielding categories.

Chromosome painting (FISH) technique, currently the most objective way to detect reciprocal translocations, now confirms that the frequency of aberrant cells by the conventional method is 70 to 80% compared with the genomic translocation frequency deduced from FISH method.

Preliminary data of electron spin resonance (ESR) measurements on tooth enamel show a close correlation with donors' chromosome aberration data.

INTRODUCTION

Except for medical purposes, radiation exposure is mostly accidental, and hence precise reconstruction of the exposure dose is usually difficult to achieve. Estimation of the dose based on some biological materials from the exposed individuals is helpful in this regard to substantiate the physical dose estimation. Such a biodosimetric technique first became feasible with the discovery of mitogens, which stimulate blood lymphocytes to proliferate in culture and provide a means to examine a large number of metaphases with minimal discomfort to subjects. This was a fantastic breakthrough not only in biodosimetry but also in immunology, because blood lymphocytes had been considered to be terminally differentiated and incapable of undergoing further cell cycle progression.

At the Atomic Bomb Casualty Commission (ABCC)/Radiation Effects Research Foundation (RERF), cytogenetic examination has been conducted since the late 1960s. Unfortunately, however, more than 20 years had already passed since the bombings. Consequently, most of the easily detectable unstable-type aberrations, represented by dicentrics, rings, and acentric fragments, had disappeared, and stable-type aberrations such as translocations and inversions were the only choice of analysis.

The term "unstable-type aberration" means that cells undergoing mitosis with the aberration meet a problem, mechanically and genetically (Figure 4.1). A dicentric chromosome is defined as a chromosome carrying two centromeres instead of the usual one; one acentric fragment accompanies such a chromosome. During late metaphase, in 50% of the cases, each of the two centromeres of the dicentric chromosome are pulled toward opposite daughter nuclei, and nuclear division cannot be accomplished because of the chromosome bridge formation between the two nuclei, which later results in fusion of the two daughter nuclei to give rise to a tetraploid cell. In the other 50% of the cases, both centromeres segregate into one of the two daughter nuclei, which does not cause any mechanical problem for cell division unless they are tangled. However, the acentric fragment is distributed randomly between the daughter cells because of the lack of a centromere, thus giving rise to partial duplication of the genome in one daughter cell and partial deletion in the other. The latter causes various adverse effects on cell viability. At each mitosis of subsequent cell divisions, the dicentric chromosome continues to cause the mechanical problem in 50% of the cases as described above, even though they bear partial duplication, which is much less adverse than the counterpart, partial deletion.

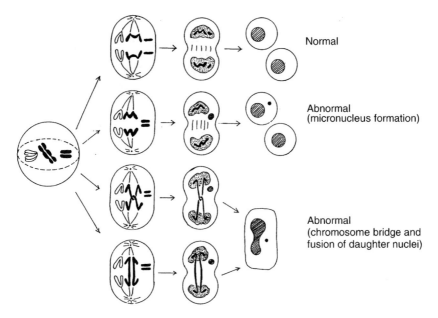

FIGURE 4.1 Schematic presentation of the fate of unstable-type aberrations.

The stable-type counterpart of the dicentric chromosome is reciprocal translocation. Such a chromosome bears a single centromere in each chromosome and hence meets no mechanical problem at mitosis. Further, there is usually no gross gain or loss of chromosome materials. Thus, such cells should have no growth disadvantage and persist in the population.

Detection of stable-type aberration requires precise knowledge of human karyotype. Because of the difficulty in standardizing the aberration detectability for each observer, it had not been recommended for biodosimetric purposes until the advent of the chromosome painting method that detects stable-type aberrations unequivocally.

The present report summarizes the cytogenetic studies conducted at ABCC/RERF for more than 20 years and the recent results from the more physical dosimetric measurements on tooth enamel.

CONVENTIONAL STAINING METHOD

General Information

A simple Giemsa staining method is commonly used for detecting unstable-type aberrations such as dicentrics and rings. Unstable-type aberrations can be identified by the abnormal structure of a chromosome bearing two centromeres in the case of a dicentric chromosome or by the lack of a centromere in the case of a fragment. In contrast, most of the chromosomes bearing a stable aberration (i.e., reciprocal translocation or inversion) appear normal by themselves because each chromosome carries a single centromere, and the abnormality shows up only when the 46 chromosomes are carefully ordered according to their length and arm ratio. Thus, detection of stable aberration requires precise knowledge of human karyotype. Although beginners require taking photographs of every metaphase for subsequent analysis, skilled people can exclude normal metaphases under the microscope and take photographs of only metaphases suspected to be abnormal. Then, the final decision can be made following the agreement of multiple observers. To avoid false negative scoring, observers are encouraged to take photographs of any suspected metaphases. The most difficult task is to standardize the aberration detectability among the observers. For this purpose, the same slides are periodically cross-checked by different observers.

Dose-Response Relationship

Results for 1,703 survivors were published by Stram et al. (1993) and Figure 4.2 shows the dose-response relationships. Two features are evident. First, the Hiroshima dose-response curve is close to linear, whereas the Nagasaki curve is more curvilinear. Second, the Hiroshima curve is about a factor of two higher at middle- to low-dose ranges. These two characteristics have been known since the 1970s (Awa, 1975), when the T65D was used to assign individual estimated doses.

Major differences between the previous T65D and current DS86 doses are twofold: the absorbed doses decreased to nearly one-half in DS86, and the neutron contribution was much reduced in Hiroshima, giving rise to only a small difference between Hiroshima and Nagasaki. (Neutron dose may be revised in the future, however.) Consequently, it is difficult to attribute the more linear dose response in Hiroshima to a larger neutron contribution.

Recently, we conducted a validation study of the dose-response relationship for a group of 2,500 survivors. The two characteristics mentioned above remain unchanged. However, a new insight was obtained after examining the average dose-response relationships according to shielding categories of each survivor. In Hiroshima, the dose responses are very similar among three groups, i.e., those who were in Japanese houses, tenement houses, and other buildings at the time of the bombings. In Nagasaki, however, the average dose response is distinctively steeper for the Japanese-house group compared with the other two groups. The

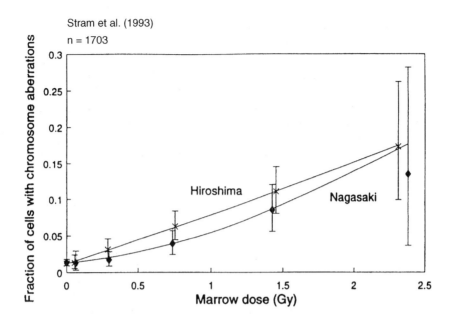

FIGURE 4.2 DS86 dose-response relationships for the frequency of lymphocytes bearing stable-type aberrations (Stram et al., 1993).

results suggest that the the dose for tenement and miscellaneous groups of the Nagasaki survivors was considerably overestimated.

Recent cancer mortality data (LSS Report 12) showed an increased city difference compared with the previous report. The increase comes rather equally from two sources: i.e., five more years of follow-up period, and extension to include a new group of survivors, mainly factory workers whose shielding conditions have been difficult to assess (Pierce et al., 1996). The cytogenetic data need to be confirmed by the new FISH technique and, if true, it would be necessary to find a way to incorporate the cytogenetic information in the epidemiologic data.

Clonal Aberrations

Clonal chromosome aberrations are defined as the same karyotypic change found in three or more metaphases in one blood sample. An identical aberration in two cells in one blood sample is not considered as clonal because a fraction of the cell population enters second mitosis in vitro in the currently used 48-hr culture protocol.

Such clonal cases have long been recognized (Awa, 1974). To date, 23 clonal cases are registered. Most of these survivors have assigned DS86 doses above 1 Gy. Such clonal expansion of lymphocytes can be derived either from a mature T cell in the periphery as a result of antigen stimulation, or from a bone marrow stem cell that proliferated extensively to produce an extremely large number of progenies. In the former case, the clonal aberration must be restricted to T lymphocytes. In addition, the patterns of T-cell-receptor gene rearrangement, which takes place during T-cell maturation in thymus and determines antigen specificity, are identical among the clonal derivatives. In the latter case, the aberration is expected to be present in different lineages of blood cells, i.e., T, B, erythroid and granulocytic lineages.

We have recently examined one survivor in detail to determine the origin of a clonal aberration (Kusunoki et al., 1995). The examinee was exposed to the bomb at the age of 20 years and is assigned a DS86 dose of 1.95 Gy. Results of repeated cytogenetic examinations showed that, in addition to nearly 30 independent aberrations among 100 cells, 3 to 8 cells bearing clonal double translocation [i.e., t(4;6),t(5;13)] have been consistently observed. Further, among 71 T-cell colonies established in vitro, 6 carried the double translocation. Similarly, among B-cell colonies established after Epstein-Barr virus transformation, 7 out of 58 carried the same double translocation. The rearrangement patterns of the T-cell-receptor gene and immunoglobulin gene were all different among the T-cell and B-cell colonies carrying the double translocation, respectively. Finally, stem cell cultures using methylcellulose medium for mainly BFU-E erythroid colonies also revealed that the same aberration was found in about 10% of the cells grown in mass culture. These results clearly demonstrate that the origin of this double translocation is in a multipotent stem cell in bone marrow, and a single stem cell in an adult could proliferate extensively to constitute as much as nearly 10% of total blood cells.

One might argue that the examinee was mosaic with regard to the aberration. Although the possibility cannot be totally excluded due to lack of pre-radiation-exposure information, it seems more likely that the two translocations were induced by radiation exposure. That most of the clonal cases are found among high-dose exposed people supports the notion.

Random Errors Associated With DS86

Calculation of DS86 dose uses parameters from interview records taken more than five years after the bombings. Thus, inaccurate memory regarding the distance from the hypocenter, or the type of structure of shielding results in random errors in estimating the dose.

Sposto et al. (1991) examined the chromosome aberration data in relation to the presence or absence of severe epilation, one of the acute radiation symptoms, defined as loss of more than two-thirds of scalp hair within 60 days after the irradiation. The authors found that the slope of dose response is roughly twice as steep in the severe epilation group as in the remainder, as shown in Figure 4.3.

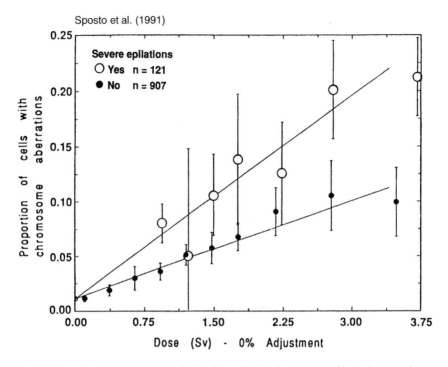

FIGURE 4.3 Dose-response relationships for the frequency of lymphocytes bearing stable-type aberrations among those who reported severe epilation (loss of two-thirds of scalp hair) and those who did not (Sposto et al., 1991).

Assuming that the random dose errors are the only source for the different slopes, they reached a conclusion that "dosimetry errors in the range of 45% to 50% of true dose are necessary to explain completely the difference in dose response between the two epilation groups."

Interestingly, the severe epilation group also showed 2.5 and 2.3 times higher risks for leukemia mortality (Neriishi et al., 1991) and incidence of cataract (i.e., lens opacity) (Neriishi et al., 1995), respectively. Although it is possible that the severe epilator group contains a larger fraction of radiosensitive individuals who show very similar elevated risks for induction of chromosome aberration, leukemia, and cataract, it seems more likely that the severe epilator group simply consists of individuals whose true doses are, on the average, nearly twice as high when compared with non-epilators at each DS86 dose level. As far as the in vitro dose-survival study using lymphocyte colony assay is concerned, no evidence for the presence of a large heterogeneity in individual radiosensitivity could be obtained (Nakamura et al., 1991, 1993). Recent electron spin resonance data on tooth enamel are also in favor of the latter hypothesis and we will come back to this issue later.

G-BANDING METHOD

General information

Precise localization of breakpoints involved in any chromosome aberration requires a banding method that gives chromosome-specific bar-code-type patterns. Among several techniques to produce the band patterns, the G-banding method is the most commonly used.

The banding method is very useful for detailed characterization of abnormalities, but its critical disadvantage is the extensive labor required for the analyses.

Comparison Between the Conventional and G-Banding Results

The conventional staining method has long been used for biodosimetric purposes with the A-bomb survivors. To determine the limitation of the assay—i.e., whether and how correctly the aberrations were detected—two studies have been performed (Ohtaki et al., 1982; Ohtaki, 1992).

In the first study, the same metaphases were examined by the two methods. Metaphases were photographed first after the conventional staining, the x-y location of each metaphase on the microscopic stage was recorded, and the slides were processed further for G-banding.

From 23 proximally exposed survivors, 896 cells were examined, and 342 had aberrations by the G-banding method (i.e., aberrant cell frequency was 38%). The conventional staining method, on the other hand, could detect 83% of the translocations (194 of 234), 59% of the inversions (32/54), 62% of the deletions (23/37) and 94% of complex aberrations (34/36), although the classifications were not always correct. The overall detection rate was 78%.

In the second study, G-band analyses were performed for 62 survivors of various doses who had been examined previously by conventional method. Linear regression analysis of conventional data (i.e., frequency of cells bearing stable aberration) plotted against G-banded data revealed a slope of 0.70, slightly smaller than the results from the first study. Thus, the conventional assay can detect 70 to 80% of the aberrations detected by the G-banding method. The results are similar to the observations by Buckton et al. (1978).

Clonal Aberrations

Previously, we presented results for clonal chromosome aberrations observed by the conventional method. Because the conventional method cannot definitively identify the chromosomes involved in an aberration in many instances, the same aberrations may be recorded as different, or different aberrations as the same. How-ever, with the G-band method, each translocation can be characterized in detail, and hence even one identical aberration each in two separate blood samples can be regarded as clonal derivatives. In other words, G-band analysis has a potential to detect clones of much smaller sizes than those detectable by the conventional method.

Our recent data after four repeated G-band tests of 16 proximally exposed sur-vivors did show numerous such small clones (173 different aberrations) among 11 survivors whose aberrant cell frequency exceeded 20% (Ohtaki, unpublished observation).

In addition to the cells bearing identical aberration(s), a new class of clonal derivatives was observed. These cells share one common aberration and some of these cells have an additional aberration. Two explanations, although not mutually exclusive, are possible. One is that genetic instability was induced by radiation exposure in a progenitor cell and, during subsequent cell division, a fraction of the progeny cell population acquired an additional aberration. Another explana-tion is that one chromosome-type aberration and one chromatid-type aberration were induced simultaneously in a progenitor cell at S phase. In this case, a bone marrow stem cell is the most likely target, because mature T cells in the periphery are mostly in G_0 phase, and chromatid-type aberrations are not usually induced in those cells. Subsequent cell divisions produced two populations sharing one aber-ration common to 100% and another common to 50% of the progenies. Detailed karyotype analyses are currently in progress to distinguish the two hypotheses.

These results suggest that the current lymphocyte pool of the survivors consists of clonal derivatives of various sizes. Determining the origin of these small clones, either from a bone marrow stem cell or from a mature T cell in the periphery, would provide valuable information in understanding lymphocyte kinetics in humans.

5q- Clones

Distribution of deletions among the chromosomes was recently summarized for 114 proximally exposed and 24 control groups. The frequencies of both appar-ent terminal deletions and interstitial deletions were about 10 times higher than the frequency of cells bearing exchange-type aberrations in the exposed group. Therefore, the deletions were most likely related to radiation exposure.

Because simple terminal deletions cause loss of the telomere structure of a chromosome, cells bearing such a deletion are unlikely to survive for nearly 50 years unless either telomerase activity was induced in those cells to add telomere structure to the broken ends, or they are in fact interstitial deletions, but one break occurred so close to the end of a chromosome that the remaining portion is too small to detect.

Whereas the apparent terminal deletions distributed widely to different chro-mosomes, nearly one-third of interstitial deletions (115 of 336) clustered on the long arm of chromosome 5 (i.e., 5q). Deletion of chromosome 5q has been well known among patients suffering from myelodysplastic syndrome (MDS) or acute myelogenous leukemia (AML), both in de novo and therapy-related cases (LeBeau et al., 1993). Interestingly, average deletion size is considerably smaller in the A-bomb survivors compared with that of MDS or AML patients, where it covers 5q14 to 5q33 in most cases (LeBeau et al., 1993). Further, scrutiny of deletions among cells from each survivor showed that the deletions tend to be identical for each survivor, suggesting that they are clonal derivatives, although the deletions distribute to wider regions, from 5q11 to 5q31, among different survivors. Thus, it seems likely that different 5q- cells were positively selected in vivo among different survivors to form clones. It is suggested that chromosome 5q bears multiple tumor-suppressor genes, which favors the hypothesis of positive selection. Hematologic data of six survivors, all of the high dose group, bearing 5q- cells at a frequency of 2% or more, did not show any consistent abnormalities, however. Thus, there is* no association between appearance of 5q- cells and any clinical indices. Careful follow-up of the frequency of 5q- cells for the six survivors will be performed to provide information on the nature of cells bearing 5q deletion.

FLUORESCENCE IN SITU HYBRIDIZATION (FISH) METHOD

General Information

The conventional method may detect as much as 70 to 80% of aberrations detected by G-banding, but extensive training of observers is indispensable, and only a few laboratories can accomplish it. On the other hand, the G-banding method is labor-intensive and is not suited for routine examinations. Flowcytometric sorting of human chromosomes and subsequent establishment of chromosome-specific DNA libraries opened a new era in overcoming these problems in detecting reciprocal translocations.

With the development of the FISH assay, reciprocal translocations can now be easily detected as a bicolored chromosome (usually in a pair) when several target chromosomes are "painted" yellow after in situ hybridization using fluorescein-tagged DNA probes, while the rest of the DNA is "painted" red by propidium iodide.

Obviously, any exchanges among the painted or unpainted chromosomes are not readily detectable. Thus, the observed exchange frequency needs to be corrected to a genome equivalent value.

For such a correction to be valid, any chromosome should have an equal chance in proportion to the length of undergoing exchange after irradiation. Breakpoint distribution of translocations in lymphocytes from A-bomb survivors by G-banding analysis showed that the frequency of breaks is linearly proportional to DNA content of each chromosome, except for chromosome 1, which was significantly more susceptible per unit DNA content (Lucas et al., 1992).

Currently, we paint chromosomes 1, 2, and 4, which cover 22% of the total genome and the observed translocation frequency is multiplied by a factor of 2.84

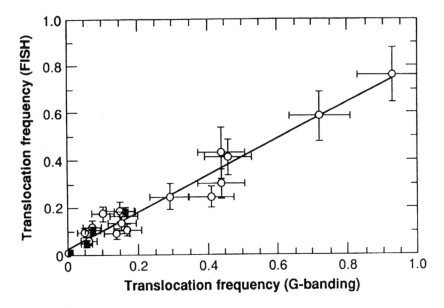

FIGURE 4.4 Correlation between FISH and G-banding data. The slope of the linear regression is 0.75, but is not significantly different from 1 (Lucas et al., 1992).

to estimate the genome equivalent value (for the details, see Lucas et al., 1992). In vitro irradiation experiments using FISH with probes for chromosomes 1, 2, and 4 showed that the deduced genomic translocation frequency agreed closely with dicentric data, an unstable-type counterpart of a two-break aberration (Lucas et al., 1992).

FISH Versus G-Banding or Conventional Method

Blood samples from 20 A-bomb survivors were examined concurrently by both the FISH and G-banding methods. The slope of the linear regression between the G-banding and FISH measurements was 0.75, although not significantly different from 1 as shown in Figure 4.4 (Lucas et al., 1992). Thus, FISH is as effective as G-banding in detecting translocations.

As for FISH versus conventional staining, we already noticed that the aberration frequency by the conventional method is 20 to 30% lower than that by G-banding. Thus, we would expect that the frequency of translocation is 20 to 30% lower in the conventional method than in FISH, assuming FISH and G-banding are equally effective. In fact, results for 97 survivors examined on different occasions by the two methods showed that the aberration frequency by the conventional method was about 80% on the average, compared with the genomic translocation frequency by FISH.

Detection Limit of a Translocated Segment by FISH

Reciprocal translocation between painted and unpainted chromosomes usually results in a pair of bicolored chromosomes, yellow (painted) chromosome with a red (unpainted) tip and red chromosome with a yellow tip. However, occasionally one of the two counterparts is undetected in a metaphase. Such an abnormality may be derived from incomplete rejoining of two breaks and subsequent loss of the acentric unrejoined fragment. Alternatively, one of the two breaks may have been so close to the end of a chromosome arm that the translocated segment was too small to detect. As far as A-bomb survivors are concerned, however, it is very unlikely that the cells bearing an unrejoined chromosome break survived several decades, because we no longer see any dose effect for the unstable-type aberrations. Therefore, the second hypothesis is more likely. Then, assuming that the translocation breakpoints distribute randomly in the genome, we could calculate the most likely minimum size of a translocated segment detectable by FISH. It turns out to be about 11 Mb for the painted chromosomes and 14 Mb for unpainted chromosomes (Kodama et al., 1997). These values are close to that estimated from G-banding data, and correspond to an average size of one band at a 400-band level.

ELECTRON SPIN RESONANCE METHOD

General Information

Electron spin resonance (ESR) is a phenomenon wherein radicals (i.e., atoms or molecules bearing an unpaired electron) absorb microwaves at a magnetic field strength characteristic of the radical species. Whereas radicals disappear very quickly in aqueous solution, they remain for years in solid materials. Calcified materials such as teeth, bones, or cave deposits are used for archeological dating by ESR method. Compared with the thermoluminescence method, ESR is nondestructive and hence repeated measurements are possible (Ikeya, 1993).

Enamel covers the crown of a tooth and can be thought of as an inorganic component of the human body. The major constituent of enamel is hydroxyapatite $[Ca_{10}(PO_4)_6(OH)_2]$, which forms a crystalline structure. About 2.5% (w/w) of the enamel consists of CO_3^{2-}, which is believed to be incorporated in place of PO_4^{2-} to terminate crystal elongation during dental enamel development. Radiation exposure to dental enamel produces carbon radicals. Originally, the carbon radicals were thought to be CO_3^{3-}, but recent work now suggests that the radicals are CO_2^{-}, which are detected by ESR. Among human tissues, tooth enamel is by far the most sensitive for biodosimetric purposes compared with tooth dentin or bones.

We have been collecting teeth extracted for medical reasons from Hiroshima A-bomb survivors participating in the RERF Adult Health Study, and preliminary results are presented here. The ESR signal on tooth enamel consists of two components, a broad background signal most likely derived from organic materials in enamel, and a sharp radiation-related signal. The background signal saturates at low microwave strength, whereas the radiation-related signal continues to increase along with the microwave power. Thus, in order to isolate the radiation-related signal from the background signal, we used two different microwave powers, 16 mW and 0.4 mW, for subtraction analysis. It should be mentioned that even at the lower microwave power, the radiation-related signal still remains, and hence the current subtraction does not isolate 100% of the radiation-related signal. Nonetheless, we believe that this is the most objective and easiest method for estimating radiation effects. Preliminary results for 11 teeth have been described elsewhere (Nakamura et al., 1994).

Results for 100 Teeth

One hundred enamel samples were measured at 16 mW and 0.4 mW for the subtraction. Among these, one failed to provide the results because of its unusually wavy pattern at 16 mW. The subtracted signal intensity was corrected first by the signal intensity of manganese located in the cavity (internal control), followed by the total amount of enamel examined to estimate the radiation signal per milligram of enamel.

The results for the relationship between ESR signal intensity thus obtained and chromosome aberration frequency of blood lymphocytes from the tooth donor showed a good correlation, with a correlation coefficient of R=0.87. Different teeth from the same donors gave reasonably close results among themselves. The frequency of chromosome aberration is usually determined after examination of 100 metaphases by the conventional Giemsa staining method. Thus, the frequency includes sampling errors. Among the 70 donors, we did not receive blood from 9 survivors, so the results are for 61 survivors.

One exception was found, however, for a man who was exposed at 1.1 km from the hypocenter and who carries stable chromosome aberrations as high as 35% of his lymphocytes, but whose tooth enamel failed to show any sign of radiation exposure. He was exposed at 15 years of age and reported various acute radiation symptoms. It turned out that the sample was a wisdom tooth that is known to develop much later than other permanent teeth, and we tentatively concluded that the enamel was most likely underdeveloped at the time of bombing.

The results of chromosome aberration frequency plotted against the physically estimated DS86 dose gave a correlation coefficient of R=0.73. As has been known for many other survivors, a good correlation predominates, but some outliers can be seen and will be discussed below.

In the present study, we did not attempt to calculate the dose from the ESR signal intensity. For this calculation, we need to irradiate the enamel with graded doses of gamma rays. If the radiosensitivity of tooth enamel in producing the ESR signal varies both among teeth and among individuals, such an in vitro irradiation procedure is critical in estimating individual dose. However, a recent study by Iwasaki et al. (1995) demonstrated that the variation is rather small. They showed that after 5 Gy of irradiation of teeth from cadavers, the coefficient of variation for the mean of the ESR signal was less than 4% for 10 repeated measurements of 4 samples, and 5.3% to 11.4% (average 6.1%) for a single measurement of 7 to 26 teeth from each donor. (Iwasaki et al. did encounter one exceptional tooth out of 72 examined that was hyposensitive for the ESR signal production, however.)

Issues Related to Cytogenetic Outliers and Severe Epilation

In the dose-response analysis of chromosome aberration frequency, two survivors were found to be outliers whose aberration frequencies were close to the background level although their estimated DS86 doses were 1.6 and 3.3 Gy, respectively. The interview records showed that they were exposed outdoors without apparent shielding. Importantly, their ESR data for tooth enamel failed to suggest any high-dose exposures, in agreement with the aberration data. Therefore, for these two survivors, it seems most likely that their memory regarding location at the time of the bombing was inaccurate. We feel it reasonable to say that lymphocyte aberration frequency and tooth enamel measurements are both manifestations of physical exposure, and are more accurate reflections of true dose than individual memory of precise location at the time of the bombings.

As mentioned earlier, the slope of the chromosome aberration dose response is about twice as steep in the severe epilation group when compared with non-epilators. Are the survivors who reported severe epilations more radiosensitive? Present ESR versus chromosome aberration data were reanalyzed according to the presence or absence of severe epilation. It was found that severe epilators were found only among middle- to high-dose exposed people (i.e., from the standpoint of ESR signal intensity), and there was no sign of their hyper-radiosensitivity for induction of chromosome aberration. Because tooth enamel ESR is a physical measurement, it is very unlikely that cellular radiosensitivity is coupled with radiosensitivity for radical induction in tooth enamel. Consequently, it could be concluded that the severe epilators' issue is merely a reflection of biased dose estimation and not of individual radiosensitivity.

FUTURE PROSPECTS

Usefulness of the Conventional Method

Recently developed FISH is excellent in detecting stable aberrations, but the necessary laboratory reagents and equipment such as fluorescence microscopes are expensive. In contrast, the conventional staining method is not perfect in detecting stable aberrations but is less expensive. Further, it has the potential to detect as much as 70 to 80% of the aberrations detectable by FISH or G-banding, provided that the observers are well trained in karyotyping. Concurrent use of FISH would be helpful in estimating the aberration detectability by the conventional method.

City Difference

Current results for the city difference between Hiroshima and Nagasaki require caution to interpret. Because most of the blood samples were processed and examined in the Hiroshima and Nagasaki laboratories separately, the difference might be a result of laboratory artifact. We plan to examine survivors from both cities in the Hiroshima laboratory using FISH to resolve the issue.

Shielding Categories

As for the difference in the slope of dose-response relationships among different shielding categories in Nagasaki, the possibility of laboratory artifact is very unlikely. Nonetheless, it would be worth examining Nagasaki survivors systematically by FISH. If the difference were confirmed, we would need to incorporate the information into epidemiologic studies.

ESR Versus Chromosome Aberrations

We plan to compare ESR results with chromosome aberration frequency of lymphocytes for 100 Hiroshima survivors (donors) to provide a definite answer to the feasibility of ESR dose estimation. We need to find a way to exclude contribution of dental x-ray exposure.

Clonal Aberrations

For several decades after the irradiation, lymphocytes or the progenitor cells bearing certain chromosome aberrations might have undergone positive selection. Currently, our knowledge of the origin of such clonal derivatives is very limited. Theoretically, clonal expansion may occur in both progenitor cells in bone marrow and mature lymphocytes in the periphery. Characterizing such clonal cells would provide unique information regarding not only long-term lymphocyte kinetics in humans, but also possible bias in biodosimetry.

Cytogenetic Epidemiology

Because FISH is suitable for automated image analysis, several laboratories are developing the programs. If such a system may become available in the future, we wish to conduct a large-scale FISH study on all the AHS participants exposed within 2.5 km from the hypocenter so that a chromosome-aberration-based epidemiologic survey may be performed. Until the introduction of an automated system, the current FISH study will continue by examining about 200 survivors a year manually.

Biodosimetry of the Exposed Parent(s) for Genetic Study

Currently, effort has been made to establish permanent B-cell lines from both parents and children as one of the tasks of the molecular genetic study. Five hundred families each for the exposed (one or both parents are exposed) and control (neither of the parents are exposed) groups are to be included. Because DS86 is biased by error to some degree, evaluation of genetic risk requires accuracy of the dose. In this regard, it is of interest to test chromosome aberration frequency in the exposed parent(s).

ACKNOWLEDGMENTS

The authors are grateful to Dr. Keisuke S. Iwamoto for his careful reading of the manuscript, and to Ms. Mayami Utaka for manuscript preparation.

PART 2
CANCER STATISTICS
AND EPIDEMIOLOGY

5

Statistical Aspects of RERF Cancer Epidemiology

DONALD A. PIERCE

SUMMARY

Here I will trace the major steps in the evolution of statistical methods used for the RERF cancer data since about 1960, and then indicate the general nature of our current descriptions of the excess risks for solid cancer as a class. From about 1960 to 1975 "contingency table" methods were the primary basis for analyses. These methods were remarkably modern and sophisticated for their time, and served the needs well. However, these provided mainly for significance tests for the *existence* of radiation effects, and did not extend well to the developing needs for *estimation* of these effects. During about 1975 to 1980, rather different methods were employed for estimation of radiation effects, largely in the course of work for the BEIR III report. These methods were well suited for estimation of the risk per unit dose, for making inferences about the shape of the dose-response curve, and for making simple comparisons such as between sexes or broad age-at-exposure groups. However, the methods did not extend well to analysis of temporal patterns in the excess risk: effects of time since exposure and age at risk. Beginning in about 1980 major advances were made in adapting for RERF needs revolutionary recent developments in analysis of survival data—methods that came to be called "relative risk regression." These were highly suitable for investigation of temporal patterns of excess *relative* risk: age-specific excess risks relative to natural background risks. Since about 1992 it has become increasingly apparent that the focus on relative risks was excessive, and that it is at least equally important to describe temporal patterns of *absolute* excess risk, and their relation to age at exposure and sex. Today there is much more importance attached to this description, and I suggest that it will become increasingly valuable in understanding the patterns in the data. In conclusion, some comparison is given of the actual nature of relative risk and absolute risk descriptions for the data on solid cancers as a class.

INTRODUCTION

Reviewing the evolution of the statistical methods used may be helpful in understanding current analyses of the RERF data on excess cancer risk.

In the next section I will present my view of this evolution of methods, beginning with an overview. The statistical methods used in each of the time periods discussed—1960–1975, 1975–1980, 1980–1992, and 1992–present—were statistically modern for their times and in each period were well suited to the needs at that time in view of what could be learned from the existing data. The evolution of the methods from one era to the next reflected both new possibilities for fuller understanding of the cancer risks brought about by the increasing accumulation of data, as well as major progress in relevant statistical methods and computer software.

In the final section I will give a brief overview of recent analyses of the data, taken largely from Life Span Study Report 12 (Pierce et al., 1996). It is my hope that these recent results will clarify some of my meanings in the discussion of the evolution of the statistical methods, and that placing these current results in that perspective will make them clearer.

EVOLUTION OF STATISTICAL METHODS

From roughly 1960 to 1975 the primary methods were those called "contingency table methods." See, for example, Beebe et al. (1978). These statistical methods, developed by Beebe, Jablon, and Charles Land, were thoroughly modern and sophisticated for their time. However, as attention turned from *testing* for a dose effect to quantitative *estimation* of its extent, these methods became inadequate. From about 1975 to 1980 estimation methods were developed and used by Land and Beebe, largely for the NAS BEIR III report (NRC, 1980). However, as the follow-up lengthened, these methods in turn became inadequate due to lack of systematic consideration of age-time patterns of the background and excess cancer risks. At about that time there were remarkable advances in biostatistics relevant to this need, and in the early 1980s further methods were developed at RERF along lines of what is now called *relative risk regression*. Again RERF was in step with the most modern approaches. Preston developed remarkable computer software for implementation of this (Preston et al., 1993), which is not only the basis now for all cancer analyses at RERF, but is widely used around the world for such work. Since about 1985 this relative risk regression has been used at RERF to obtain far more effective and detailed inferences from the cancer data, especially in regard to *temporal patterns* of the excess risk. See, for example, Preston and Pierce (1988), Shimizu et al. (1990), and many other RERF reports of that era.

Finally, in recent years there has been another major development—the realization that clearer understanding of the excess cancer risks may come from less emphasis on excess *relative* risks, and more on the excess *absolute* risks. This is discussed in the most recent of the periodic reports on these data: Life Span Study Report 12 (Pierce et al., 1996). I will now outline all these developments in a little more detail.

The basic idea of the contingency table approach was to divide the cohort into strata defined by city, sex, and age-at-exposure categories. Within each of these strata the "expected numbers" of cancer deaths in dose categories were obtained by allocating the total observed number according to the dose-specific person years (PYR) at risk. This corresponds to the expected numbers of cancer deaths in dose categories, under the hypothesis that the radiation exposure had no effect. Then, still within each stratum, one could consider by dose category the differences $(O_d - E_d)$ between observed and expected numbers of cancer deaths. If there is a dose effect, these should display an increasing trend with dose, which can be summarized by the trend statistic: $\sum d(O_d - E_d)$. The essence of the contingency table approach is to then sum these stratum-specific trend statistics over all the strata. The theoretical details of this are statistically interesting, and it is a very good way to test for a dose effect. The problem is that it does not extend well to estimation of the effect once it is clear that there is one. The primary reason for this limitation is that the expected numbers E_d as defined above are not what would be expected *in the absence* of radiation exposure, but rather if the exposure *had no effect*. Consequently, the $(O_d - E_d)$ do not really indicate the excess cancer due to radiation.

After some valiant but unsuccessful attempts to work around this limitation within the contingency table approach, a quite different approach for estimation was taken, largely for analyses for the NAS BEIR III report. This was in principle quite simple, based on fitting by weighted least-squares regression-type models of the form

$$[Cases/PYR]_d = \alpha + \beta d + \epsilon \qquad (5.1)$$

where the left-hand side represents rates within dose categories, α represents the intercept term, β denotes the contribution of doses, d, is the excess risk, and ϵ is the error term. This was done both for the entire cohort, and within strata of sex and age at exposure. To my knowledge no consideration was given to following the contingency table approach of estimating parameters within such strata, and then combining these results over strata. Rather, when within-stratum estimates were made, this was more for the purpose of *comparing* them than *combining* them. Models quadratic rather than linear in dose were also used, especially for leukemia. There were two limitations to this approach, one of which was just mentioned regarding strata. The other, also present in the contingency table approach but becoming more important with the lengthening follow-up, has to do with age-time variations in risk. There are two issues involved in this second limitation:

1. the rapid increase of background cancer rates with age (or follow-up time for this fixed cohort), making it statistically important to estimate the excess risk fundamentally by comparing dose groups among persons of about the same age;

2. the specific need to investigate the age-time patterns in the excess cancer risk.

I should say that the extent to which (1) is a serious issue in a cohort study where everyone is aging together involves delicate statistical issues which cannot be addressed here, and in particular whether one is estimating the absolute or relative excess risks. However, regardless of the importance of point (1), the method just described did not lend itself well to study of the temporal patterns of excess risk, which was becoming a more important issue as the follow-up lengthened.

As mentioned earlier, there was in the late 1970s a revolutionary advance in biostatistics bearing directly on these considerations. By 1980 RERF statisticians were adapting the new methods for application to the Life Span Study, and what follows is closer to their adaptation than to the original advance. The fundamental idea can be seen by extending the regression approach indicated above to a stratification based on both dose and age. By "age" here is meant attained age, rather than age at exposure. Thus, stratification on this variable involves stratifying the cohort *experience* rather than its subjects. Then we may consider extended regression models such as

$$[Cases/PYR]_{ad} = B(a)[1 + \beta d] + \epsilon. \tag{5.2}$$

The remarkable thing is that the parameter β can be estimated without considering specifics of the age-specific background rate function $B(a)$. Essentially this is because an estimate of β for each age stratum can be obtained from ratios of the left-hand side at various doses but for that fixed age, and then these estimates can be averaged over the age categories, since β does not depend on age. This is in fact an extension of the stratification principles of the contingency table methods, an extension to stratification on age at risk. A primary reason why the development of relative risk regression was considered so elegant and important by the statistical community is that, as with most important ideas, it extended some well-established principles and formalized methods which had already seen some use by the most insightful workers.

Now what is estimated from the above model is the *excess relative risk* (ERR), rather than the *excess absolute risk* (EAR) as in the previous model. That this can be done so elegantly and simply was, and is a primary reason for the subsequent emphasis placed on relative risks, not only at RERF but in all of epidemiology. Extending this approach to models quadratic in dose presents no difficulties. More importantly, extending it to allow the parameter beta to depend on sex and age at exposure does not interfere with the basic idea, since with a suitable mathematical model or further stratification, inferences may still be based on ratios within fixed age groups, thereby "eliminating" the function $B(a)$.

In fact, although with some loss of elegance and ideal statistical properties, it is possible to allow the parameter β to depend to some extent on age as well. Thus what is actually used at RERF are regression models of the form

$$[Cases/PYR]_{sad} = B(s,a)[1 + ERR(s,a,d)] + \epsilon, \qquad (5.3)$$

where s denotes strata related to city, sex, and age at exposure. $\mathrm{ERR}(s, a, d)$ refers to a mathematical model for the ERR which may vary with strata, age, and dose, with the dose dependence often but not always taken as linear. The function $B(a)$ cannot be entirely eliminated when the ERR is modeled as depending on age, but as long as the age effects are modestly formulated this does not cause serious difficulties. Moreover, for solid cancers the models ordinarily used do not require allowing the ERR to vary with age, and the formulation with ERR depending on this serves mainly for testing the age-constancy of the ERR. Various parametric models are used for the excess relative risk function $\mathrm{ERR}(s, a, d)$. It has become routine in this way to study the effects of "modifying factors" in relation to dose, namely city, sex, age at exposure, and attained age. These are very powerful statistical methods—perhaps too powerful, in that they increase the chance of overinterpreting the data. Asking the "right questions" becomes the critical issue when analytical methods rather outrun the extent of actual information in the data.

This brings us to developments of the most recent era, about 1992 to present. It has become increasingly clear that there are serious limitations in the focus on the ERR as opposed to the excess *absolute* risk (EAR). One that has been realized for quite some time pertains to the solid cancer ERR for women being about twice that for men, which is potentially misleading. That is, it should be realized that women have about half the background cancer rate as men, and indeed the EAR is about the same for the sexes. It is clearly wrong, even though common in epidemiology in general, to simply think of the ERR as "the risk." In a more subtle and complex way, this issue arises in interpreting the effect of age at exposure. Generally speaking, those exposed as children have shown a much higher ERR over the follow-up than those exposed as adults. This is commonly interpreted as meaning that children are "more sensitive" to radiation in regard to induction of cancer. But this large ERR for children is actually the result of dividing a quite small EAR by the even smaller background cancer risk for young ages. However, as noted, this issue is far more subtle than that involved in the sex-specific ERRs. Describing the EAR involves fitting regression models of the form

$$[Cases/PYR]_{sad} = B(s,a) + EAR(s,a,d) + \epsilon, \qquad (5.4)$$

which can be done, but with a substantially different statistical approach that involves formulating parametric models for the background rate $B(s, a)$. A most important point to understand is that the issues do not involve whether one can fit the data better with Equation 5.4 than with Equation 5.3. When the specific models used for ERR(s, a, d) and EAR(s, a, d) are sufficiently rich and well-selected, the fits to the data will be about the same, and in fact the resulting product $B(s, a)$ ERR(s, a, d) will be similar to the function EAR(s, a, d). What really differs in the choice of models Equation 5.3 or Equation 5.4 is *what is being described*, rather than how well the data are being fitted.

To say that either approach is better would mean that the description it provides is more useful. An important issue in evaluating this is that there are at least two rather different uses of the cancer data: (1) in making radiation protection decisions, and (2) in providing a more fundamental understanding of the nature of radiation-induced cancer. Of course, the latter understanding is important in radiation protection decisions as well, but it is useful to separate a sort of empirical assessment of radiation risks from actually understanding them. In this sense some issues in (1) involve projection of risks for this cohort beyond current follow-up, estimating lifetime risks for specific ages at exposure and sex, transporting these risk estimates to other cultures where background cancer risks are different, and so forth.

Simplicity is always an important aspect of a model's usefulness. Description of the ERR has a simplicity which is particularly important for use (1) above. Aside from those exposed as children, we can quite adequately describe the ERR for solid cancers as a constant value for all remaining lifetime (after some minimal latent period), depending on age at exposure and sex. This simplicity is very convenient, both for subsequent calculations and simply for summarization purposes. However, we believe that such a description may be very poor indeed for understanding what is really going on.

The dominant feature of an EAR description for solid cancer is a very strong increase with age. There is some question whether this increase is really due to age or to time since exposure, and trying to make such a distinction is a large part of what is challenging about the nature of the data. For the moment, let us say that the increase is best thought of in terms of age. The other important feature of the EAR description is that there is relatively little dependence on sex or age at exposure. But even when these two factors are ignored, there is no doubt that for descriptive purposes the EAR model is rather less convenient than the age-constant ERR description. That is, to say for example that for solid cancers the ERR per Sv following exposure is about 38% for males exposed at age 30 is far simpler than to describe how the EAR increases with age.

SOME RECENT RESULTS

I will finish with a brief presentation and discussion of recent analyses of the excess risk for solid cancer mortality in the RERF cohort, given in more detail in Life Span Study Report 12 (Pierce et al., 1996). There are certainly problems involved in considering together all solid cancers, but as the data are much weaker for specific cancer sites, much of what can be learned about age-time patterns of the excess risk

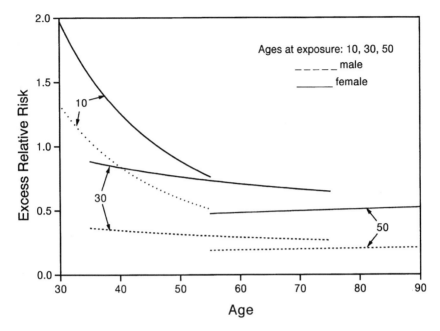

FIGURE 5.1   Solid cancers: excess relative risk per Sv.

must at least begin with such oversimplification. The excess risk for solid cancers is very linear in dose up to about 3 Sv organ dose, and so describing risk per unit dose is adequate.

Figure 5.1 portrays the ERR for solid cancers as a function of age. These risks could be plotted against time since exposure, but the use of age has the advantage of emphasizing that the follow-up for different ages at exposure has been for different stages of lifetime. There are 3 pairs of curves, one each for ages at exposure of 10, 30, and 50 years, and the distinction within pairs being by sex. As noted earlier, the ERR for women is about twice that for men, but this should probably be thought of only as reflecting that the women have about half the background cancer mortality. For ages at exposure 30 and 50, the ERR has been very constant in age and time throughout the follow-up to date, although at somewhat different levels. For those exposed as children the ERR has decreased sharply over the follow-up period. It should be understood, however, that the evidence pertaining to this decrease is rather weaker than Figure 5.1 might imply. The decrease is only marginally statistically significant ($p = .06$ for a two-sided test). Certainly, it is too soon to draw firm enough conclusions about the nature of the decrease to be able to project what the future may hold.

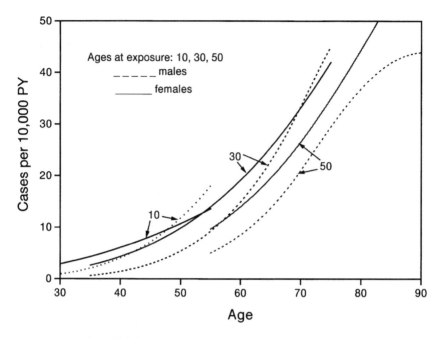

FIGURE 5.2 Solid cancers: excess absolute risk per Sv.

Figure 5.2 portrays the EAR in the same manner. There is at best very weak evidence that the different curves are estimating any differential true effects by sex or age at exposure. The important aspects of this description are that the EAR increases with age for all ages at exposure and both sexes, and that one would not be very seriously misled to consider all the curves in the figure to be estimating a single curve that depends only on age. If one estimated such a single curve for the EAR and then computed from it the corresponding ERR curves, the result would be roughly a pair of decreasing curves for each sex, decreasing in age but with no (or small) age-at-exposure effect. Such a pattern can indeed be seen, at least roughly, in Figure 5.1. This corresponds to an interesting hypothesis raised by Kellerer and Barclay (1992), in a paper that is important for its highly sensible discussion of issues in interpreting the RERF data. Their suggestion is that what is usually considered an age-at-exposure-dependent ERR effect, constant in age (excluding children), could also be considered an ERR decreasing with attained age, with no age-at-exposure effect.

It will almost certainly develop in time, whether it is an issue now or not, that there are statistically significant distinctions between the curves shown in Figure 5.2. However, it may become more important as the study progresses not to let such statistically significant differences interfere inordinately with understanding the data. Science is not primarily the analysis of complexity, but rather trying to understand phenomena through intentionally simplified hypotheses and descriptions of observational data. This is most important in very large studies involving complicated phenomena, where there will always be statistically significant departures from simple models due simply to the size of the study. Recognition of this seems likely to become increasingly important regarding the RERF data.

ACKNOWLEDGMENTS

I would like to thank Gilbert Beebe, Seymour Jablon, Charles Land, and Jack Schull for introducing me to this work, and for their contributions leading to the state of what has been discussed here. I also want to thank Dale Preston and Michael Vaeth for invaluable collaborations with me in this area. Permission to use Figure 5.1 and 5.2 here has been granted by Radiation Research. The Radiation Effects Research Foundation (formerly ABCC) was established in 1975 as a private nonprofit Japanese foundation, supported equally by the government of Japan through the Ministry of Health, and the government of the United States through the National Academy of Sciences, under contract with the Department of Energy.

6

A Historical Review of Leukemia Risks in Atomic Bomb Survivors

DALE L. PRESTON

SUMMARY

An excess risk of leukemia was one of the earliest delayed effects of radiation exposure seen in the victims of the atomic bombs dropped on Hiroshima and Nagasaki in August of 1945. Now, more than 50 years after the bombs were dropped, this excess is widely seen as the most apparent long-term effect of radiation. Indeed, death from leukemia is, for many people, a primary symbol of the terrible price paid by the victims of the bombings. Leukemia risks have been a major concern of Atomic Bomb Casualty Commission (ABCC) and Radiation Effects Research Foundation (RERF) researchers. With more than 50 citations in the topical index of the ABCC/RERF Technical Report series (RERF 1992), leukemia is considered in more ABCC/RERF reports than any other single subject. In this paper, I review the history of efforts by scientists of the ABCC and RERF to identify and quantify the excess risks of leukemia in the atomic bomb survivors. The report begins with a discussion of efforts to identify leukemia cases. I then discuss how the description and understanding of the leukemia risks in the survivors has evolved over time. I will conclude with a summary of our current understanding of the nature of these risks. This summary is based upon results in recent reports on leukemia incidence (Preston et al., 1994) and mortality (Pierce et al., 1996). In his recent book Jack Schull (1995) provides additional information on leukemia studies at ABCC and RERF.

ASCERTAINMENT OF LEUKEMIA CASES

As expected based on the limited knowledge available in 1945, medical observers who examined the victims in the immediate aftermath of the bombs noted severe acute effects on the hematopoietic system of heavily exposed survivors. At that time it was not clear whether or not survivors would experience an increased risk of leukemia, but there was concern that this might be the case. The documents dealing with the establishment of ABCC in 1946 and 1947 recognize, among other things, the need to look for cases of leukemia among the survivors. The first major non-genetic study conducted at ABCC was a clinical survey of the hematological status of survivors and controls carried out in 1947 and 1948. The survey included about 925 exposed persons, primarily children, who reported epilation and a similar number of unexposed controls sampled from the nearby city of Kure. In the report on this survey (Snell et al., 1959) no prevalent cases of leukemia were found among those examined.

However, by the late 1940s Japanese physicians—principally Yamawaki Takuso, a pediatrician at the Hiroshima Red Cross Hospital, and Tomonaga Masanobu, a hematologist at Nagasaki University—were becoming concerned about an apparent excess of leukemia cases among the survivors. They felt that evidence for an elevated risk was especially strong for those who were within 1,000 m of the hypocenter at the time of the bomb. Dr. Yamawaki reported his observations to Wayne Borges. These initial observations led to the commencement of a series of ABCC leukemia surveys. The first of these surveys was carried out under the direction of Drs. Borges and Yamawaki, together with Jarrett Folley, Yamasowa Yamamichi, and William Valentine in Hiroshima and, in Nagasaki by Dr. James Yamazaki with the cooperation of Dr. Tomonaga and others at Nagasaki University. The first ABCC report dealing with the elevated leukemia risks is a memorandum written by Valentine (1951). In addition to the ABCC efforts to record leukemia cases, Dr. Watanabe Susumu of Hiroshima University undertook an independent effort.

The ABCC leukemia surveys continued with varying levels of intensity through-out the 1950s. Published reports (Folley et al., 1952; Moloney and Lange, 1954; and Brill et al., 1962) suggest that major efforts to update the leukemia data were made in 1951, 1953, and 1957. These surveys attempted to ascertain all cases of leukemia and related disorders (including lymphoma, aplastic anemia, and myeloma) among survivors resident in the areas around Hiroshima and Nagasaki. Cases were identified on the basis of diagnoses at or referrals to ABCC, records from local hospitals, examination of Hiroshima and Nagasaki death certificates, autopsy reports, and even, in the early years, newspaper reports of survivor deaths. Until the 1960s there appear to have been no consistent standards for case ascertainment or review. In addition, the surveys were conducted and, in some cases, the reports were presented separately for Hiroshima and Nagasaki. There was no precise definition of the population of interest or even of whom to consider as a survivor. (These problems were common to virtually all of the epidemiological studies undertaken by ABCC during its first decade.) The poorly defined population from which cases ascertained in the ABCC Leukemia Surveys were drawn is often referred to as the "open-city" population.

From about 1960, Stuart Finch led an effort to improve the quality of the leukemia survey data by standardizing data collection and review procedures. This lead to the formal establishment of the ABCC Leukemia Registry (Finch, 1965a, 1965b; Belsky, 1972). The Leukemia Registry operations manual prescribed that cases be collected for all "survivors of the atomic bombs with Honseki [place of legal family registration] in Hiroshima and Nagasaki regardless of their present place of actual residence" and for all others seeking care for these diseases in either of the cities. An important aspect of the procedures adopted for the Leukemia Registry was the requirement that detailed shielding histories be obtained and doses assigned for all cases among survivors who were within 2,000 m of the bomb. Data collected for the ABCC leukemia surveys were reviewed and incorporated into the Leukemia Registry. In addition, the Hiroshima data collected by Dr. Susumu of Hiroshima University were also reviewed and added to the registry (Finch, personal communication). The Leukemia Registry was quite active from its inception through the late 1980s.

Between 1985 and 1990 a special effort was undertaken to classify cases in the Leukemia Registry using the French-American-British (FAB) system for the detailed characterization of acute leukemia subtypes, and modern classifications for other leukemia subtypes (Matsuo, 1988). At the same time, efforts were made to identify all cases of adult T-cell (ATL) leukemia in the registry. This latter effort was thought to be important because of the fairly recent discovery that the virus (HTLV-1) associated with ATL was relatively common in the Nagasaki area. It was possible to classify about 64% of the cases in the registry using modern criteria.

Since 1958, the Hiroshima and Nagasaki Tumor Registries (Mabuchi et al., 1994) have been collecting data on the incidence of leukemia, lymphoma, myeloma, and other cancers. For various reasons, until recently the Tumor Registry data (especially in Hiroshima) were not considered adequate for analyses of cancer incidence in the survivors. However, over the past decade, the efforts of Dr. Mabuchi and his colleagues have led to improvements in the quality and completeness of the Tumor Registry data (Mabuchi, 1997). At this time, the Tumor Registries are the primary source of data on new cases of leukemia and related diseases among the A-bomb survivors. While the recent incidence analysis (Preston et al., 1994) used data from both the Leukemia and Tumor Registries, there has been no formal merging of the data from the two sources.

ANALYSES OF LEUKEMIA RISKS IN THE A-BOMB SURVIVORS

As noted above, the first ABCC report on leukemia was the 1951 memo by Valentine (Valentine, 1951). This memo summarized the findings of the first ABCC

TABLE 6.1 Leukemia cases in Atomic Bomb Survivors 1948–50 (based on data from Folley et al., 1952).

Distance from hypocenter (m)	Population (Est.)	Observed cases	Expected cases	Excess cases
0-999	2,071	4	0.1	3.9
1,000-1,499	13,823	12	0.6	11.4
1,500-1,999	23,363	6	1.0	5.0
2,000 +	155,970	7	7.0	0.0
Total	195,227	29	8.7	20.3

leukemia survey covering the years from 1948 through 1950. In 1952, Folley et al. published a report on the data presented in the Valentine memo. Table 6.1 shows the distribution of the 29 leukemia cases (19 from Hiroshima, including 4 cases among the 925 people examined during the 1947 hematological survey) considered in this paper, by distance categories. The table also includes estimates of the number of expected and excess (observed minus expected) cases computed by application of the crude rate for survivors in the 2,000 m+ category to survivors in the other distance groups. Despite uncertainties about the completeness of case ascertainment or the accuracy of the population size estimates, evidence for an excess risk of leukemia associated with the atomic bomb exposure is compelling.

Folley et al. (1952) note that the cases were primarily acute leukemias or chronic myelogenous leukemias, and that acute cases seemed most common in the proximal group. While Valentine had concluded that there was some suggestion that the youngest survivors might be at greater risk than those exposed later in life, Folley et al. were careful not to emphasize this possibility.

In 1953 William Moloney, a hematologist, became chief of medicine and organized a series of reports (Lange et al., 1954; Moloney, 1955; Moloney and Lange, 1954; Moloney and Kastenbaum, 1955) describing leukemia incidence through 1952. Based on 75 cases (52 in Hiroshima) that had been found among the survivors, it was concluded that there was clearly an excess risk of leukemia and that this excess appeared to decrease in a manner that was roughly linear in the logarithm of distance. As in the earlier reports, it was noted that the cases were predominately acute types and chronic myelogenous leukemia, while chronic lymphocytic leukemia was extremely rare in the survivors regardless of distance. The authors indicated that age at exposure and sex had no apparent effects on the risks. In a summary of the results that appeared in the New England Journal of

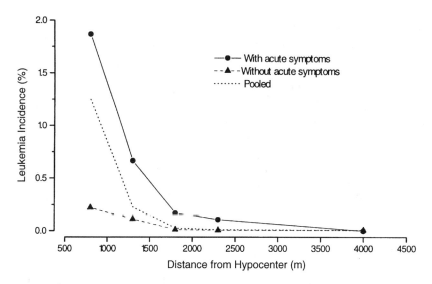

FIGURE 6.1   Leukemia incidence in A-bomb survivors (1948–55) based on Moloney and Kastenbaum (1955).

Medicine, Moloney (1955) emphasized that "whereas a large single dose of ioniz-ing radiation was required for leukemogenesis in survivors, repeated smaller doses under other circumstances may have a greater leukemogenic effect." Moloney and Kastenbaum (1955) carried out an analysis of leukemia incidence by distance for Hiroshima survivors with and without acute radiation symptoms (epilation, oropharyngeal lesions, or purpura). Figure 6.1 summarizes these data (a similar figure is given by Schull, 1995).

Table 6.2 presents the estimated numbers of survivors with and without acute symptoms, as reported by Moloney and Kastenbaum. The population size esti-mates in this table are considerably lower, especially for the most proximal and distal groups, than in the initial reports (Table 6.1). These changes reflect the re-sults of ongoing surveys of the survivor populations and the uncertainties inherent in any effort to carry out analyses of the ill-defined "open-city" population from which cases were being drawn.

Rising public concern over the possible health effects of atmospheric nuclear testing led to a paper in *Science* (Lewis, 1957) that used leukemia data from ABCC and data on the dose-versus-distance estimates from the York air-dose curves, as presented by Neel and Schull (1956), to arrive at an estimate of 2 excess leukemia cases per million person-years. These were the first published quantitative cancer risk estimates based on the ABCC data. In response to the *Science* article, Wald (1958) published a short updated summary of the Hiroshima leukemia data, stress-ing the point that uncertainties in dose estimates meant that efforts to quantify the

TABLE 6.2 Estimated numbers of survivors and survivors' acute symptoms by distance from hypocenter (from Moloney and Kastenbaum, 1955).

Distance from hypocenter (m)	Population (Est.)	Acute Symptoms	%
0 - 999	1,200	750	63
1,000-1,499	10,500	2,250	21
1,500-1,999	18,700	1,750	9
2,000 +	50,500	850	2
Total	98,100	6,550	7

leukemia risks were premature. A review of hematological findings, including the leukemia results, during the first decade after the bombs was published by Wald et al. (1958).

The third ABCC leukemia survey covered the period through 1958. The results were described in a series of papers (Heyssel et al., 1959; Tomonaga et al., 1959; Brill et al., 1962) that appeared in the first year of the ABCC Technical Report series. By this time a total of 209 leukemia cases had been reported in the survivors. However, for their primary summary, the investigators decided to limit attention to those cases that had occurred among members of the newly created Life Span Study (LSS) cohort. (In the papers mentioned above, the LSS is called the Master Sample, a term that is currently used to refer to the larger population from which the LSS was selected.) Limiting the data to cases among LSS cohort members reduced the number of usable cases to 95. These reports also presented some of the first dose-response curves to appear in ABCC publications. Figure 6.2 is a reproduction of the Hiroshima dose-response curve from Heyssel et al. (1959). It was found that the dose response was consistent with linearity in dose.

The reports on the ABCC leukemia survey noted that excess risks were higher among those exposed under the age of 10, and that men had greater excess risks than women. Observed risks were also observed to have been increased from 18 months to two years after exposure. Although risks were still elevated, the excess seemed to have peaked at some time between 5 to 10 years after exposure. Although the papers presented data on the dose response, it was noted that the risks in Nagasaki were higher than those in Hiroshima, suggesting that there were some problems with the York-curve-based T57D dose estimates. Brill et al. (1962) noted that by the late 1950s more leukemia cases had already been seen among proximal survivors than would be expected in their lifetimes and argued that this could serve to refute the notion that the excess could be explained in terms of accelerated aging

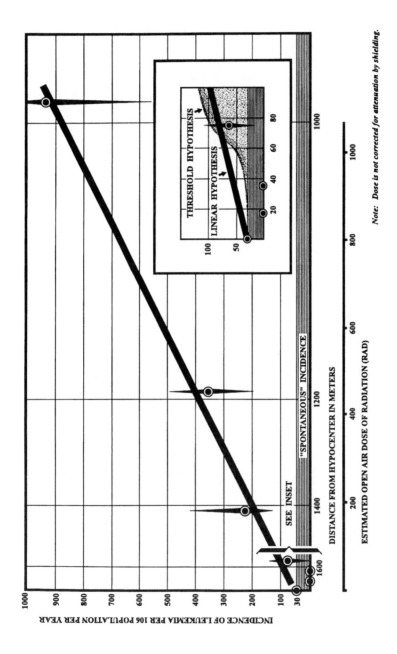

FIGURE 6.2 Hiroshima leukemia dose response based on 1946–1957 leukemia cases and T65D dosimetry (from Heyssel et al., 1959).

(i.e., that the excess cases were simply cases that would have been seen later in life, but exposure had simply hastened their development).

The next comprehensive report on the leukemia data (Ishimaru et al., 1971) considered leukemia cases in the LSS for the 1950 to 1966 period. This was one of the first major ABCC reports to make use of the T65D dosimetry. At this time there were 279 cases in the Leukemia Registry, but only 117 of these people were members of the LSS. It was reported that while the excess risk had peaked in the early 1950s, risks among the survivors were elevated throughout the study period. The results appeared to support earlier observations that the magnitude of the excess depends on age at exposure and sex. Excess risks were higher in Hiroshima than in Nagasaki, and the authors suggest that this difference might be due to neutrons. The deficit of CML cases in Nagasaki was also noted, along with a suggestion that a neutron effect was the "most tenable explanation" for this difference. LSS follow-up was extended to 1971 in a report by Ichimaru et al. (1978). This report presents an extensive analysis of age-and-time patterns in the excess risk and makes the point that excess risks appeared to have peaked earliest and at the highest levels for those exposed early in life, while for those who were older at the time of exposure the peak was less sudden, and the excess seems to have continued over a longer period of time.

A series of reports on leukemia incidence was published in 1981. These reports considered leukemia incidence in the LSS for the period from 1950 through 1978 (Ishimaru et al., 1981a; Ichimaru et al., 1981) and in the "open-city" population for the period from 1946 through 1978 (Ishimaru et al., 1981b). By this time the number of cases among LSS cohort members had increased to 181 (149 in Hiroshima), while there had been 509 cases (297 in Hiroshima) in the "open-city" population. The conclusions reached in these analyses were similar to those of the earlier reports. However, it was emphasized that the dose response for acute leukemias appeared to be non-linear, with a large limiting value for the RBE. The report stated that no excess had been seen in Nagasaki since 1970. It also noted that chronic leukemia cases were twice as likely to be seen in Hiroshima (47 of 149 cases) as in Nagasaki (6 of 40 cases). The nature of the dependence of the excess leukemia risks on age at exposure, time, and city were illustrated by Ichimaru et al. (1981) using the schematic diagram shown in Figure 6.3.

The LSS leukemia incidence data of Ichimaru et al. (1981) were also analyzed by Pierce et al. (1983). They used the data to illustrate how new regression-based methods could be used to make inferences about excess absolute or relative risks in the LSS. The methods described and developed in this paper have become the standard methods for virtually all subsequent analyses of mortality and incidence in the LSS and are widely used by researchers concerned with risk estimation in other populations.

Recently Tomonaga et al. (1991) presented an analysis of the data on the "open-city" population for the period from 1946 through 1980. A total of 766 leukemia cases were used in these analyses. Tomonaga and his colleagues discussed the

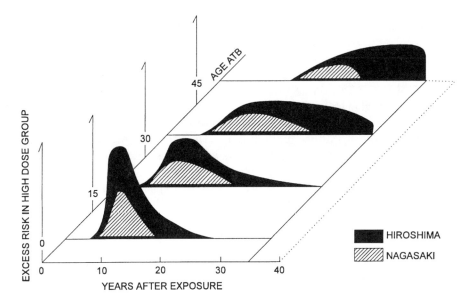

FIGURE 6.3 Summary of the pattern of LSS excess leukemia risks from Ichimaru et al. (1981).

results of the FAB reclassification and used complex statistical methods based on ratios of relative risks to deal with problems caused by the inadequate information about the size of the population covered by the Leukemia Registry. Although the FAB system provides detailed subclassifications for acute myeloid leukemias (AML) and acute lymphocytic leukemias (ALL), the primary analyses were based on the broader classifications of ALL, AML, and chronic myeloid leukemia (CML), supplemented with a group of other leukemias that included adult T-cell leukemia (ATL). The results of this analysis are generally consistent with earlier findings. However, it was suggested that the CML dose response might be relatively more linear at low doses than that for AML.

CURRENT RESULTS

Analyses of the LSS leukemia incidence for the period from 1950 to 1987 were published in 1994 (Preston et al., 1994) as one of a series of reports on cancer incidence in the LSS. In this section I will summarize the results of these analyses. For these analyses, descriptions of the excess risks were given primarily in terms of the excess absolute risk as a function of dose, time since exposure, age at exposure and sex. Preston et al. (1994) present detailed models for the excess absolute risk for all leukemias as a group and for ALL, AML, CML, ATL, and other leukemias. The analyses focused on leukemia cases among survivors in the LSS cohort who had DS86 shielded kerma estimates less than 4 Gy and who were living in Hiroshima, Nagasaki, or nearby areas at the time of diagnosis. A total of 231 leukemia cases met these criteria. This total includes 32 ALLs, 103 AMLs, 57 CMLs, 23 ATLs, and 16 leukemias of other types. All except one of the ATL cases occurred in Nagasaki survivors.

Table 6.3 presents the number of leukemia cases observed in bone marrow dose categories, by city, for the full follow-up. The table also gives estimates of the number of excess cases for each period in each dose group. These estimates were computed as the difference between the observed and the expected numbers of cases. The expected numbers of cases were computed using the background (zero dose) model for leukemias of all types given by Preston et al. (1994, p. S94). Overall, about one-third of the 231 cases considered here are associated with radiation exposure. There were six additional leukemia cases among the 263 survivors who had DS86 kerma estimates over 4 Gy. It seems likely that all of these cases could be attributed to radiation exposure. There were also 24 leukemia cases among the 7,190 LSS survivors who had an unknown dose. Since most of these survivors were located within 2,000 m of the hypocenter ATB and are likely to have received fairly large doses, it is probable that 50% or more of the leukemia cases can be attributed to the exposure. This would suggest that there were about 100 extra leukemia deaths in the LSS over the 37 years of follow-up covered by this study.

Table 6.4 provides a breakdown of the observed and estimated excess leukemia cases by dose and decade of follow-up. The results indicate that there were appreciable excess risks in all decades except possibly the 1980s. Even in the most recent periods there is a suggestion of a small excess in the highest dose groups.

Figure 6.4 summarizes the temporal pattern of the fitted excess rates by sex for three age-at-exposure groups. This plot is based on the excess absolute risk model in Preston et al. (1994). The pattern of risks is not simple. Generally, risks tended to be largest in the early years of follow-up. Within each age-at-exposure group the peak was higher and subsequent rate of decrease greater for men than for women. The fitted model suggests an increase with time in the EAR for women exposed at age 50. This apparent increase should not be overinterpreted. While the point estimate of the slope is positive, the estimate has a large standard error. Confidence levels for this slope at the 90% level or even lower levels include 0 or even negative values.

Table 6.5 presents observed numbers of cases for five subgroups of leukemia cases in the very low dose (< 0.01 Sv) group and the exposed group, together with the estimated number of excess cases for each type. The excess cases were computed as the difference between the observed cases and the fitted type-specific background models given in Preston et al. (1994). Because these results were based on separate models, the total number of excess cases differed slightly from that in

TABLE 6.3 Observed and excess LSS leukemia cases by dose category and city, 1950–87.

Marrow Dose (Sv)	Hiroshima			Nagasaki			Total		
	Subjects	Obs.	Excess	Subjects	Obs.	Excess	Subjects	Obs.	Excess
< 0.01	26,711	59	8.2	18,481	31	1.3	45,192	90	9.4
0.01- 0.1	17,727	26	-7.9	5,589	12	3.5	23,316	38	-4.4
0.1 - 0.2	4,946	7	-2.6	940	1	-0.5	5,886	8	-3.1
0.2 - 0.5	5,127	26	16.3	1,265	1	-1.1	6,392	27	15.2
0.5 - 1	2,309	20	15.7	998	4	2.4	3,307	24	18.1
1 - 2	1,157	20	17.8	637	7	5.9	1,794	27	23.7
2+	339	15	14.5	105	2	1.9	444	17	16.3
Total	58,316	173	61.8	28,015	58	13.4	86,331	231	75.2

TABLE 6.4 Observed and excess LSS leukemia cases by dose category and time period, 1950–87.

Marrow Dose (Sv)	1950–60		1961–70		1971–80		1981–87	
	Obs.	Excess	Obs.	Excess	Obs.	Excess	Obs.	Excess
<0.01	21	5.3	18	-1.5	30	5.4	21	0.2
0.01- 0.1	5	-3.3	8	-2.2	15	2.0	10	-0.9
0.1 - 0.2	4	1.8	1	-1.7	3	-0.4	0	-2.8
0.2 - 0.5	10	7.7	9	6.1	6	2.4	2	-1.0
0.5 - 1	14	12.9	5	3.5	4	2.2	1	-0.5
1 - 2	13	12.4	9	8.2	3	2.0	2	1.2
2+	11	10.9	1	0.8	3	2.8	2	1.8
Total	78	47.7	51	13.2	64	16.3	38	-2.0

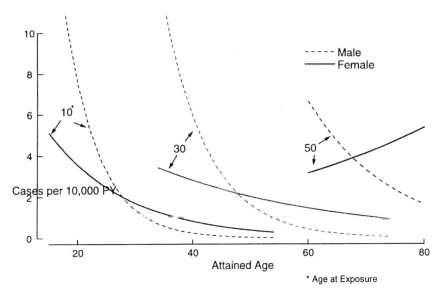

FIGURE 6.4 Temporal patterns of excess leukemia at (ERR_{1Sv}) by sex and age at exposure (Preston et al., 1994).

the earlier tables. The only group for which there is no evidence of an excess risk is the ATL leukemia group. For the other specific types and the "Other" category, radiation appears to account from 50 to 75% of all of the cases among those with significant exposures.

Based on a joint analysis of the ALL, AML, and CML data, Preston et al. (1994) concluded that while the CML dose response seems more linear than that for the other types, the evidence for this is weak, and it is not possible to reject a the hypothesis of a non-linear dose response with the same shape for these three groups. However, they also found that there are significant subtype differences in the sex, age at exposure, and time-dependence of the excess risks.

CONCLUSIONS

It has been more than 45 years since it became clear that the Hiroshima and Nagasaki survivors had an increased risk of developing leukemia. During this time the scientists and staff at ABCC and RERF, with the cooperation and support of the survivors and the local medical and medical research communities, have carried out extensive investigations of the nature of this risk. While excess leukemia risks have been seen in other radiation-exposed populations, the ABCC/RERF data provide the most detailed description and serve as the primary source of quantitative

TABLE 6.5 Observed and excess LSS leukemia by exposure category and subtype, 1950–87.

	Observed cases		
Subtype	<0.01 Sv	0.01-4 Sv	Excess
ALL	9	23	17
AML	43	60	30
CML	17	40	26
ATL	14	9	−2
Other	7	9	6

estimates of these risks. In what follows, I will offer some brief comments and opinions on several issues of ongoing concern regarding leukemia in the A-bomb survivors.

While there is clear evidence of a radiation dose response for leukemia, there are a number of questions about the nature of these risks at low doses. The first of these concerns whether or not there is a threshold exposure below which no excess leukemia risk is seen. With the possible exception of CML (for which the dose response may be somewhat more linear than for other types), the leukemia dose response in the survivors is non-linear, with low doses having a smaller effect than would be predicted by linear extrapolation from high doses. However the LSS data do not offer convincing evidence of a threshold. Indeed these data can rule out thresholds of 0.5 Gy or more, and threshold models do not offer substantial improvements in fit over a simple linear-quadratic model. A related question concerns whether or not low doses, e.g., below 0.1 Sv, have a protective effect. The models developed by Preston et al. (1994) included no explicit constraints for non-negative low dose risks but the dose-response parameter estimates were all positive. Thus, while the LSS data cannot rule out this possibility, they also offer no real evidence of the existence of such an effect. An important question regarding radiation effects on leukemia that cannot be answered by the survivor studies concerns the effects of protracted, low-dose rate exposures. Recent combined analyses of data on nuclear workers in several countries (Cardis et al., 1995a) suggest that there may be elevated risks associated with such exposures, as do preliminary analyses of workers (Koshurnikova et al., 1994) and the general population (Kossenko and Degteva, 1994) with chronic radiation exposures resulting from plutonium production at the Mayak facility in the former Soviet Union.

Another question that arises in most discussion of the LSS leukemia data concerns the nature of differences in the dose response for Hiroshima and Nagasaki. There are important differences in background leukemia rates in the two cities. CML background rates in Nagasaki are about one third of those in Hiroshima, and ATL accounts for about 40% of the Nagasaki cases but almost none of the Hiroshima cases. In addition there are a number of unresolved questions about the accuracy of the Hiroshima neutron dose estimates (Straume et al., 1992). However, despite these issues, current analyses fail to demonstrate significant differences in the excess risks for the two cities for all leukemias as a group. There is however, a significant difference in absolute risks for CML considered by itself, which is proportional to the large difference in CML background rates.

The primary question that arises with respect to temporal patterns is whether or not the excess risk has disappeared. Since about 1970 it has often been stated that the excess is no longer apparent, especially in Nagasaki. The data presented in Table 6.4 suggest that excess risks were apparent at least until the end of the 1970s and, at least for heavily exposed survivors, into the 1980s. The LSS data also suggest that for those exposed as adults, risks have continued throughout their lifetimes. While it is certainly true that risks have fallen dramatically with time, especially for those exposed as children, continued follow-up of this population remains important for our understanding of the complex nature of the leukemia risks. As for the question of whether or not Nagasaki risks have disappeared, as noted above there is no convincing evidence that the temporal patterns or the level of the excess risks (except possibly for CML) in Nagasaki differ from those in Hiroshima.

As data on the experience of the survivors accrue, increasing attention is being focused on radiation effects on cancer incidence and on non-cancer effects. However, it will also be important to continue to monitor and analyze leukemia incidence in the LSS, as there is undoubtedly more that can be learned from this population about this important long-term effect of radiation exposure.

7

Tumor Registries and Cancer Incidence Studies

KIYOHIKO MABUCHI

SUMMARY

It is now widely recognized that the induction of solid cancer, in addition to that of leukemia, is the most important late health consequence in the atomic bomb survivors and other populations exposed to ionizing radiation. The current risk estimates for radiological protection are in fact dictated by the pattern and size of radiation-induced solid cancer risk. Until recently, the assessment of cancer risk carried out at ABCC/RERF had primarily relied on death-certificate-based mortality data. Cancer incidence data provide information needed to assess a broader spectrum of the effects of radiation exposure. However, unlike routinely collected mortality data, the acquisition of cancer incidence data requires a systematic, large-scale effort to ascertain and register all cancer cases occurring in a defined population on an ongoing basis. Recognizing the importance of incidence data, ABCC in the early years collaborated with the local medical community in establishing tumor registries in Hiroshima and Nagasaki (Ishida et al., 1961). As a result, the Hiroshima Tumor Registry was inaugurated in May 1957, and in April of the following year the Nagasaki Tumor Registry was started. These were the first population-based registries established in Japan.

THE EARLY YEARS

Although the excess number of leukemia cases in the atomic bomb survivors became clear several years after the exposure, it is quite remarkable that the question on the possible effect on solid cancers was still in much debate as late as in the mid-1950s. In 1956, Oho reported increased mortality from various types of malignant

117

TABLE 7.1 Cancer incidence by distance from hypocenter: Hiroshima, May 1, 1957–December 30, 1958.*

Site	Distance (m)	Population	Cases	Incidence*
All malignancies	500 - 999	973	16	1,287.1
	1,000 - 1,499	8,688	71	528.6
	1,500 - 1,999	15,318	91	392.6
	2,000 - 2,499	13,915	68	297.3
	Non-irradiated	202,727	464	280.5
Malignancies	500 - 999	973	13	1,023.1
excluding	1,000 - 1,499	8,688	66	479.0
leukemias and	1,500 - 1,999	15,318	89	380.1
lymphomas	2,000 - 2,499	13,915	66	283.4
	Non-irradiated	202,727	447	270.1

*Incidence per 100,000, adjusted for age and sex.

neoplasms in the "exposed" persons in Hiroshima based on death certificate data for 1951 through 1955 (Oho, 1956). The early LSS mortality data also showed a similar increase for cancers other than leukemia for proximal survivors, but the interpretations were difficult because of the concern about possible biases in reporting based on death certificates (Beebe et al., 1961; Jablon et al., 1963). However, the first tumor registry report by Harada and Ishida demonstrated a clear relationship of solid cancer incidence to distance from the hypocenter (Harada and Ishida, 1960)(Table 7.1).

The report by Harada and Ishida, published in 1960, included cancer incidence data for the first 20 months of the Hiroshima Tumor Registry operation. The analysis of the tumor registry data involved: the general population of Hiroshima; open population, estimated from the 1950 National Census; as well as the fixed Life Span Study sample selected for mortality follow-up conducted by ABCC. The incidence of all malignant neoplasms increased in an exponential or linear fashion with a decreasing distance, on a logarithmic scale, from the hypocenter. The authors also noted that the age-specific rates of cancer (other than leukemia) for proximal survivors were a constant multiple of the corresponding rates for the non-exposed, suggesting the increase in absolute incidence. Subsequent incidence reports included data from both Hiroshima and Nagasaki and added several more years of follow-up (Harada et al., 1963; Ide, et al., 1965); these supported the earlier incidence results. Therefore, in the early years, the tumor registries were quite active and appeared promising.

The following decades, however, saw a rapid decline in registry activities, especially in Hiroshima. Deterioration in the relationship with the local medical community led to an increasing difficulty in sustaining collaboration from certain major hospitals in Hiroshima. To supplement case reporting, tissue registries were started in Hiroshima and Nagasaki in 1973, with initial financial support provided by the US National Cancer Institute. The tissue registries are pathology-based registries intended to collect and store pathology slides and tumor information for histologically diagnosed tumor cases in the Hiroshima and Nagasaki areas. While these registry features are quite unique and useful for pathological investigations, the lack of a population base has remained a serious shortcoming for epidemiological studies.

From the late 1960s through early 1980s, the tumor registry activities became less visible, but routine data collection work continued in both cities. During this time, the tumor registry data were primarily used as an index source for identifying cancer cases for individual researchers interested in specific sites of cancer. Several important studies of this kind were carried out, including the breast cancer series (Tokunaga et al., 1987) and cancers of the stomach (Matsuura et al., 1983), lung (Kopecky et al., 1986; Yamamoto et al., 1986), ovary (Tokuoka et al., 1987), thyroid (Akiba et al., 1991), and colon and rectum (Nakatsuka et al., 1992).

The tumor registry data were at one time analyzed as a supplement to the periodical analysis of the LSS mortality data (Beebe et al., 1961). But it was not until 1981 that the next comprehensive cancer incidence report was published and this report included only the Nagasaki portion of the LSS sample (Wakabayashi et al., 1981). This report included a total of 1,412 cancer incidence cases occurring in the Nagasaki subcohort during the period 1959–78, as compared with a total of 759 cancer mortality cases for the same period. The number of excess cases attributable to radiation exposure was estimated to be 150.5, or about 11%. The radiation dose estimates were based on T65DR, and the shape of a dose-response curve, for essentially gamma radiations resulting from the Nagasaki bomb, was linear for solid cancers and either linear or linear-quadratic for leukemia.

RECENT DEVELOPMENTS

Recent developments as described below have completely modernized the Hiroshima and Nagasaki Tumor Registries; these tumor registries, as they now exist, are fully developed population-based registries. In the late 1980s RERF began to devote increased attention and resources to rebuilding and improving the tumor registries to meet research needs, i.e., to providing high quality incidence data for risk assessment. The Hiroshima data collection lagged behind, but by this time the relationship of RERF with the major hospitals in Hiroshima had substantially improved so that retroactive case ascertainment was possible. Data collection was resumed from several hospitals, which had previously not been visited, and new hospitals were added to the hospital visitation schedule. In order to achieve consistency in data and data handling between Hiroshima and Nagasaki, the registry staff from both cities worked together to develop common procedures and manuals. All tumor and tissue registries were then assembled, together with data from other relevant records such as death certificates, RERF pathology records (autopsy and surgical), clinical records from the Adult Health Study, and, for hematological tumors, the Leukemia Registry records. Some hospitals were re-visited to follow up on cases for which previously abstracted information was incomplete or insufficient. Following the newly established procedures and rules, all records were reviewed, a standardized summary was prepared for each primary tumor, and these data were then coded and entered into a newly developed database with extensive quality control checks.

By the late 1980s, both the Hiroshima and Nagasaki cancer incidence data had a sufficiently good quality, and they were included in a series of monographs on world-wide cancer incidence data, *Cancer Incidence in Five Continents*, compiled by the International Agency for Research on Cancer (IARC; Muir et al., 1987; Parkin et al., 1992). By the early 1990s, the cancer incidence data for the LSS cohort population were updated through 1987 and were of a sufficient quality and consistency that comprehensive risk analyses were warranted. In 1994, a series of LSS cancer incidence reports were published (Mabuchi et al., 1994; Thompson et al., 1994, Preston et al., 1994, and Ron et al., 1994a).

CURRENT CANCER INCIDENCE DATA

Data Quality

Unlike mortality data that are routinely collected by established procedures throughout Japan, incidence data are generated by assembling information from multiple sources with varying degrees of diagnostic confirmation. Therefore, quality of data and consistency in collection and handling of data are the important concerns. The Hiroshima and Nagasaki Tumor Registries both employ an active case ascertainment with abstraction of medical records by hospital visitation as the primary means, augmented by data from the tissue registries. These data are linked with the RERF major samples through the Master File for identification of cancer cases in members of the samples. For members of the LSS, the data are further supplemented by death certificate data as well as data from a number of clinical and pathological programs which have been undertaken over the years at ABCC/RERF.

As measures of data quality IARC uses three numerical indices: (1) the proportion of cases registered with histologically verified diagnosis (histological verification, HV), (2) the proportion of cases registered from death certificates only (DCO), and (3) the mortality/incidence ratio (M/I) (Waterhouse et al., 1982). The DCO rate and M/I ratio provide a gauge of completeness of reporting, whereas the HV rate is considered a measure of accuracy of diagnostic data. Currently,

TABLE 7.2   Indices of quality in Hiroshima and Nagasaki Tumor Registries compared with other selected registries (males).

Registry (years)	DCO(%)[a]	M/I[b]	HV(%)[c]
Hiroshima (1978–80)	9	0.63	77
Nagasaki (1978–82)	7	0.55	67
Osaka, Japan (1978–82)	25	0.70	58
Miyagi, Japan (1978–82)	15	0.67	67
Singapore (Chinese)(1978–82)	7	-	76
Israel (All Jews) (1978–81)	6	-	86
Ontario, Canada (1978–82)	1	0.54	82
Seattle, USA (1978–82)	1	0.49	93
Denmark (1978–82)	1	-	91
Hamberg, Germany (1978–79)	30	0.76	54

[a] Death certificate only: percent of cases notified from a death certificate only.
[b] Mortality/incidence ratio: ratio of mortality to incidence cases.
[c] Histological verification: percentage of cases whose diagnosis is verified histologically.

the Hiroshima and Nagasaki Tumor Registry data show DCO rates of 7–9%, M/I ratios of about 0.50, and HV rates of about 70% or higher. As seen in Table 7.2, these are better than any other in Japan and comparable to those for many other established registries. The homogeneity of incidence data as measured by these and other indices of data quality across various substrata such as by age, time, and radiation dose revealed no indication of the presence of potential bias or confounding which may influence cancer risk estimates using the current data set (Mabuchi et al., 1994).

Solid Cancer Risk

Incidence data have several important advantages for risk assessment. Incidence data provide a complete picture of the spectrum of cancer outcomes, including both fatal and less fatal tumors. Even for fatal cancers, the survival time allows for timely identification of tumor occurrences. These advantages result in a substantial increase in the number of cancer cases ascertained, which in turn enhances statistical power for assessing the dose response as it is affected by various modifying factors. Although the collection of cancer incidence data began in 1958

TABLE 7.3 Observed and expected solid cancer incident cases, Life Span Study cohort, 1958–87.

Dose (Sv)*	Subjects	Observed	Expected	Excess
<0.01	42,702	4,286	4,267	19
0.01-0.1	21,479	2,223	2,191	32
0.1-0.2	5,307	599	574	25
0.2-0.5	5,858	759	623	136
0.5-1	2,882	418	289	129
1-2	1,444	273	140	133
2.0+	300	55	23	32
Total	79,972	8,613	8,106	507

*Dose (Sv) is the dose to the large intestine (colon).

and is limited to those residing in the tumor registry's catchment area, the number of incident cancer cases far exceeds that of mortality cases. For the period of 1958–87, the total number of solid cancer incidence cases (first primary cancers) in the Hiroshima and Nagasaki areas was 8,613, 25% more than the 6,887 mortality cases for the period of 1950–87 with no residential restriction (Thompson et al., 1994, and Ron et al., 1994a). Of these, about 500 incident cases are considered an excess due to radiation exposure (Table 7.3); this is 65% more than the 304 excess mortality cases. Because of the ability to capture less fatal cancers and to provide better diagnostic accuracy, the new incidence data have, for the first time in the LSS cohort, demonstrated significant excess risk for liver and non-melanoma skin cancers. The data have also provided updated information on the risk of thyroid, breast, and salivary gland cancers. While an improvement in accuracy and precision results in a refined site-specific risk assessment, it is also important to recognize that the lack of statistical significance for specific cancer sites does not necessarily mean the absence of radiogenic tumor response for those sites. It is remarkable that increased risk, albeit statistically insignificant for some sites, is seen for almost all types, perhaps with the exception of uterine tissues (Table 7.4).

One of the most striking findings from the solid cancer data is the linearity of the dose-response curve (Figure 7.1). The LSS cancer mortality data have also supported the linear dose response, but the linearity shown by the incidence data is remarkably unequivocal. This linearity is contrasted to the non-linear dose response for leukemia (Preston et al., 1994; Vaeth et al., 1992). The difference may reflect different underlying biological mechanisms for carcinogenesis and

TABLE 7.4 Observed and excess incident cancer cases, by site.

Cancer site	Observed	Excess
Oral cavity and pharynx	64	6
Esophagus	84	5
Stomach	1,305	85
Colon	223	32
Rectum	179	8
Liver	283	31
Gallbladder	143	3
Pancreas	122	4
Lung	449	85
Nonmelanoma skin	91	22
Female breast	289	92
Uterus	349	-12
Ovary	66	12
Prostate	61	4
Bladder	115	19
Kidney	34	5
Nervous system	69	4
Thyroid	129	33

leukemogenesis, the former being a multistage, multifactorial process, while the latter is more closely associated with chromosomal rearrangement.

The temporal patterns of cancer risk are of special importance in risk assessment. Table 7.5 shows the numbers of observed and excess solid cancer cases by age at exposure for four time periods. It is striking that almost half of the excess cases in the entire follow-up period were seen in the last 7 years, 1981–87, of the current follow-up. When examined by age at exposure, it is clear that most of the excess cases in the earlier periods were in those exposed at ages 20 years or older, whereas in the later periods more excess cases have been seen in those exposed at younger ages. Particularly noteworthy is the rapidly increasing number of excess cases for those exposed at very young ages, less than 10 years, emphasizing the importance of incidence-based follow-up for the young survivors as they enter ages where increased background cancer incidence rates are expected.

The risk for solid cancer based on incidence data demonstrates the well-known dependence on sex and age at exposure, as illustrated in the left panel in Figure 7.2 (Preston, 1995). This figure shows that the excess relative risk (ERR) for solid cancer among those exposed as adults has been nearly constant over the current

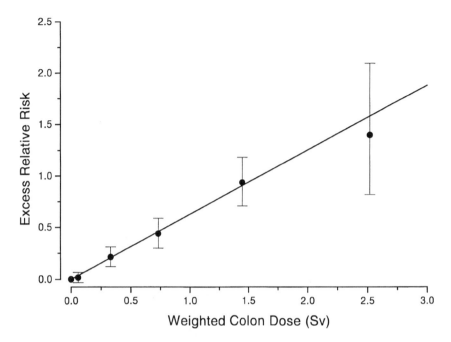

FIGURE 7.1 Dose-response for solid cancers. LSS incidence data, 1958–87.

follow-up period, whereas the ERR for those exposed as children has been decreasing in the recent years following the very high level in the early years. The
ERR for women is about twice as high as for men. In terms of excess absolute rate
(EAR) per unit dose, the sex difference is less striking for those exposed as adults
(right panel in Figure 7.2), as has been shown by the mortality data. However, the
sex difference in EAR remains large for those exposed as children; this is different from the mortality results and may reflect the inclusion of a large number of
incident cases of thyroid, breast, and skin cancers, the risk of which is especially increased among those exposed at young ages. In general, it is important to note that
although the ERR for those exposed as children shows a declining trend, the EAR
for these survivors continues to increase with time, as background rates increase
with advancing age. These patterns are essentially comparable to those for cancer
mortality risk, and emphasize the importance of further follow-up, especially for
those exposed during childhood.

   Through the tumor registry multiple primary tumors can also be ascertained
in a systematic manner. The analysis of the latest cancer incidence data was
restricted to first primary tumors. The main reasons for this restriction were (1)
that patients who developed tumors may be more likely than others to be followed

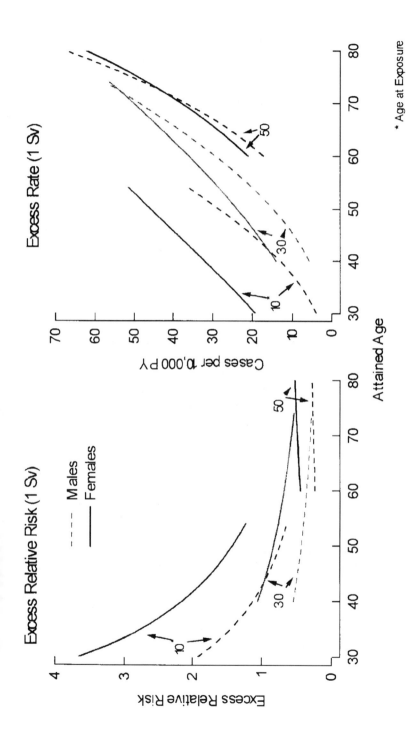

FIGURE 7.2 Excess relative risk (ERR$_{1Sv}$) and absolute risks (cases/10^{4}PYSv) for solid cancers as a function of age ATB and attained age.

TABLE 7.5 Observed and excess incident solid cancer cases, 1957–87.

Age ATB*	Calendar period							
	1957–65		1966–75		1976–80		1981–87	
	Obs.	Excess	Obs.	Excess	Obs.	Excess	Obs.	Excess
0 - 9	19	1.7	63	-1.4	75	6.9	208	40.3
10 - 19	76	11.7	245	54.2	203	27.7	469	84.9
20 - 29	137	19.0	290	3.0	237	10.4	462	31.1
30 - 29	326	42.6	575	-29.0	418	-0.7	707	50.4
40 - 49	573	13.7	982	38.8	473	15.0	489	30.4
50+	793	31.5	611	25.6	133	9.9	49	-11.2
Total	1924	120.2	2766	91.3	1539	69.2	2384	225.9

*ATB denotes age at time of bombing.

closely and thus to have better chances of second tumors being detected, and (2) the possible confounding and/or modifying effect of some cancer treatment modalities. Currently analyses are undertaken to assess the risk of second primary tumors in the LSS sample. There is much evidence of an elevated cancer risk in patients who have received radiation treatment, sometimes in combination with chemotherapy, for various types of cancer. These data have been useful for risk assessment, but a potential bias due to inclusion of susceptible individuals is some concern. The second primary cancer study in the LSS sample should provide much needed population-based data on background second cancer risk and the magnitude of second cancer risk associated with radiation exposure.

The tumor registry data have several limitations, the most important of which concerns migration of the study subjects. Over the years, a considerable number of subjects in the LSS sample (estimated to be more than 20%) have moved out of the catchment areas of the Hiroshima and Nagasaki Tumor Registries. It is therefore necessary to adjust for migration when using the tumor registry data for assessing cancer risk in the LSS. In a previous study of breast cancer incidence (Tokunaga et al., 1987), all cases including those diagnosed outside the Hiroshima and Nagasaki areas were included, but adjustment was made for incomplete reporting of those who migrated out of the area. In the current report, cases were restricted to those residing in the area and person-years were adjusted to allow for the migration effect, as described by Sposto and Preston (1992). An alternative and more direct approach would be to obtain residential data for each LSS subject. Work is in progress to create an address database for the LSS using information from previous mail surveys and other sources available at RERF. Such individual data can be used in the future to compute the person-years at risk. The address data should also be helpful in finding incident cases in the areas to which individuals are known to have moved and where active population-based registries are available.

Another important limitation arises from that fact that the tumor registry operation began in 1957 in Hiroshima and 1958 in Nagasaki. Currently work is under way, in some of the site-specific studies, to reconstruct cancer incidence data for the pre-registry years using early pathology and hospital documents that are still kept at University and some other hospitals. As there already was a significant excess in solid cancer in the first 20 months of the tumor registry, from May 1957 through December 1958 (Harada and Ishida, 1960), it is likely that there was also some excess solid cancer risk, albeit presumably small, before that time.

RELATED CANCER STUDIES

In addition to the role that the tumor registry incidence data play in risk assessment, the functional tumor registries have greatly facilitated the conduct of many epidemiological investigations which have led to further understanding of how cancers are induced by radiation exposure, possibly in combination with other etiologic factors. Several important new studies have been initiated and will be conducted in the future.

Among the most active projects at the present time is a series of site-specific cancer studies. The tumor registry represents the collection of diagnostic data from various sources spanned over many years. Such data reflect diagnoses made by many different physicians using diagnostic standards acceptable at the time of practice. Under the established guidelines (Tokunaga et al., 1988), several important site-specific cancer incidence studies have been undertaken. Case ascertainment has been undertaken through the Hiroshima and Nagasaki Tumor Registries, but histopathological verification of diagnoses is being achieved by a panel of designated pathologists, and tumor types are being classified using internationally accepted tumor classification systems. The pathology panel consists of pathologists who specialize in specific types of cancer in Hiroshima and Nagasaki. The Hiroshima and Nagasaki Tissue Registries facilitate these studies by allowing access to pathology slides needed for pathology review.

The major advantage of site-specific studies is the capability to study specific questions under strictly controlled conditions. A liver cancer study now in progress serves as an example of what can be achieved by site-specific studies. Since diagnostic misclassification is a major concern for liver cancer, the primary objective of the liver cancer study is to assess the relationship between atomic bomb radiation and liver cancer based on data confirmed by a panel of pathologists. Another arm of the liver cancer study is a case-control study nested in the cohort. The possible role of hepatitis B (HB) virus and hepatitis C (HC) virus infection is investigated, while the pathology review is extended to include diagnosis of any accompanying liver cirrhosis and identification of HBV, and molecular techniques are used to characterize the HB and HC viruses.

Other site-specific studies underway include such sites as breast, salivary glands, skin, thyroid, ovary, central nervous system, breast, lung, and lymphoid tissues. Some of these studies have been initiated because of the well-recognized variability in diagnoses (skin cancer, salivary tumors, thyroid cancer, lymphoid tumors), while other studies focus on specific histologic types (basal cell skin carcinoma, neurolemmoma of the neural tissue), tumor locations (lung cancer), or both benign and malignant tumors (salivary tumors, thyroid tumors, neural tumors, ovary tumors). For some sites—such as lung, skin, breast—molecular studies are also carried out to characterize the nature and spectrum of mutations at oncogenes and tumor-suppresser genes.

Among the studies of increasing importance in the future are those involving characterization of molecular changes using archived tissue samples. A pathology program, which was started at the inception of ABCC, has undergone several major revisions. The autopsy program, a mainstay of the RERF pathology program, had long been quite active but was terminated in 1988; the surgical program, another major program, was also active before it was replaced by the tissue registries starting in 1973. Through these programs, a large number of archived tissues have been collected. These archived tissues have been found to be useful for molecular biological studies which are now beginning to play an important role in addressing biological issues of current interest. Now that these pathology programs are no longer active, it is expected that the tissue registries will become an important source of tissue materials for molecular biological studies.

CONCLUSIONS

The long-term follow-up of the Life Span Study cohort has been a major source of epidemiologic data for risk assessment. The evidence based on the latest cancer incidence data and mortality data (Pierce et al., 1996) suggests that the excess solid cancer risk persists many decades after the exposure and will probably remain elevated throughout life. This means that the number of excess cases will continue to increase, possibly more rapidly than before, as the younger cohort members enter older age categories. At the present time, more than half of the LSS cohort members are still alive. Lifetime follow-up has been almost completed for those exposed to the bomb at ages over 50, but more than 90% of those exposed during childhood are alive. Therefore, risk projection based on the incomplete follow-up of the subjects, especially those exposed as children, presents the major uncertainty in risk estimates. A further follow-up, say for another two or three decades, will be critical in obtaining more definitive data on the temporal patterns of solid cancer. For the future follow-up, cancer incidence data will become an increasingly more important addition for risk estimates. Because of the advances in cancer treatment in recent years, a larger number of cancer patients survive longer. Although mortality data may become less valid as the outcome measure of cancer risk, death will continue to serve as an essential and useful endpoint.

In discussing the purposes and uses of cancer registries, Jensen and Storm stated that "the cancer registry provides a crucial basis for epidemiology since it holds information on the distribution of cancer, including non-fatal cases." Further, "the collection of records of cancer patients from a defined population facilitates the in-depth study of cancer in individuals minimizing the selection bias" (Jensen and Storm, 1991). For epidemiological studies on cancer, the functional tumor registry is essential, and through well-designed and well-conducted epidemiological studies it will be possible for us to obtain further biological insights. Rapid advances in our understanding of cancer will also demand new types of research. The tumor registries will play a pivotal role in meeting advancing research needs in a timely manner.

8

Solid Cancer Incidence and Mortality in the Life Span Study of Atomic Bomb Survivors

ELAINE RON, DALE L. PRESTON, AND
KIYOHIKO MABUCHI

SUMMARY

For several decades, mortality data have been the mainstay of cancer risk estimates derived from the Life Span Study (LSS) cohort of atomic bomb survivors. However, with the recent availability of comprehensive incidence data, the evaluation of cancer incidence is now possible. Solid tumor incidence data are obtained by matching the LSS cohort with the Hiroshima and Nagasaki Tumor Registries. These registries were established in 1958 and cover all residents of Hiroshima and Nagasaki. The LSS mortality data are obtained through the nationwide family registry system. Using the incidence data, 8,613 first primary solid cancer cases were identified between 1958 and 1987, compared with 6,887 solid cancer deaths reported on death certificates throughout Japan between 1950 and 1987. Of cancer patients identified through the tumor registries, 23% were alive in 1988, the underlying cause of death was not attributed to cancer for 14%, and the type of cancer recorded on the death certificate was different than in the tumor registries for 5%. Still, the incidence data generally confirmed the mortality findings. Significant excess risks were observed in both the incidence and mortality series for all solid cancers, for cancers of the stomach, colon, liver (when it was defined as primary liver cancer or liver cancer NOS on the death certificate), lung, breast, ovary, and urinary bladder. In general, the risk estimates were larger based on incidence data. No significant radiation effects were seen for cancers of the pharynx,

131

rectum, gallbladder, pancreas, nose, larynx, uterus, prostate or kidney in either series. A significant excess of esophageal cancers was observed in the mortality data, and significant excesses of non-melanoma skin cancers and cancers of the salivary gland and thyroid were seen in the incidence data. Incidence data allow a more precise description of radiation-induced cancer risks because the number of incident cases is considerably larger than the number of cancer deaths; incidence data provide the only reliable source of data for malignancies with relatively good survival rates, and diagnoses are more accurate and detailed. Mortality data are easy to obtain and are virtually complete for all of Japan starting in October 1950. The next 20 years will yield extremely important information on radiation exposure at early ages, and analyses of both incidence and mortality are needed to fully understand the role of radiation in carcinogenesis.

INTRODUCTION

For several decades, mortality data have been the mainstay of cancer risk estimates derived from the Life Span Study (LSS) cohort of atomic bomb survivors (Shimizu et al., 1990). This cohort has been followed up by the Atomic Bomb Casualty Commission until 1974 and by the Radiation Effects Research Foundation (RERF) subsequently. Population-based cancer registries were established in Hiroshima and Nagasaki over 30 years ago, but problems with data collection have limited their use. During the 1980s, RERF worked to improve the registries' data quality. Previously collected data were reviewed, revised, and supplemented, and new methods for the systematic collection and management of future data were instituted. As a result, comprehensive incidence data became available for study (Mabuchi et al., 1994). Last year, four reports summarizing cancer incidence in the LSS were published (Mabuchi et al., 1994; Thompson et al., 1994; Preston et al., 1994; Ron et al., 1994a). The wide use of these data (UNSCEAR, 1994) demonstrates their valuable contribution to risk estimation, and ensures that they will play a larger role in the future research of RERF.

MATERIALS AND METHODS

The LSS cohort includes 120,321 people; however, in this report, we did not include the 26,580 individuals who were not in Hiroshima or Nagasaki at the time of the bombings, or the 7,109 persons with unknown radiation doses. As described by Mabuchi et al. (1994) and Thompson et al. (1994), solid tumor incidence data are obtained by matching the LSS cohort with the Hiroshima and Nagasaki Tumor Registries. These registries were established in 1958 and cover all residents of Hiroshima and Nagasaki. The Hiroshima and Nagasaki registries use the same methods for ascertaining cases, the majority of which are identified by searching hospital records. These data are supplemented by abstracting information from the Hiroshima and Nagasaki Tissue Registries; the Nagasaki Prefectural Cancer

TABLE 8.1 Characteristics of incidence and mortality series (Life Span Study of atomic bomb survivors).

	Incidence series	Mortality series
Years covered	1958–1987	1950–1987
Catchment area	Hiroshima, Nagasaki	All Japan
Study population	79,972	86,309
Person-years of follow-up	1,950,567	2,588,874
Mean age ATB*	26.8	29.0
Number of total tumors	9,014	7,308
Number of solid tumors	8,613	6,887

* Age ATB denotes age (years) at time of bombing.

Registry; the Leukemia Registry; data from the RERF clinical, surgical and autopsy programs; notification by physicians; and review of national death certificates. Because the cancer incidence data are derived from the Hiroshima and Nagasaki Tumor Registries, follow-up for the incidence series begins January 1, 1958, and includes the 79,972 LSS members who were alive and not known to have had cancer before 1958 (Thompson et al., 1994) (Table 8.1).

The LSS mortality data are obtained, in three-year cycles, through the nationwide family registry system (*koseki*). For each deceased survivor, the underlying cause of death is abstracted from the death certificate. Mortality follow-up begins on October 1, 1950, and includes the 86,309 LSS members who were known to be alive on that date.

For the incidence series, follow-up ended at the first diagnosis of a primary cancer, or the date of death, or December 31, 1987, whichever came first. In the mortality series, follow-up ended at the date of death or December 31, 1987. There are more than 600,000 additional person years of follow-up in the mortality series than in the incidence series because follow-up began eight years earlier. At the time of the bombings (ATB), the mean age was 29 years for subjects in the mortality series, compared with 26.8 years for survivors in the incidence study, because many people who were older ATB died before the incidence follow-up began (Ron et al., 1994a).

Organ doses were computed as the sum of the gamma-shielded kerma and ten times the neutron-shielded kerma, as estimated by the 1989 version of DS86 (Roesch, 1987; Fujita, 1989). Kerma is expressed in Gy, and the weighted organ doses are expressed in sieverts. Site-specific analyses were based on estimated organ doses, except for skin cancers, for which we used kerma because no skin dose was calculated as part of DS86. Solid tumors as a group were analyzed using intestinal doses. Persons with kerma estimates over 4 Gy were excluded from the analysis because the accuracy of these dose estimates is questionable (Pierce et al., 1990).

Risk estimates were calculated using Poisson regression methods. Analyses were based on general excess relative risk (ERR) models and time-dependent excess absolute risk (EAR) models. Background rates were modeled as a function of city, gender, attained age, and year of birth. A linear dose-response function was used as the standard model, although we tested for non-linearity. Maximum likelihood methods were used to calculate parameter estimates and 95% confidence intervals (Cox and Hinckley, 1974). The statistical methods are described in more detail in Thompson et al. (1994) and Preston et al. (1994).

RESULTS

Using the incidence data, 9,014 first primary cancer cases were identified between 1958 and 1987, compared with 7,308 cancer deaths reported on death certificates throughout Japan between 1950 and 1987. Of the incidence cases, 8,613 (95.5%) were solid tumors, and of the cancer deaths, 6,887 (94.2%) were attributed to solid cancers. When the mortality series was limited to deaths occurring during the incidence series time period (i.e., 1958–1987) 2,671 more incident cancer cases than cancer deaths were identified. Further restricting the mortality series to deaths occurring between 1958 and 1987 in Hiroshima or Nagasaki would result in 3,155 more incident cancer cases than cancer deaths.

The number of deaths from cancers of the digestive and respiratory systems, occurring in all of Japan between 1950 and 1987, was not much smaller than the number of cases in the incidence series. But, for cancers of the male genital and urinary system, the incidence series was at least twice as large as the total mortality series. The number of incident cases of cancers of the skin, female breast, and thyroid was more than three times greater than the number of deaths. Cancer ascertainment was not only more complete using incidence data, but diagnoses were more precise than those recorded on death certificates. For example, specific diagnoses for cancers of the uterus were available for 88.1% of the incident uterine cancers, whereas in the mortality series, 71.6% of cancers of the uterus were recorded as not otherwise specified (NOS) (Table 8.2).

When we compared the first primary cancer incidence diagnoses with the underlying causes of death, we found 23% of cancer patients were still alive in 1988 and, therefore, would not have been included in an analysis of the mortality data. In addition, the underlying cause of death was recorded as non-cancer for 14% of the cancer patients. Another 5% had a different type of cancer listed on the death certificate than in the tumor registries, even when the comparison was made at the organ-system level. This means that only 58% of the cancer cases analyzed in the incidence series would be included in the mortality series with the cancer diagnoses listed in the cancer registries. Although statistics vary depending on the

TABLE 8.2 Distribution of uterine cancers in the incidence and mortality series (Life Span Study of atomic bomb survivors).

Cancer site	Incidence series (%)	Mortality series (%)
Cervix uteri	553 (76.4)	110 (25.0)
Uterine corpus	85 (11.7)	15 (3.4)
Uterine NOS	86 (11.9)	315 (71.6)
Total uterine cancers	724 (100)	440 (100)

*NOS denotes not otherwise specified.

cancer type, if death certificates were the only source of medical information, less than 50% of the cancer cases would have been classified as in the tumor registry for the majority of sites studied. For cancers with particularly poor survival, such as cancers of the esophagus, liver, pancreas, and lung, over 70% of the incident cancers would have been identified using death certificate diagnoses.

Even though there were considerable differences in the two data sources, the incidence data generally confirmed the mortality findings. Significant excess risks were observed in both the incidence and mortality series for all solid cancers. Based on incidence data, the excess relative risk at 1 sievert (ERR_{1Sv}) was about 40% larger and the excess absolute risk per 10,000 person-year sievert ($EAR/10^3 PYSv$) almost three times higher than the risk estimates derived from mortality data (Table 8.3). Based on the incidence series, 504 excess cancer cases were predicted, compared with 304 excess cancer deaths based on the mortality data. Even during the last follow-up period (1976–1987) when more cancer deaths would be expected because of the older age of the cohort, 128 more excess cancer cases were predicted than cancer deaths.

Using either the incidence or mortality data, excess relative risks were approximately two times greater for females than males, although the level of significance was higher using the incidence data. The estimated excess relative risks for persons under 20 years ATB were substantially larger in the incidence series than

TABLE 8.3 Comparison between solid tumor risk estimates in the incidence and mortality series by gender, age at exposure, and time since exposure (Life Span Study of atomic bomb survivors).

	Incidence			Mortality		
	Observed	ERR_{1Sv}	$EAR/10^3 PY Sv$	Observed	ERR_{1Sv}	$EAR/10^3 PY Sv$
Gender						
Female	2,467	0.81	34.4	1,836	0.63	13.8
Male	1,860	0.33	19.0	1,616	0.23	7.5
Age ATB[a]						
<20	721	1.53	22.4	360	0.70	3.8
>20	3,606	0.44	33.9	3,092	0.41	18.6
TSE[b]						
<19	991	0.69	20.8	1,111	0.43	7.7
20–29	1,367	0.37	15.0	967	0.49	13.4
30–42	1,969	0.64	35.8	1,374	0.53	20.3

[a] Age ATB denotes age (y) at time of bombing.
[b] TSE denotes time (y) since exposure.

the mortality series, whereas for persons exposed to the bombings after age 20 the risks derived from incidence and mortality data did not vary as much. Since the excess relative risks for developing cancers of the breast, thyroid, and skin, which are often not fatal, are particularly pronounced among persons exposed to the bombings during childhood, this finding would be expected. During the first 19 years of follow-up, risks were considerably higher using incidence data than mortality data. In fact, using incidence data, risks were high in the first follow-up period, decreased in the second, and increased again in the last, whereas in the mortality data risks increased with increasing time since exposure.

The EAR for females was 1.8 times greater than for males in both the incidence and mortality series. The high risk for females was observed particularly among survivors exposed to the bombings before age twenty. However, the EARs increased with attained age more rapidly for males than for females. By about age 55 years, males had a higher EAR of cancer deaths than females, whereas this gender crossover was seen at about age 68 in the incidence data.

Using either a relative or absolute risk model, statistically significant radiation effects were demonstrated in both the incidence and mortality data for cancers of the stomach, colon, liver (when it was defined as primary liver cancer or liver cancer NOS on the death certificate), lung, breast, ovary, and urinary bladder (Table 8.4). In general, the ERR point estimates were larger in the incidence series than in the mortality. Particularly notable were the high risks observed for the incidence of cancers of the stomach, colon, and lung, compared with the mortality series. The EAR estimates also were larger when based on incidence data than mortality.

No significant radiation effects were seen for cancers of the pharynx, rectum, gallbladder, pancreas, nose, larynx, uterus, prostate, or kidney in either series. For cancers of the oral cavity and pharynx, the risk estimates based on the incidence data were positive, while those based on mortality data were negative. The opposite situation was observed for cancers of the uterus: negative risk estimates based on incidence data and positive estimates based on mortality data. However, the confidence bounds around these estimates were wide and overlapped.

A significant excess of esophageal cancers was observed in the mortality but not in the incidence data (Table 8.4). This is explained partly by the extremely high risks of death from esophageal cancer seen shortly after the bombings, i.e., before incidence data were available.

Non-melanoma skin cancers, and cancers of the salivary gland and thyroid were rarely evaluated in mortality studies, but we demonstrated a significant excess of each of these cancers in the incidence data (Table 8.4). There were so few deaths due to salivary gland tumors that we couldn't evaluate them using mortality data. Although the number of incident salivary gland cancers was also small (13 exposed and 9 non-exposed cases), the ERR was very high (ERR$_{1Sv}$ = 1.77; 95% CI = 0.15–6.0) and was greater for persons exposed before age 20.

Based on 168 first primary cases of non-melanoma skin cancer, there appeared to be some evidence for non-linearity (p = 0.12 for upward curvature) in the dose

TABLE 8.4 Risk associated with incidence and mortality findings (Life Span Study of atomic bomb survivors).

	ERR_{1Sv}		$EAR/10^3\,PY\,Sv$	
	Incidence	Mortality	Incidence	Mortality
Significant excess risks in both incidence and mortality				
Stomach	0.32	0.22	4.8	1.9
Colon	0.72	0.52	1.8	0.51
Liver[a]	0.49	0.46	1.6	1.3
Lung	0.95	0.65	4.4	1.9
Breast	1.6	1.3	6.7	1.3
Ovary	0.99	1.2	1.1	0.69
Bladder	1.0	1.5	1.2	0.49
Significant excess risks in incidence only				
Salivary	1.8	No data	–	–
Skin[b]	1.0	0.31	0.84	0.034
Thyroid	1.2	0.094	1.6	0.016
Significant excess risks in mortality only				
Esophagus	0.28	0.60	0.30	0.45

[a] Includes deaths due to primary or NOS liver cancer.
[b] Non-melanoma skin cancer only.

response. A spline model appeared to fit the data best. An inverse relationship between risk and age at exposure was also demonstrated. The ERR_{1Sv} was 7.2 for persons under age 10 ATB, 2.4, 1.2, and 0.27 for those 10–19, 20–39, and 40+ ATB, respectively. Based on only 36 deaths from non-melanoma skin cancer, the ERR_{1Sv} was approximately one-third of that observed in the incidence data.

There were only 49 deaths due to thyroid cancer compared with 225 incident cases. Of the 49 deaths, only 37 were the same cases as in the incidence series; 4 had other cancers in the tumor registry, 3 were not residents of Hiroshima or Nagasaki and 5 died before the tumor registries were established. Based on the mortality data, the ERR_{1Sv} was 0.09. In contrast, even at doses as low as 25 cGy, we observed a large excess risk in the incidence data and found a dramatic difference in the effect of radiation exposure during childhood and adulthood. The ERR for persons <10 ATB was 9.5 (95% CI = 4.1–18.9) but the risk for persons >20 ATB was 0.1 (<-0.2–0.8). Although survivors who were routinely examined

TABLE 8.5 Excess relative risks for breast, non-melanoma skin, thyroid and other solid cancers in incidence and total solid cancers in mortality series (Life Span Study of atomic bomb survivors).

Population group	Incidence ERR$_{1Sv}$		Mortality ERR$_{1Sv}$
	Breast, skin, thyroid cancers (n=935)	Other solid cancers (n=7,677)	All solid cancers (n=6,887)
All survivors	1.9	0.45	0.43
Female	1.9	0.61	0.63
Age ATB <20*	5.5	0.75	0.70

* Age ATB denotes age (y) at-time-of-bombings.

clinically as part of the Adult Health Study had 2.5 times more thyroid cancers than other LSS members, the slope of the dose response did not differ significantly between the examined and non-examined groups.

The radiation effects observed for non-fatal cancers strongly influenced the incidence-based risk estimates. Breast, thyroid, and non-melanoma skin cancers comprised 10.7% of the solid cancer cases in the incidence series, but only 3.8% in the mortality series. Using incidence data, the ERR$_{1Sv}$ for these three cancer types was 1.86, i.e., about four times larger than for other incident cancers combined (Table 8.5). For breast, thyroid, and non-melanoma skin cancers, all females and individuals (males and females) exposed to the bombings before age 20 had excess relative risks that were considerably greater than for males or people above age 20 ATB. The ERR$_{1Sv}$ for survivors under age 20 ATB was 5.53. When breast, thyroid, and non-melanoma skin cancers were not included in the incidence data, the excess relative risk estimates based on incidence or mortality were virtually the same.

Because the number of deaths specifically reported as due to primary liver cancer was small, a significant ERR was found only when deaths from primary liver cancers were combined with liver cancer deaths not designated as either primary or secondary. Using incidence data, we found that the ERR was higher when good diagnostic data were available. The ERR$_{1Sv}$ was 0.49 (95% CI = 0.16–0.50) when all liver cancers were evaluated. But when the analysis was restricted to histologically confirmed cancers, the ERR$_{1Sv}$ rose to 0.66 (95% CI = 0.11–1.44). The ERR$_{1Sv}$ was 0.41 (95% CI = -0.14–1.27) for cases diagnosed clinically and 0.36 (95% CI= -0.11–1.11) for cases ascertained through death certificates.

Histologic type was available from the tumor registries for 69% of the lung cancer cases. Approximately 45% were adenocarcinomas, 31% squamous-cell

TABLE 8.6 Lung cancer excess relative risk estimates by histologic type (Life Span Study of atomic bomb survivors).

Histologic type	Number of cases		ERR_{1Sv}
	<0.01 Sv[a]	≥ 0.01 Sv[a]	(95% CI)
Adenocarcinoma	115	139	1.30 (0.6-2.2)
Squamous-cell carcinoma	90	88	0.81 (0.2-1.8)
Small-cell carcinoma	26	41	2.07 (0.6-4.6)
Other and NOS	39	38	0.43 (-0.23-1.71)
Total lung cancer[b]	416	456	0.95 (0.60-1.36)

[a] Lung dose equivalent.
[b] The number of lung cancer cases is greater than the sum of the histologic types because no histology was available for 31% of the cases.

carcinomas, and 12% small-cell carcinomas. Although we found no significant difference in the slope of the dose response by histological diagnosis, the ERR for small-cell carcinomas was extremely large (Table 8.6). Females had a higher ERR for adenocarcinomas than males, and the ERR for squamous-cell carcinomas increased with increasing age at exposure and decreased with increasing time since exposure.

DISCUSSION

Incidence data allow a more precise description of radiation-induced cancer risks because the number of incident cases is considerably larger than the number of cancer deaths, they provide the only reliable source of data for malignancies with relatively good survival rates, and diagnoses are more accurate and detailed because they are based on hospital, clinic or pathology records. For example, 75% of the cancers identified among the LSS members were verified histologically, and another 4.4% were diagnosed based on direct observation, e.g., endoscopy, bronchoscopy, or surgery. For several cancer types, over 90% of the tumors were histologically confirmed. Information on method of diagnosis, tumor location and size, and treatment also may be available.

The cancer incidence results have generated questions and hypotheses about risk patterns for specific cancers. To further study these questions, detailed site-specific incidence studies including histologic review are underway. A salivary gland cancer incidence study, which included special searches for salivary gland tumors occurring outside the parotid gland, has just been completed (Land et al., 1996), and a paper on non-melanoma skin cancers will be submitted shortly. Studies of cancers of the nervous system, liver, ovary, and thyroid are in the data collection phase, and plans for other site-specific studies are currently being developed.

The primary weaknesses of the incidence data are that coverage is restricted to LSS members who reside in the Hiroshima and Nagasaki Tumor Registry catchment areas, and that solid tumors cannot be ascertained from before 1957, i.e., for the first 12.5 years after the bombings. Approximately 8% of the cancer deaths between 1958 and 1987 occurred outside Hiroshima and Nagasaki. To take this into account, a migration adjustment, based on information from the Adult Health Study, was made in the incidence analysis (Sposto and Preston, 1992). Thirteen percent of cancer deaths occurred between 1950 and 1958, before the tumor registries were established. These missed incident cases result in an underestimate of the EAR and could affect excess relative risk estimates if the distribution of cancer types differed from later years. Although a small excess risk of solid tumors has been observed in the LSS and the ankylosing spondylitis study five to ten years after the bombings, a latency period of ten or more years is generally accepted (UNSCEAR, 1994). Thus, while the lack of follow-up before 1958 is clearly a drawback, the problem may not be severe.

Mortality data are easy to obtain and are virtually complete for all of Japan starting in October 1950 (Shimizu et al., 1990). Mortality data are unavailable for 1945 to 1949 because the LSS cohort was selected from the 1950 Japanese census data (Beebe and Usagawa, 1968). Mortality data, however, only provide information on fatal cases, and since cancer may be detected long before death, time factors cannot be evaluated directly using mortality data. Furthermore, inaccurate or non-specific death certificate diagnoses (cancers erroneously reported as non-cancers, cancer site misclassified or reported as "ill-defined" or "not otherwise specified," and non-cancers recorded as cancers) limit results regarding specific cancer sites (Glasser, 1981; Cameron and McGoogan, 1981; Mollo et al., 1986).

In a study comparing 5,000 autopsy and death certificate diagnoses in the LSS, 9% of deaths reported as due to cancers by the death certificates were not cancer deaths as judged by autopsy, and 25% of cancers diagnosed at autopsy were missed on death certificates (Ron et al., 1994b). Sposto et al. (1992) report that these errors result in a 12% underestimate of the ERR and a 16% underestimate of the EAR. Clinical information often is not available at the time a death certificate is written.

The underlying cause of death as listed on death certificates often overrepresents the number of liver cancers, probably because metastatic cancers were misclassified. The failure to distinguish between primary and secondary liver cancers also makes reliance on mortality data questionable for this site (Ron et al., 1994b).

The finding of a lower ERR in the mortality data, when compared with incidence data before 1965, may be a result of a latency period for developing radiation-induced cancer. Since only mortality data are available before 1958, the mortality risk estimates would be diluted by a high proportion of spontaneous cancers. Furthermore, because people generally are diagnosed with cancer years before dying from it, radiation risks will be detected earlier when incidence data are analyzed.

Because individuals who were exposed to the bombings during childhood are only now entering the ages for spontaneous cancers to develop, the next 20 years will yield extremely important information on radiation carcinogenesis. Although much has been learned from mortality data over the last few decades, analyses of both incidence and mortality data are needed in the future.

9

Studies of Workers Exposed to Low Levels of External Radiation

ETHEL S. GILBERT

SUMMARY

Several epidemiologic studies of workers who have been exposed occupationally to low levels of external radiation have been conducted in the United States, Great Britain, and Canada. These studies provide a direct assessment of risk and thus serve as a check on the validity of risk estimates obtained through extrapolation from the Japanese atomic bomb survivor studies. This paper reviews results of the worker studies, giving particular attention to national and international efforts to reduce uncertainties by combining data from several studies. Estimates based on the international analyses, which provide the most comprehensive evaluation of the worker data, are lower than those obtained from linear extrapolation from the atomic bomb survivors, but confidence limits indicate that risks are compatible both with a reduction of risks at low doses and with risks that are about double the linear atomic bomb survivor estimates. Overall, the worker studies confirm the appropriateness of basing radiation protection standards on the atomic bomb survivor data. In addition, the studies illustrate the difficulties in obtaining direct estimates that are sufficiently precise for risk assessment, and thus affirm the continued importance of the atomic bomb survivor data for this purpose.

INTRODUCTION

Although this chapter does not directly address the A-bomb survivor studies, these studies have served as the basis for establishing radiation protection standards for the workers who are the subject of this paper. Because the A-bomb survivor studies have clearly established that exposure at relatively high doses and dose rates increases the risk of cancer, occupational exposures have been deliberately limited, with the expectation that risks for most workers would be very small and that observable health effects would be unlikely.

Many nuclear facilities began operations in the 1940s and 1950s, and thus the worker populations have matured to a point where the appropriateness of risk estimates based on extrapolation from A-bomb survivors can be evaluated. Obviously there are uncertainties in risk estimates that are extrapolated not only from high to low doses and dose rates, but also from a Japanese population exposed under special circumstances in 1945 to modern-day populations of other races and nationalities. Some have claimed that this extrapolation process has underestimated risks, perhaps by an order of magnitude or more, while others argue that extrapolation from high-dose data has overestimated risks.

This paper reviews the findings of the worker studies, emphasizing results of the recently completed international combined analyses. Special attention is given to what the worker studies tell us about the usefulness of the A-bomb survivor data for estimating risks of exposures at low doses and dose rates.

INDIVIDUAL STUDIES OF WORKERS EXPOSED
TO EXTERNAL RADIATION

Several studies of workers in the United States, the United Kingdom, and Canada have addressed the effects of external exposure to radiation (Beral et al., 1988; Douglas et al., 1994; Fraser et al., 1993; Gilbert et al., 1993b; Gribbin et al., 1993; Wiggs et al., 1991; Wiggs et al., 1994; Wilkinson et al., 1987; Wing et al., 1991). The characteristics of the populations included in these studies are given in Table 9.1. Cause-specific mortality was the health endpoint of primary interest in all the studies, although cancer incidence was also evaluated in two of the United Kingdom studies (Douglas et al., 1994; Fraser et al., 1993). Additional information on these studies can be found in the individual papers, in a review by Gilbert (1995), in Cardis et al. (1995b), and in a recent report of the United Nations Scientific Committee on the Effects of Atomic Radiation (UNSCEAR, 1994).

In addition to the studies listed in Table 9.1, workers at the Savannah River Plant have also been studied (Cragle et al., 1988), but results of analyses making use of radiation exposure data have not yet been reported. Studies of workers at the Oak Ridge Y-12 plant (Checkoway et al., 1988), the Linde facility (Dupree et al., 1987), and Pantex (Hadjimichael et al., 1983) are excluded from consideration in this paper because dose-response analyses regarding external radiation exposure were not conducted, either because this was not a major exposure and/or exposure data were inadequate for such analyses.

A major strength of the studies listed in Table 9.1 is that estimates of the whole-body penetrating dose, obtained from personal dosimeters worn by the workers, were available for each worker for each year of employment at the facility of

TABLE 9.1 Number of workers, number of deaths, and collective dose equivalent from several epidemiologic studies of workers monitored for external radiation.

Study population	Number of workers[a]	Number of deaths[a]	Collective dose equivalent (Sv)
Hanford	32,643	6,200	854
Oak Ridge National Laboratory (ORNL)	8,318	1,524	144
Rocky Flats Weapons Plant (RFP)	5,313	409	N/A[b]
Mound Laboratory	3,229	304	96
Los Alamos National Laboratory (LANL)	15,727	3,196	188
UK Atomic Energy Authority (UKAEA)	21,545	3,021	862
UK Atomic Weapons Establishment (UKAWE)	9,389	972	73
Sellafield Plant	10,276	2,144	1,317
Atomic Energy of Canada Limited (AECL)	8,977	878	315

[a]These are the numbers of workers and deaths included in dose-response analyses.
[b]Not available in Wilkinson et al. (1987), but the total person-Sv for workers included in combined US analyses was 241.

interest. The information in Table 9.1 is based on workers included in dose-response analyses; in most cases, these analyses were limited to workers who were monitored for external radiation, and at Hanford, they were also restricted to workers employed at least six months at the site. The United Kingdom studies and the Hanford study included both male and female workers, but most of the exposure in all studies was received by males.

The dose distributions in these studies were generally highly skewed to the right, with the majority of the population receiving relatively little exposure. For example, in the recent international study, about 60 percent of all workers monitored had cumulative doses (added up over all years of employment) less than 10 mSv; only 20 percent of monitored workers had accumulated doses of 50 mSv or more, and less than 2 percent (1,752 workers) had accumulated doses of 500 mSv or more (Cardis et al., 1995b). By contrast, risk estimates based on the Life Span Study of A-bomb survivors are strongly dependent on survivors with doses exceeding 1 Sv (1000 mSv).

STATISTICAL METHODS USED TO ANALYZE
DATA FROM WORKER STUDIES

External Comparisons

A first step in analyzing data from the worker studies has been to compare death rates of the study population to those of the general population using national vital statistics. For all the studies listed in Table 9.1 death rates were lower than those of the general population, in most cases substantially lower. With the exception of Sellafield, the all-cause standardized mortality ratios (SMRs) ranged from 0.62 to 0.82, and the all-cancer SMRs ranged from 0.71 to 0.86; in all cases (except Sellafield), these SMRs differed significantly from unity. For Sellafield, both the all-cause and all-cancer SMRs were 0.96, and did not differ significantly from unity.

The low SMRs exhibited in most of the nuclear worker studies are typical of groups employed in industries that are free from serious hazard and reflect the "healthy worker effect" (McMichael, 1976), a phenomenon that is not fully understood but probably results at least in part because those too ill to be employed have been selected out of the population. It is not clear why Sellafield differs from the other studies in not exhibiting such an effect.

Dose-Response Analysis

Factors related to the healthy worker effect have made SMRs difficult to interpret. Because of these difficulties, and also because overall SMRs that do not specifically examine more highly exposed workers may be greatly diluted, this approach is not ideal for examining the effects of occupational radiation exposure. For this latter purpose, dose-response analyses, which make use of available data on radiation dose, provide a less biased and more sensitive method of analysis. With this approach, death rates are compared by level of radiation exposure, and, in a sense, the very large number of workers with very little exposure serve as the control group.

The statistical methods used for dose-response analyses in the worker studies are similar to those that have been used to analyze data from the A-bomb survivor studies. In both the worker and A-bomb survivor studies, observed and internally based expected deaths by dose category have been calculated, trend tests that are sensitive to an increase in death rates with increasing radiation dose have been conducted for many different disease categories, and estimates that express the risk per unit of exposure have been obtained. A difference in the worker data and the A-bomb survivor data is that dose is protracted over time in the worker studies. This complicates analyses in that it is necessary to account for changes in cumulative dose as workers are followed over time, and makes it especially difficult to examine time-related factors such as the modifying effects of time since exposure, age at risk, and age at exposure.

Because an objective of the worker studies is to compare risk estimates with those that have been obtained through extrapolation from the A-bomb survivor data, it is important to use comparable models for estimating risks. Recent analyses of the A-bomb survivor data, especially analyses used to derive the risk models used by BEIR V (NRC, 1990) and UNSCEAR (UNSCEAR, 1988), have estimated the excess relative risk (ERR) based on a model in which the relative risk is given by $1 + \beta z$, where z is the dose and β is the ERR.

In analyses of the A-bomb survivor data, modification of the ERR by sex, time since exposure, age at exposure, and age at risk has been evaluated. Although some attempts have been made to evaluate these factors with the worker data, power was usually insufficient to distinguish among models. For this reason, analyses of worker data have drawn on findings from the A-bomb survivor studies for treatment of these factors. For example, doses in most analyses of worker data have been lagged for 2 years for leukemia and 10 years in analyses of other types of cancer because these were the choices used in recent risk assessments based on the A-bomb survivors. Also, most worker analyses have been based on a model in which the ERR remains constant as workers age, again an assumption that is reasonably consistent with A-bomb survivor data.

RESULTS OF DOSE-RESPONSE ANALYSES IN INDIVIDUAL STUDIES

Dose-response analyses have been conducted for all studies listed in Table 9.1, and have generally included analyses of all cancers (or all cancers excluding leukemia), leukemia (usually excluding chronic lymphocytic leukemia), and several other specific types of cancer. Most studies found no evidence of a statistically significant positive association of all-cancer mortality and radiation dose, but such associations were found for both Oak Ridge National Laboratory (ORNL) workers and for

United Kingdom Atomic Weapons Establishment (UKAWE) workers. Leukemia exhibited a statistically significant positive correlation in Mound Laboratory workers, Sellafield workers, and Atomic Energy of Canada Limited (AECL) workers.

A few other diseases showed statistically significant associations in single studies, but these can reasonably be explained as chance findings given the number of statistical tests conducted. Diseases showing statistically significant associations in more than one study include Hodgkin's disease in Hanford and Los Alamos National Laboratory (LANL) workers, lung cancer in ORNL and UKAWE workers, and cancer of the esophagus in ORNL (Gilbert et al., 1993a) and LANL workers. Earlier analyses of both Hanford and Sellafield workers had indicated a statistically significant association with multiple myeloma, but the association was no longer significant in the most recent analyses of data from either study.

Table 9.2 shows estimates of the ERR for all cancers excluding leukemia and for leukemia excluding chronic lymphocytic leukemia (CLL) for those studies where such estimates were calculated. These estimates were those presented in the papers on the individual studies listed in Table 9.1. For comparison, it is noted that linear estimates for male A-bomb survivors exposed between the ages of 20 and 60 were 0.18 per Sv for all cancers excluding leukemia and 3.7 per Sv for leukemia excluding CLL; these estimates were calculated at IARC as described in Cardis et al. (1995b). For leukemia excluding CLL, a linear estimate from a linear-quadratic dose-response function of 1.4 per Sv was also calculated.

For all cancers, the confidence limits from most studies included the possibility of zero or no risk, and also included estimates several times the linear estimates from male A-bomb survivors exposed as adults. For two of the smaller studies, ORNL and UKAWE, statistically significant correlations were identified, and the confidence limits excluded the A-bomb survivor-based estimates. The UKAWE correlation resulted primarily from a correlation for lung cancer, and nonmalignant respiratory disease showed a similar correlation; thus, smoking-related bias may explain this finding.

The ORNL estimates shown in Table 9.1 are those calculated by Wing et al. (1991). Gilbert et al. (1993a) also analyzed the ORNL data and obtained a smaller all-cancer ERR with confidence limits that included zero. The difference in results can be attributed to several differences in the statistical methods employed, particularly the choice of dose categories used in the analyses; a detailed discussion of these differences is found in Gilbert et al. (1993a). Also, the ORNL correlation was strongly influenced by deaths from cancers of the esophagus and larynx; both diseases have been linked with smoking (Doll and Peto, 1981).

As noted above, leukemia showed a significant positive correlation with radiation dose in Sellafield, Mound, and AECL workers. It can be seen that the confidence limits for all studies in Table 9.2, including both Sellafield and AECL, included both the linear and linear-quadratic extrapolations from the A-bomb survivor estimates. Although no risk estimates were presented for Mound workers, the correlation for

TABLE 9.2 Estimates of the excess relative risk (ERR) from several epidemiologic studies of workers monitored for external radiation.[a]

Study population	ERR$_{1Sv}$ (95% CI)	
	All cancers excluding leukemia[b]	Leukemia excluding CLL[c]
Hanford	-0.00 (<0.00, 1.00)	-1.10 (<0.00, 3.00)
Oak Ridge National Laboratory	3.30 (0.90, 5.70)	6.40 (<0, 27.00)
UK Atomic Energy Authority	0.80 (-1.00, 3.10)	-4.20 (-5.7, 2.60)
UK Atomic Weapons Establishment	7.60 (0.40, 15.00)	– –
Sellafield	0.10 (-0.40, 0.80)	14.00 (1.90, 71.00)
Atomic Energy of Canada Limited	0.05 (-0.70, 2.20)[d]	19.00 (0.10, 113.00)[d]

[a] These results are taken from the papers describing the individual studies.
[b] Results for ORNL and UKAWE include leukemia.
[c] Results for ORNL and UKAEA include chronic lymphocytic leukemia.
[d] 90% confidence intervals.

this study resulted from two deaths with doses exceeding 50 mSv, one of which was CLL.

COMBINED ANALYSES OF DATA ON NUCLEAR WORKERS

Description of National and International Combined Analyses

A broad assessment based on the totality of evidence from all studies is obviously needed, and, for this reason, analyses based on combined data from several studies have been carried out at both national and international levels. A major objective of these combined analyses is to obtain more precise estimates of risk than those obtained from individual studies, which have generally exhibited very wide confidence limits (Table 9.2).

In the United States, analyses of data on workers at Hanford, ORNL, and Rocky Flats were published in 1989 (Gilbert et al., 1989) and later updated (Gilbert et al., 1993b). Data on Mound and LANL workers were not available at the time these combined analyses were conducted. In the United Kingdom, results of combined analyses of the three facilities listed in Table 9.1 were published in 1994 (Carpenter et al., 1994), and these analyses are subsequently referred to as "combined UK analyses."

In addition, a study of National Registry of Radiation Workers (NRRW) was published in 1992 (Kendall et al., 1992). This is in a sense a combined analyses in that it includes workers in the facilities listed in Table 9.1, as well as workers in several smaller facilities. The total number of workers in the NRRW study was more than double the number of workers in the combined UK study. However, because many of the additional workers in the NRRW study were recent entrants to the industry, and because of the longer period of follow-up for some workers in the combined UK study, the number of deaths in the two studies was very similar. Results from the two studies were also very similar, and subsequently only the combined UK analyses as reported by Carpenter et al. (1994) are considered in this paper.

International combined analyses have also been carried out, which included the studies listed in Table 9.1 with the exception of Mound and LANL (data from Mound and LANL were not available at the time the international analyses were initiated). This effort was coordinated by the International Agency for Research on Cancer (IARC), but all investigators for the individual studies participated in the planning and interpretation of the data. Comparability of dosimetry for different studies and in different time periods received special attention from a dosimetry subcommittee comprised of dosimetry experts from the three countries. The main findings from the international analyses were published in 1994 (IARC, 1994), and a more detailed description of these results was published recently (Cardis et al., 1995b).

Table 9.3 shows characteristics of the study populations included in the national and international combined analyses; the single Canadian study is also shown for completeness. The numbers from the three countries do not add up exactly because criteria for inclusion of subjects were not identical. For example, international analyses included only subjects that had been employed at least six months at the facility, while the UK and Canadian studies did not have such a restriction.

Results of Combined Analyses of All Cancers Excluding Leukemia and Leukemia Excluding Chronic Lymphocytic Leukemia (CLL)

Table 9.4 shows estimates and confidence limits for the ERR based on the national and international analyses. For comparison, estimates for male A-bomb survivors exposed between the ages of 20 and 60 are also shown; as noted above, these estimates were calculated at IARC as described in Cardis et al. (1995b).

TABLE 9.3 Number of workers, number of deaths, and collective dose equivalent from national and international analyses of workers monitored for external radiation.

Study population	Number of workers[a]	Number of deaths[a]	Collective dose equivalent (Sv)
United States[b]	44,943	7,863	1,237
United Kingdom[c]	40,761	6,900	2,303
Canada[d]	8,977	878	315
International analyses	95,673	15,825	3,843

[a]These are the numbers of workers and deaths included in dose-response analyses.
[b]Includes workers at Hanford, ORNL, and Rocky Flats.
[c]Includes workers at UKAEA, UKAWE, and Sellafield.
[d]Includes workers at AECL.

For both the national and international combined analyses, the estimates for all cancers excluding leukemia were very close to zero. However, the upper confidence limit based on the international analyses was lower than those for any of the individual studies. That is, in this case, the objective of obtaining more precise estimates and tighter confidence limits was achieved. These confidence limits included both negative values and values larger than the linear A-bomb survivor-based estimate.

For leukemia excluding CLL, the United States estimate was negative, but both the United Kingdom and Canadian studies showed significant positive correlations. Given the wide confidence limits, the findings were not necessarily incompatible. A test for homogeneity of the leukemia ERR estimates among the individual facilities included in the international analyses yielded a p-value of 0.08; such tests, however, have limited statistical power.

The estimate for leukemia excluding CLL, based on the international data, was intermediate between the combined US and combined UK estimates, with confidence limits extending from very close to zero to about twice the linear A-bomb survivor-based estimates. The leukemia ERR estimate differed significantly from zero, with a one-tailed p-value of almost exactly 0.05. The result was strongly driven by six leukemia deaths with cumulative doses exceeding 400 mSv, compared with 2.3 deaths expected in this category; four of these deaths occurred in Sellafield workers. Statistical significance was not achieved when analyses were restricted to doses below 400 mSv.

The lengths of the confidence limits from the A-bomb survivors may be contrasted with those from the international worker study. Risks have been much more

TABLE 9.4 Estimates of the excess relative risk (ERR) from national and international analyses of data from studies of workers monitored for external radiation.

	ERR_{1Sv} (90% CI)	
Study population	All cancers excluding leukemia	Leukemia excluding CLL
United States	0.00 (<0.00, 0.80)	-1.00 (<0.00, 2.20)
United Kingdom	-0.00 (-0.50, 0.60)[a]	4.20 (0.40, 13.00)[a]
Canada	0.05 (-0.70, 2.20)	19.00 (0.10, 113.00)
International	-0.07 (-0.40, 0.30)	2.20 (0.10, 5.70)
Japanese atomic bomb survivors[b]	0.18 (0.05, 0.34)	3.70 (2.00, 6.50)

[a] 95% confidence intervals.
[b] These estimates and confidence intervals were calculated at IARC, and were based on male survivors exposed between the ages of 20 and 60.

clearly demonstrated in the A-bomb survivor study, and the A-bomb survivor estimates are more precise than those from the workers. Also, A-bomb survivor estimates based on all the data including females and those exposed in childhood are of course more precise that those shown in Table 9.4. Nevertheless, the precision of the worker-based estimates is beginning to approach that for the A-bomb survivors. Of course, uncertainties resulting from dosimetry and confounding are not included in these confidence limits, and the latter are likely to be more severe for low-dose data than for high-dose data.

Results of Combined Analyses of Specific Cancer Types

Both national and international analyses included statistical tests for the association between radiation exposure and development of cancer. In the US combined analyses, these tests yielded statistically significant results (one-tailed p<0.05) for cancers of the esophagus, cancer of the larynx, and Hodgkin's disease. In the

UK combined analyses, melanoma and other skin cancers, and ill-defined and secondary cancers showed statistically significant associations with radiation dose. For none of these disease categories is there any a priori reason based on results from high-dose studies to expect an especially strong correlation with radiation dose.

In the international analyses, 30 specific cancer types (subcategories of the all-cancer-excluding-leukemia category) were tested for an association with radiation dose. Of these, multiple myeloma was the only type of cancer to exhibit a statistically significant result, with a p-value of 0.037. With the standard ten-year lag period, this disease had not exhibited a significant correlation with radiation dose in any of the national analyses, but had shown a significant correlation in earlier combined US analyses (Gilbert et al., 1989) and in the combined UK study with a 20-year lag. Given the number of tests conducted, obtaining such a correlation for a single type of cancer is not inconsistent with chance. Recent analyses of cancer incidence data on A-bomb survivors have not shown a statistically significant correlation of multiple myeloma and radiation dose. Thus, there is no a priori reason to expect a correlation for multiple myeloma that is stronger than that expected for other types of cancer. Further discussion of the multiple myeloma correlation is found in Cardis et al. (1995b) and in Gilbert et al. (1993b).

ADDITIONAL UNCERTAINTIES IN RISK ESTIMATES DERIVED FROM NUCLEAR WORKER STUDIES

The confidence limits presented above include only uncertainty from sampling variation. Additional uncertainty derives because of potential bias resulting from confounding and from the fact that dose estimates are subject to errors.

Confounding

The national and international combined analyses, and analyses of data from individual studies, were adjusted for age, calendar year period, sex, and sometimes additional variables. This was usually accomplished through stratification on these variables.

For the international analyses, a special effort was made to develop a measure of socioeconomic status for each contributing study, although the measure differed from study to study depending on the available data. It was hoped that this socioeconomic measure might serve as a surrogate measure for smoking, diet, and other lifestyle factors that might confound analyses. For the Hanford study, job category data were used to assign socioeconomic status in a manner that was reasonably comparable to that used to assigned social class for the UKAEA and UKAWE studies. For ORNL, socioeconomic status was determined by paycode (monthly, weekly, hourly); for Rocky Flats by education; and for Sellafield by industrial or non-industrial job category. For AECL, no measure of socioeconomic status was available. In addition, the international analyses explored the role of variables such as length of employment and time since initial employment, although the final analyses did not include adjustment for these variables.

In the international analyses, supplementary analyses were performed with various alternative treatments of confounding factors. The overall conclusion—that results were consistent with a range of estimates from zero to estimates a few times larger than linear A-bomb survivor estimates—was unaltered by these alternative analyses; however, the precise quantification of confidence limits was affected. For example, adding adjustment for duration of employment increased the upper confidence limit on the all-cancer-excluding-leukemia estimate from 0.3 to 0.5 per Sv and led to a negative lower confidence limit for leukemia excluding CLL.

Unfortunately, adequate data on every factor that might influence cancer mortality analyses were not available. Thus, results might be biased, positively or negatively, because of inadequately measured or unidentified confounders. Bias is a concern in all epidemiologic studies, but is less important in studies in which exposures of interest increase the risk of disease by a factor of two or more. In nuclear-worker studies, where for most workers and diseases the increase is much less than a factor of two, biases may be larger than the expected effects, and this must be kept in mind in interpreting results of these studies.

Dosimetry

A major strength of the worker studies is the availability of objective quantitative measurements of exposure, and the quality of these measurements greatly exceeds exposure data usually available in occupational studies. Nevertheless, the recorded estimates obtained from personal dosimeters are subject to several sources of potential bias and uncertainty.

For the purpose of comparison of worker-based estimates with those obtained from the A-bomb survivor data, estimates should ideally be expressed in terms of absorbed dose to various organs. The objective of current dosimetry systems, however, is to estimate deep dose (energy absorbed at a depth of 1 cm in tissue), and in earlier years the objective dose may not have been as clearly defined. The bias in recorded dose as an estimate of bone marrow dose or doses to other organs depends on the dosimetry system in use, the energy of the radiation, and the geometry (direction from which the radiation was received). A difficulty in estimating the bias is that the energy and geometry are not known for any individual exposure, and even average values for a facility are difficult to determine.

At Hanford, extensive efforts have been made to document historical dosimetry practices (Wilson et al., 1990) and, recently, to estimate the bias in recorded doses as estimates of bone marrow dose and lung dose (as a typical internal organ for use in evaluating risks from all cancer excluding leukemia), and also the uncertainty associated with this estimated bias (Gilbert et al., 1996). This latter effort estimated that the recorded dose overestimates bone marrow dose by about 50%, and overestimates lung dose by about 10%, but with large uncertainty for both factors.

In addition, a variety of sensitivity analyses have been carried out on the Hanford data, addressing potential dosimetry biases for several different sources (Gilbert and Fix, 1995a). Overall, the results of these sensitivity analyses were reassuring and did not greatly modify overall conclusions of the study. Nevertheless, as was the case with different approaches to adjusting for confounders, the precise quantification of confidence limits was affected.

As part of the international analyses, dosimetry experts from each of the three countries met several times to evaluate the comparability of recorded doses from different time periods and from different facilities (Cardis and Esteve, 1991). For the most part, dose estimates were judged to be reasonably comparable and reliable. However, it was concluded that doses from low-energy photons and from neutrons were not reliably measured, especially in earlier time periods. Also, external dose estimates do not reflect doses that might have been received from internal depositions. Fortunately, for most workers, most exposure is from high-energy photons, and such exposure was judged to be reliably estimated. Some efforts were made in the international analyses to identify workers with potential dosimetry problems and to conduct supplementary analyses with these workers excluded; these analyses did not greatly modify the estimated ERR, but did increase the upper confidence limits, primarily because of the smaller sample size. The dosimetry experts also estimated that recorded dose would overestimate bone marrow dose by about 20%; however, this evaluation was not as detailed or extensive as that noted above for the Hanford study.

CONCLUSIONS

The radiation worker studies are probably the most informative of the various low-dose studies that have been conducted. These studies have a wide range of doses that have been objectively measured through the use of personal dosimeters. The recent international analyses included nearly 100,000 workers and a collective dose of about 3800 Sv. What do these studies tell us about the importance of the A-bomb survivor data for estimating risks of radiation exposures at low doses and dose rates?

First, the worker studies have clearly demonstrated the continued need for the use of high-dose data to obtain estimates that are sufficiently precise for risk assessment purposes. Although combining data from studies of workers in three countries provides more precise risk estimates than individual studies, the confidence intervals for these combined estimates were still large. For cancer excluding leukemia, the lower confidence limits were negative, while for leukemia excluding CLL, the lower bound was very close to zero; for both disease categories, the upper confidence limits were about twice as high as estimates obtained from linear extrapolation from the A-bomb survivor data.

The precision of estimates obtained from worker studies can be expected to improve in the future. Only about 15% of the workers in the international study were dead by the end of the follow-up period included in this study, and, thus, there is a great deal more that can be learned from these data. Also, a collaborative study of additional nuclear workers in several countries is being conducted, with IARC serving as the coordinating agency (Cardis and Esteve, 1992). To realize the full potential of these new studies, these workers will need to be followed for many years, but they may eventually help us to overcome some of the sample size problems, and thus reduce the statistical uncertainty in risk estimates. Uncertainty resulting from confounding is not so easily overcome and is especially troubling in low-dose studies. Although the worker studies can contribute important supplemental information, it seems likely that the A-bomb survivor studies will continue to be the main source of information on low-level radiation exposure.

A second way that the worker studies have demonstrated the usefulness of the A-bomb survivor data is by confirming that risk estimates obtained through extrapolation are reasonably appropriate. The direct study of workers, especially the combined international analysis, has shown that it is unlikely that linear extrapolation has seriously underestimated risks. This is important because claims have been made by some that risks might be underestimated by an order of magnitude or more (Stewart and Kneale, 1990). The upper confidence limits from the international analyses fairly effectively rule out this possibility.

There is of course interest in whether risks at low doses and dose rates might be less than those predicted by extrapolation from the A-bomb survivor data, and whether or not there might be a threshold below which there is no risk. The statistically significant correlation of leukemia risk and radiation dose observed in the international study might be interpreted as providing direct evidence of risk at low doses and low dose rates. However, the fact that the result is just barely significant (one-tailed $p = 0.046$) and the possibility of bias resulting from either confounding or dosimetry mean that this conclusion should be drawn somewhat tentatively.

Unfortunately, the worker studies are uninformative on the issue of whether extrapolation might have overestimated risks. Although worker results are compatible with reduced risks at low doses, they are also compatible with risks that are higher than those predicted by high-dose studies. Sampling uncertainty alone makes it extremely difficult to distinguish between no risk and the very small risks that would be predicted in workers if linear extrapolation were correct. Additional uncertainty from confounding and dosimetry add to the difficulty.

To conclude, results of studies of workers thus far demonstrate both the importance and appropriateness of the A-bomb survivor data for the purpose of estimating risks of radiation exposure at low doses and dose rates. Worker studies have already provided useful upper confidence limits on risks. Although there will always be limitations to the worker data, continued follow-up, new studies of nuclear power workers, and combined analyses should lead to more precise estimates of risk based on a direct assessment at low doses and dose rates, and thus provide important information supplemental to that obtained from the A-bomb survivors.

PART 3
MUTAGENESIS AND CARCINOGENESIS

10

Studies on the Genetic Effects of the Atomic Bombs: Past, Present, and Future

JAMES V. NEEL, JUN-ICHI ASAKAWA, RORK KUICK,
SAMIR M. HANASH, AND CHIYOKO SATOH

SUMMARY

The results of some 48 years of studies on the potential genetic effects of the atomic bombs are reviewed. There are no statistically significant differences between the children of the exposed survivors and a suitable comparison group of children with respect to frequency of congenital defects, sex ratio, survival, cancer, growth and development, cytological abnormality, and mutation affecting a variety of biological markers. However, there is a small deviation in the direction of a genetic effect, and this is used to develop an estimate of the mutational doubling dose of acute ionizing radiation of between 1.7 and 2.2 Sv equivalents. It is suggested that a review of the mouse data on the genetic effects of radiation is in satisfactory agreement with this estimate. Possible future studies at the DNA level on immortalized cell lines are discussed. The possibility that an increase in tumors of all types can be used as a "pit canary" for genetic damage is also discussed.

INTRODUCTION

This program provides an unusual opportunity not only to review the past studies at Atomic Bomb Casualty Commission/Radiation Effects Research Foundation (ABCC/RERF) on the genetic effects of the atomic bombs, but also to consider what, if any, future studies are indicated, and I shall attempt to do both. Past studies to be summarized are the work of many minds and hands, all recognized in our various publications. It is impractical in this brief presentation to mention the names of all investigators with significant involvement over the past 48 years as I proceed, but I note especially the major role Jack Schull has played in these studies. I will also in this presentation mention some recent unpublished work and calculations, and note that my co-authors on this paper are Dr. J. Asakawa, Dr. C. Satoh, Mr. R. Kuick, and Dr. S. Hanash.

BRIEF SUMMARY OF PAST STUDIES ON THE GENETIC EFFECTS OF ATOMIC BOMBS

The genetic studies of the children of survivors of the atomic bombings have been summarized several times in the past five years (Neel et al., 1990; Neel and Schull, 1991; Neel, 1995), and I propose to be very brief in the present summary. As of now, the reproductive performance of atomic bomb survivors living in Hiroshima and Nagasaki is complete. Most of the studies to be mentioned have been based on a cohort (or subset thereof) consisting of 31,150 children born to parents one or both of whom were within 2,000 meters of the hypocenter at the time of the bombing (ATB) and a matched comparison cohort of 41,066 children born to survivors beyond this distance, or to parents now living in either city but not there ATB. Over the years, data have been collected on these children (F_1) concerning untoward pregnancy outcomes (major congenital malformation and/or stillbirth and/or neonatal death); sex of child; malignant tumors with onset prior to age 20; death of liveborn infants through an average life expectancy of 26.2 years, exclusive of death from malignancy; growth and development of liveborn infants; cytogenetic abnormalities; and mutations altering the electrophoretic behavior or function of a selected battery of erythrocyte and blood plasma proteins. Most of the parents have been assigned radiation exposures based on the DS86 schedule; where this was impossible, DS86-type doses have been approximated. All of the data have been analyzed on the basis of these exposures. In a series of analyses that involved fitting a linear dose-response regression model to the occurrence of the various indicators of radiation damage, the model including up to six concomitant variables depending on the indicators in question, there was no statistically significant effect of combined parental exposure on any of these indicators. I wish to emphasize that, relatively limited though our knowledge of mammalian radiation genetics was in 1947, this possible outcome was anticipated (Genetics Conference, 1947), but the decision was made to proceed nevertheless with what has become the largest study in genetic epidemiology ever undertaken on the basis that, whatever the outcome, it was of profound interest.

In the planning stages of the study, inasmuch as most births were midwife-attended home deliveries, we were gravely concerned over the possibility of concealment of congenital abnormalities, since such births, in a society with open *koseki* records, might be concealed because they might stigmatize the family. We went to great pains to enlist the cooperation of the midwives and obstetricians, who readily grasped the significance of the study. The attendant at delivery was urged to report at once any abnormality, whereupon a physician employed by ABCC examined the baby as soon as possible; in any event, all newborns were examined by an ABCC physician within two or three weeks of birth, and a 30% sample reexamined at approximately nine months of age. For what it is worth, the data on congenital defects in the children of unexposed parents were very similar to world figures (Neel, 1958). This clinical program was terminated in 1953.

A fact that we believe went far to ensure the quality of the data was that for some years after the end of WWII, the Japanese government maintained the wartime system of special rations for Japanese women who registered their pregnancies at the completion of the fifth lunar month. By incorporating a registration for the genetic studies into this ration system, we had the basis for a prospective study of pregnancy outcomes, an aspect of the study design whose importance we cannot overemphasize. The registrations for the genetic studies were initiated in March 1948; these children would have been conceived in October 1947. There is thus a hiatus of some two years between the bombings and the first direct collection of data (but births in those first two years were incorporated into certain of the study groups).

A convenient metric for measuring the genetic effects of ionizing radiation is the estimated dose of radiation necessary to equal the impact of spontaneous mutation on the indicator in question, i.e., the doubling dose. If we assume that despite the absence of a statistically significant regression of any of the indicators on parental exposure, the regressions that were derived are nonetheless the best current indicators of the genetic effects of radiation on humans, then these regressions can be employed to generate a doubling dose. In order to make such an estimate, one must have an estimate of the contribution of spontaneous mutation in the parents to each of the indicators. This is sometimes relatively easy; sometimes, as in the case of major congenital defect, it is more difficult. Also, the estimate is restricted to the effects of the mutation of single genes with discrete effects. Thus, it was not useful to calculate a doubling dose for such indicators as physical growth and development and the sex ratio (although neither of these yielded findings suggestive of a radiation effect). For the remaining five indicators, the findings were as shown in Table 10.1. Since all the regressions evaluate the effects of essentially non-overlapping indicators (i.e., those that are independent of each other) and are based on the same cohorts or subsets thereof, it is legitimate to combine these additively. The sum of the five individual regressions is 0.00375/Sv equivalent, with a considerable but difficult-to-define error. We estimated, as shown in Table 10.1, that in each generation of newborns, something like 6.3 to 8.4 infants per 1,000 would ultimately exhibit some one of these five indicators because of mutation in the parental generation. The estimate of the doubling dose thus falls between 0.00632 and 0.00375 Sv equivalents and 0.00835 and 0.00375 Sv equivalents, or 1.68 to 2.22 Sv equivalents, with a poorly defined error term.

TABLE 10.1 A summary of the regression of the various indicators on parental radiation exposure and the impact of spontaneous mutation on the indicator, after Neel et al. (1990).

Trait	Regression coefficient, β, for combined parental dose (Sv)	s.e.(β)	Contribution of spontaneous mutation
UPO	+0.00264	±0.00277	
F_1 mortality	+0.00076	±0.00154	0.0033-0.0053
Protein mutations	-0.00001	±0.00001	
Sex-chromosome aneuploids	+0.00044	±0.00069	0.0030
F_1 cancer	-0.00008	±0.00028	0.00002-0.00005
Total	0.00375		0.00632-0.00835

Most of the gonadal ionizing radiation humans will experience will be low-level, chronic, or intermittent, rather than the acute (single dose), relatively high exposures of the atomic bombs. I say "relatively" because the average gonadal dose for both parents combined varied in the various different studies between 0.32 and 0.60 Sv equivalents; in the calculation of these doses the relatively low neutron component was assigned a relative biological effectiveness (RBE) of 20. In the mouse experiments, the gonadal doses were usually 3.0 or 6.0 Gy. Acute radiation at relatively high doses is on a per-unit basis genetically more mutagenic than chronic radiation, the dose-rate factor ranging from 3 to 10, depending on the precise indicator. The best studied system in mice, the Russell 7-locus test, yielded one of the lowest dose-rate factors, a factor of 3. Because of the linear-quadratic nature of the radiation dose–genetic damage curve, we suggest that the dose-rate factor to be used in extrapolating from the effects of acute to chronic radiation on the basis of the Japanese data should be 2, so that the doubling dose for low-level chronic ionizing radiation in humans becomes 3.4 to 4.4 Sv equivalents.

A COMPARISON OF THE FINDINGS IN HIROSHIMA–NAGASAKI WITH STUDIES OF THE GENETIC EFFECTS OF IONIZING RADIATION ON MICE

This doubling dose estimate for humans is substantially higher than the extrapolation to humans from the results of experiments on mice that dominated thinking about genetic risks from approximately 1950 to 1985. The figure most often quoted for acute radiation was about 0.4 Gy. Accordingly, in 1989, together with Susan Lewis (Neel and Lewis, 1990), I undertook a review of the accumulated data on mouse radiation genetics, including deriving, for each indicator that had been studied, an estimate of the doubling dose. The results with respect to specific locus tests are shown in Table 10.2. Note the wide range in the various estimates, to which we found it impossible to attach errors in the usual statistical sense. Not shown there (because the data do not lend themselves to the calculation of a doubling dose) are the important results of Roderick (1983), who estimated for mice a recessive lethal mutation rate in post-spermatogonial cells per locus from ionizing radiation of only $0.35 \times 10^{-8}/0.01$ Gy, whereas for the Russell 7-locus system, the corresponding rate for all mutations was $45.32 \times 10^{-8}/0.01$ Gy, approximately 80% of these mutations being homozygous lethal. As Roderick pointed out, this was about one hundredfold difference, although the error term to be attached to his estimate was large but difficult to calculate. The simple average of all the estimates in Table 10.2, unweighted because we could think of no good way to weight the individual studies, was 1.35 Gy, with an indeterminate error. Given the relatively high doses at which the mouse experiments had been performed, we felt that in extrapolating to the effects of low-level, chronic, or small intermittent exposures, a dose-rate factor of 3 was appropriate (Russell et al., 1958). The estimate of the genetic doubling dose of chronic ionizing radiation thus became 4.05 Gy, in surprising agreement with the estimate for human exposures. But note that the estimate for mice is gametic, whereas that for humans is zygotic.

TABLE 10.2 A summary of the various indicators of genetic radiation damage pursued in murine experiments and the doubling dose they yield (Neel and Lewis, 1990).

System	Doubling dose (Gy)	Strain of origin of treated males
Russell 7-locus	0.44	101 x C3H
Dominant visibles	0.16	various
Dominant cataract	1.57	101/El x C3H/El
Skeletal malformations	0.26	101
Histocompatibility loci	>2.60	C57Bl/6JN
Recessive lethals	0.51 ⎫ 0.80 ⎬ 1.77 4.00 ⎭	DBA C3H/HeHx101/H CBA, C3H
Loci encoding for proteins	0.11	various
Recessive visibles	3.89	C3H/HeH x 101/H
Average	1.35	

In retrospect, it is instructive to consider why the results of the Russell system so dominated the thinking of 20 and 30 years ago. It was a simple system yielding clear-cut results for which a great deal of data became available. When other studies suggested lesser genetic effects of ionizing radiation, the results were often dismissed as reflecting the use of "less sensitive" systems. Furthermore, the desire to be conservative in risk setting led to reliance on the systems that yielded the most striking results. However, I suggest that—for reasons Russell could not know at the time, and which will now be enumerated—the loci he selected probably had higher spontaneous and induced mutation rates than the average locus.

Russell in his very first papers recognized that the key assumption was that his loci were representative of the genome. There are now data for the mouse indicating a sevenfold range in the rate per locus with which spontaneous mutation results in phenotypic effects (Green et al., 1965; Schlager and Dickie, 1967). With respect to humans, Chakraborty and Neel (1989) have suggested from population data a tenfold range in the spontaneous mutability of human genes. In Russell's data, radiation produced 18 times more mutations at the $\underline{s}$ locus than at the $\underline{a}$ locus, surely a signal to extrapolate with caution (reviewed in Searle, 1974). Finally, in a somatic cell mutagenization experiment in our laboratory, utilizing the TK6 line of human lymphocytoid cells and employing ethylnitrosourea as mutagen, the protein products of 263 loci were scored for the occurrence of mutants resulting in electrophoretic variants, employing a two-dimensional polyacrylamide gel system (Hanash et al., 1988). Ten of the 263 loci whose protein products were being scored were known to be associated with genetic polymorphisms, on the basis of family-oriented studies on gels derived from human peripheral lymphocytes. The induced mutation rate at these 10 loci was 3.6 times greater than at the monomorphic loci, an observation with a probability <0.004. The relevance of all these observations to the possible bias in the Russell system is that to set the system up, Russell drew on loci characterized by genetic variation. There had to be at least two alleles known for each of the loci in his system, and it helped in creating the optimum phenotype for scoring if there were even more alleles available to choose among. This use of loci for which variants were readily available introduced the bias.

There is an additional reason why the previous extrapolation from mouse to man was conservative. The mouse doubling-dose estimate was male based. The demonstration (Russell, 1965) that although in the first few litters post-treatment the offspring of radiated female mice exhibited about the same amount of genetic damage as the offspring of radiated males, there was no apparent damage in the later litters of these females, created a dilemma for risk setting. Was the human female similar to the mouse female in this respect? The estimates of Table 10.2 are male based. To be conservative, in extrapolating to the human situation, the mouse male-derived risks were applied to both sexes. In the Japanese data, however, radiated females contribute about half the dose.

As a result of the studies on the genetic effects of the A-bombs, plus this evaluation of the totality of the murine data, the case for a major revision downward in the previous estimates of the genetic risks of radiation for humans must be very seriously considered.

SOME CURRENT USES OF THE RESULTS OF THE ABCC/RERF GENETIC STUDIES

The results of studies on radiation-induced oncogenesis at RERF have provided worldwide standards. The results of the genetic studies are beginning to do likewise. Without doubt, the most spectacular use of these data has been in connection with the suit brought against British Nuclear Fuel, with respect to the allegation that the "hot spot" for childhood leukemia in Seascale, West Cumbria, England, was related to the employment of four of the fathers of these children in the nearby Sellafield Nuclear Reprocessing Facility, i.e., that the high number of cases was a result of leukemogenic germ-line mutations induced in these fathers (Gardner et al., 1987). The radiation received by the fathers in question was well within the occupational guidelines, i.e., no father over a working period of 6-13 years received more than an estimated total exposure of 200 mSv. Thus, in effect this legal action, which became the most expensive civil suit in the history of the English courts, was a challenge to existing guidelines. The data on leukemia in the F_1 of Hiroshima and Nagasaki, including a subset of the data on malignancy in the F_1 (Yoshimoto et al., 1990), became pivotal to the trial. A precise calculation is impossible, but to a first approximation a genetic interpretation of the Seascale leukemia cluster implied genetic sensitivities some 4,000 to 6,000 times greater than suggested by the Japanese data. The trial thus challenged existing genetic guidelines. I was amazed at the lengths to which plaintiff's counsel went to attempt to discredit the Japanese study in their efforts to legitimize their own case. In the end, the data emerged unscathed, and the judge found resoundingly for the defense. The study that precipitated this trial is probably the most egregious example of a false positive in genetic epidemiology on record (Neel, 1994a).

The analysis of the F_1 data that we just summarized was not completed in time to figure strongly in the results of the BEIR V report but did receive prominent consideration in the most recent (1993) update of the United Nations Scientific Committee on the Effects of Atomic Radiation (UNSCEAR), entitled, "Sources and Effects of Ionizing Radiation."

NEW TECHNIQUES FOR EVALUATING THE GENETIC EFFECTS OF ATOMIC BOMBS

In the 1980s, it became apparent that the techniques of molecular genetics might be brought to bear on the question of the genetic effects of the bombs. Accordingly, in 1985, the RERF began the task of establishing cell lines appropriate to such studies, the goal being approximately 500 mother/father/child trios in which one or both parents had been exposed to the atomic bombs, and a similar set of comparison constellations in which neither parent had been exposed. Where more than one child was available, he or she was included in the study, and the mean number of children per parental set was 1.45. The task of establishing the cell lines is almost completed. Meanwhile, the genetics staff launched on an exploration of the technologies that might be employed. To date, three systems utilizing the DNA approach have been piloted by RERF staff and their collaborators.

A DGGE System

The first of these employed the denaturing gradient gel electrophoresis (DGGE) technique of Myers et al. (1985) and Lerman et al. (1986) primarily in combination with other techniques to detect unknown single nucleotide substitutions (Hiyama et al., 1990; Takahashi et al., 1990; Satoh et al., 1993). In terms of nucleotides examined per unit of technician time, the most efficient of the several variations of this approach which were explored involved amplification of target sequences with PCR, digestion of these sequences to fragments of approximately 500 bp, followed by DGGE of the fragments (Satoh et al., 1993). A total of 6,724 bp were examined per individual, using the human coagulation factor IX gene as substrate. In this pilot study, samples from 63 couples and 100 of their children were examined. Half of the children were born to parents one or both of whom had been exposed to the atomic bombs. Eleven previously undescribed nucleotide substitutions were detected in the parents. No mutations were detected in the approximately 672,000 nucleotides examined in the 100 children in the study, not surprising in view of the estimated spontaneous mutation rate per nucleotide per generation of 1×10^{-8} (Neel et al., 1986). This technique is most efficient in the detection of nucleotide substitutions.

The Use of Minisatellites

The second approach, published under the authorship of Kodaira et al. (1995), concentrates on studying mutation that alters the length of minisatellite loci, also known as VNTR loci (Variable Number of Tandem Repeats). Employing Southern blot analysis of a battery of six minisatellites (*Pc-1, 1λM-18, ChdTC-15, pλg3, λMS-1, and CEB-1*), the investigators identified 6 mutations in 390 alleles from 65 parents whose gametes represented an average exposure of 1.9 Sv equivalents, a mutation rate of 1.5%; and 22 mutations in 1,098 alleles of the 183 gametes from the unexposed parents, a mutation rate of 2.0%. The difference is not statistically significant and, in any event, in a direction opposite to hypothesis. The authors calculate that, given the observed spontaneous mutation rate and using standard power function statistics (a type I error of 0.05 and a type II error of 0.20), it would be necessary to survey two samples (exposed and unexposed) of 1,188 germ cells each to detect a significant difference at the 0.05 level. Certainly, given the need to extend our knowledge of the genetic effects of radiation, the series should be extended at least that far. Furthermore, given the general acceptance of the fact that ionizing radiation produces mutations, even without a significant difference between the two datasets, these data, like the data from the previous studies, can be taken at face value and used to produce a doubling dose estimate for this phenomenon; but this can only be done if, unlike the present situation, there is an excess of mutations in the children of exposed parents. However, even if the present deficiency of mutations in the children of exposed parents were to persist in an expanded series, the data can be used, at stated probability levels, to place a lower limit on the doubling dose.

Because the function of these minisatellites is so poorly understood, they are not a very satisfactory marker of radiation damage for a public interested in the phenotypic impact of an increased mutation rate on its children. However, it occurred to me, in writing an invited editorial to accompany the paper of Kodaira et al. (Neel, 1995), that these studies might have a value that could not have been anticipated a few years ago. Since 1991, eleven diseases have been recognized for which the mutational basis is an increase in the numbers of a specific trinucleotide repeat embedded in the gene (reviewed in Ashley and Warren, 1995). Some of these are quite well-known causes of mutational morbidity (e.g., the fragile X syndrome, myotonic dystrophy, Huntington's chorea). The basic repeat unit in the minisatellites employed in the study of Kodaira et al. varies from 5 to 43 nucleotides. If mutation at these loci can be shown to follow the same principles as mutation affecting the length of minisatellites, then the latter may in the future serve as surrogates for the former.

Two-Dimensional DNA Gels

We come now to a third approach, in which our group in Ann Arbor has been collaborating with Dr. Asakawa of RERF. While RERF was establishing the cell lines, we in Ann Arbor devoted considerable effort to exploring the utility of two-dimensional polyacrylamide gel electrophoresis of protein solutions in the study of mutation (Neel et al., 1984; Neel et al., 1989). Figure 10.1 illustrates a typical preparation based on peripheral lymphocytes, in which approximately 200 proteins can be visualized with the clarity necessary for unequivocal classification. We obtained good identification of electrophoretic variants of these proteins, but, as a result of the many steps between gene and protein quantity, only a small subset of these proteins were quantitatively so reproducible that there was satisfactory discrimination between those 50% and those 100% in protein quantity—a vital prerequisite to the study of null mutations. However, we were successful in developing an algorithm that, with a minimum of operator intervention, would compare the gels of a father, mother, and child, with the objective of identifying characteristics of the child's gel not present in either parent, and so putative mutations (Skolnick and Neel, 1986; Kuick et al., 1991). This approach to mutation requires that gels be run on the child and both parents.

Beginning in 1979, techniques became available for visualizing the DNA fragments resulting from genomic digests on a two-dimensional gel (Fischer and

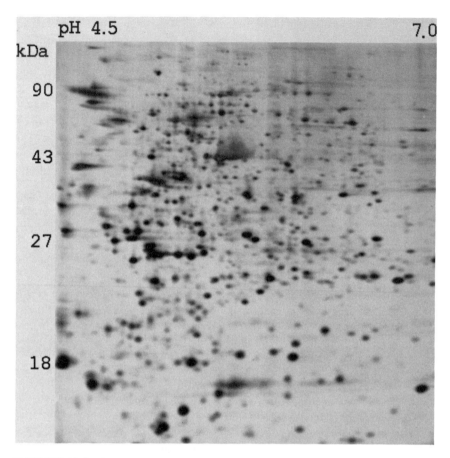

FIGURE 10.1   A computer-processed image of a silver-stained two-dimensional polyacrylamide gel of the protein contents of peripheral lymphocytes.

Lerman, 1979; Uitterlinden et al., 1989; Yi et al., 1990; Hatada et al., 1991). We are now exploring the applicability of this approach to studying the genetic effects of the atomic bombs, employing a modified version of the technique described by Hatada et al. (1991) and developed by Asakawa (1995). Hatada et al. (1991) have termed this technique "restriction landmark genomic scanning" (RLGS). A gel based on a lymphocytoid cell line in the RERF collection in shown in Figure 10.2.   For a diploid organism such as our species, in the absence of sex-linkage or genetic variation, each spot is the product of two homologous fragments. For these preparations, genomic DNA was digested with *Not*I and *Eco*RV restriction enzymes and the *Not*I-derived 5' protruding ends were α-^{32}P labeled. These fragments were electrophoretically separated in an agarose disc gel, which was subsequently treated with *Hinf*I to further cleave the fragments in situ. The resulting fragments are separated perpendicularly in a 5.25% polyacrylamide gel

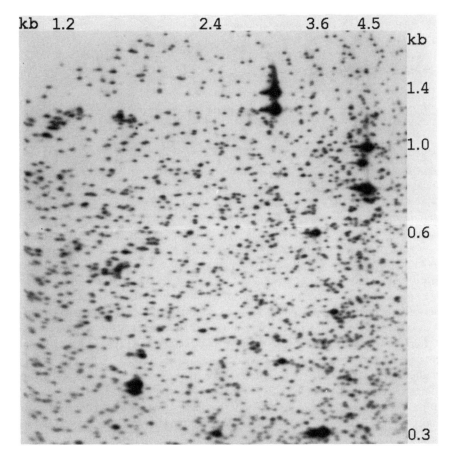

FIGURE 10.2 A computer-processed image of a two-dimensional gel of enzyme-digested genomic DNA obtained from one of the lymphocytoid cell lines established at RERF.

(33 cm × 46 cm × 0.05 cm). Autoradiograms are then obtained (Asakawa et al., 1994, 1995; Kuick et al., 1995).

The visual comparison of the gel of a child with those of its parents to detect attributes of a child's gel not present in either parent (i.e., a potential mutation) would be extremely demanding, the type of activity guaranteed to lead to a high turnover rate in technicians. Fortunately, the computer algorithm developed for the analysis of protein gels did just as well with these complex DNA images (Asakawa et al., 1994). Among the approximately 2,000 DNA fragments to be visualized on these preparations, we initially identified a subset of approximately 500 for which the coefficient of variation (CV) of spot intensity is ≤0.12; this reproducibility permitted distinguishing between spots of normal intensity and spots with 50% intensity with high accuracy (i.e., two-fragment or one-fragment spots). Already we have identified mendelizing genetic variation involving some 10% of these fragments. Figure 10.3 indicates a comparable area from the gels of two different trios in which segregating variation involving two different polymorphic fragments is present. We believe that with impending technical developments, the battery of fragments suitable for quantitative scoring may increase to 600 or 700. Other enzyme combinations can be used for the genomic digests that precede the gel runs, and, by altering the electrophoretic conditions, larger DNA fragments can be visualized. Currently we are working on three different types of gels for which we believe there is little overlap in the DNA fragments visualized. Furthermore, we have demonstrated the feasibility of recovering (and characterizing the nucleotide sequence of) specific fragments from the gels (Asakawa et al., 1994). Thus, a mutant fragment can be precisely studied. As in the study of proteins, this approach requires running gels on both parents as well as the child.

We would not like to lead you to believe that there is something magic in these new approaches. Their implementation will require a major effort. For each of them, RERF staff and ourselves have attempted to calculate the magnitude of the effort required to reach a significant difference between controls and the children of exposed parents, based on a variety of assumptions considering the doubling dose for the mutational endpoint. There is no time in this presentation to go into the excruciating details of these calculations, but they are available upon request.

Earlier, we made the point that the past genetic studies at RERF examined different aspects of the bomb's potential genetic damage, so that the results of different studies could be combined into a composite picture. These DNA studies will stand alone, since the DNA damage that will be detected could be manifest in a variety of the previous endpoints, varying from impaired survival or a predisposition to cancer to an electrophoretic variant, so that the findings from this study would in principle overlap with the previous findings.

CAN A BRIDGE BE BUILT BETWEEN SOMATIC CELL GENETIC STUDIES IN A-BOMB SURVIVORS AND GERM-LINE STUDIES IN THEIR OFFSPRING?

For many years now, the desirability of a somatic cell indicator of genetic damage has been obvious, and several possible systems have been explored at RERF. Four such systems deserve mention: (1) frequency of cytogenetic damage in cultured lymphocytes, (2) frequency of mutations in the glycophorin system, (3) frequency of mutations in the HGPRT system, and (4) frequency of mutant T lymphocytes defective in the expression of the T-cell antigen receptor gene. Time does not permit a discussion of the pros and cons of each of these systems, and in any event, two will be discussed later in this book. Each of these systems has been shown to provide evidence of radiation effects, but each has its limitations as a barometer of germ-line damage. For instance, one of the standbys in cytogenetic studies, the dicentric chromosome, is unstable and either would not be transmitted at gametogenesis or,

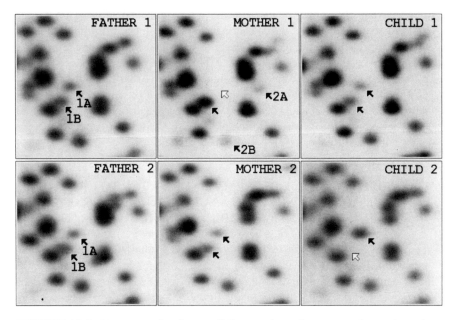

FIGURE 10.3 Two examples, in two different trios, of one type of genetic variation encountered in computer images of the two-dimensional DNA gels prepared
from the RERF cell lines. In the first example, the father is heterozygous for a
segregating variant, the mother is homozygous for the normal fragment, and the
child has received a normal fragment from the mother and the variant from the
father. In the second example, both parents are heterozygous for a variant but the
child has received the normal fragment from both parents.

if transmitted, would probably be incompatible with fetal survival. The nature of
the genetic variation revealed by the glycophorin system cannot be studied because
the erythrocyte is enucleate, but the phenotypic findings suggest that many of the
changes detected are the result of somatic cell crossing over rather than a mutation
in the usual sense of the term. For each of these indicators, there is the question
of how well it represents the genome as a whole. There is also the question of
the nature of the damage decay curve following the initial genetic damage. For
these and other reasons, the investigators working with these systems have been

properly cautious in suggesting doubling dose estimates, albeit these are the type of estimates needed for comparison with germinal rates and for guidelines regarding permissible exposures. We would like to suggest a fresh approach to the matter of A-bomb-induced somatic cell damage. Figure 10.4 is a two-dimensional DNA preparation from an individually and directly cloned B lymphocyte transformed by the Epstein-Barr virus, prepared in the Ann Arbor laboratory. The detailed similarity to the preparation based on the Japanese cell lines is striking. Although our studies are still preliminary, we suggest that this technique may represent a new and powerful multi-locus approach to the study of radiation-induced somatic cell damage. This technique has the additional advantage that the results of a study of A-bomb survivors would be directly comparable to the results of a study of their children using the same technique. We remind you that just as is possible for the studies in the F_1 cell lines, in principle any apparent somatic mutation that is detected in this system can be precisely characterized.

CANCER PREVALENCE IN PERSONS SUBJECTED TO INCREASED RADIATION EXPOSURES AS A SURROGATE FOR GENETIC STUDIES ON THEIR OFFSPRING

There is, in addition to the indicators mentioned in the last section, one final "somatic cell" indicator of the genetic effects of the atomic bombs: namely, the increase in benign and malignant tumors in the survivors of the atomic bombs. The question is, Can cancer be a "pit canary" for germ-line damage? The best current estimate of the amount of acute ionizing radiation that will double the frequency of solid tissue benign and malignant tumors is 1.66 Sv equivalents (Shigematsu and Mendelsohn, 1995). We regard the similarity of this estimate to the genetic estimate as, at the very least, a useful coincidence. However, Mendelsohn (see Chapter 13) has suggested that the linear relationship between exposure and cancer excess in the survivors is consistent with the radiation having added one additional hit (i.e., mutation) to the complex multi-hit process that characterizes oncogenesis. If this thesis can be sustained, then the cancer doubling dose of 1.66 Sv equivalents is indeed a genetic doubling dose, albeit based on the genes involved in oncogenesis rather than on the genes involved in the indicators we have previously discussed. From the standpoint of risk-setting, there are many important philosophical differences between somatic cell genetic damage and germ-line genetic damage. Most obviously, somatic cell genetic damage runs its course with the death of the affected person, whereas germ-line damage may result in multiple affected persons in multiple generations. Furthermore, the cancers resulting from radiation exposure on average occur some 20–25 years after the exposure; if the exposures are occupational, then the cancers appear in the later years of life. On the other hand, genetic defects resulting from germ-line mutation will generally be apparent at birth and pursue a lifelong course. Nevertheless, an argument can be made to the effect that any public health measures that protect "adequately" against

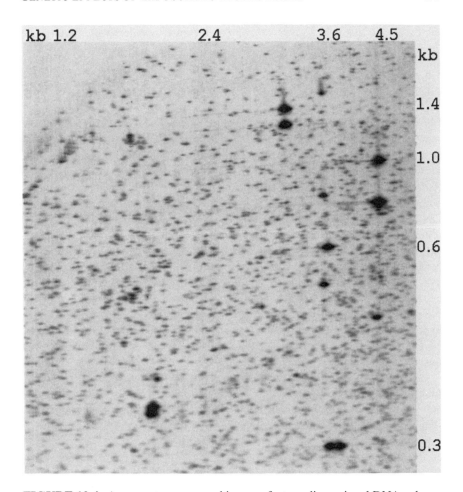

FIGURE 10.4 A computer-processed image of a two-dimensional DNA gel prepared from an EBV-transformed single-cell lymphocytoid clone derived from an American Caucasian.

an increased incidence of tumors following whole-body radiation exposures will protect adequately against germ-line damage. If accepted, then monitoring for genetic effects becomes somewhat easier, since most populations one might wish to monitor for genetic effects are populations subject to cancer reporting, the populations often supporting cancer registries. Conversely, a population that shows no (or an "acceptable") increase in cancer probably has an "acceptable" level of germ-line genetic damage, which is certainly true of radiation exposures, less certainly true of chemical exposures. This philosophy would have to be applied judiciously: an increase in thyroid malignancy in an area subject to a high fallout of radioactive iodine does not readily translate to genetic damage. The greatest single drawback to adopting this philosophy is the 20–25 year lag between the exposure and the cancer, so that if in a given situation an increase in cancer were found to be "unacceptable," the germ-line genetic damage has already been done.

"IF WE HAD IT TO DO OVER AGAIN"

Looking back on the long-running complex study of the genetic effects of the atomic bombs, it is important to ask, given all the amazing developments in genetic science these past 50 years, how differently should such a study be designed if it were beginning today? The most obvious change in the research design would be to include studies at the DNA level from the outset. However, it will be some years before it is possible to extrapolate with the desired precision from damage at the DNA level to gross phenotypic effects, and these latter are what the public which ultimately supports such studies really wants to know. Accordingly, we would suggest that any future study should still include most of the components of the study in Japan: frequency of congenital malformations and stillbirths, death rates among liveborn children, growth and development of surviving children, cancer and chromosomal abnormalities in children of exposees. The one study that would probably not be repeated would be a search for electrophoretic and activity variants in proteins. At the time, these studies were the best available approach to detecting nucleotide substitutions and small deletions in DNA, which comprise roughly one-third of the spectrum of radiation-induced damage in DNA. Now the ability to examine DNA directly offers the possibility of much more efficient detection of such lesions. Otherwise, however, looking both backward and forward, we suggest that any future study of the genetic effects of an ionizing or chemical agent ideally should include all but one of the components of the Japanese study plus, now, a DNA component.

There is another aspect to the question of how we would proceed if we had it to do over again based on administrative and psychosocial considerations. With respect to the former, recently an unidentified official at the Department of Energy has said that the early mission of the ABCC had been "absolutely without question" to perform secret research to benefit the weapons program (Macilwain and Swinbanks, 1995). While there was indeed material stamped "secret" in the files of the ABCC in the early days, since any material that touched in any way on the atomic bomb was then classified, there was to my knowledge no research project ever undertaken in secret specifically for the benefit of the military, and I challenge that official to document his statement. Likewise, these are times when the revisionist historians are showing great interest in the atomic bombings and related matters. Thus, I draw your attention to the recent book by Susan Lindee, *Suffering Made Real*, which depicts the ABCC in its early studies as riding roughshod over Japanese sensibilities in its efforts to collect the data demanded by the Atomic Energy Commission. Lindee has never understood that the ABCC program was set up to be fiercely independent by the Committee on Atomic Casualties of the National Academy of Sciences and the National Research Council. Furthermore, the recent publications of Schull (1990, 1995), Yamazaki and Fleming (1995), and Neel (1994b) document the very real efforts made to respect Japanese sensibilities and interact with the appropriate local groups.

For appropriate evaluations of these distortions, the reader is referred to the reviews of Putnam (1995) and Finch (1995). Unfortunately, when the uninstructed and naive review Lindee's book, they seem quite willing to accept her distortions. A recent series in the *Chugoku Shimbun*, Hiroshima's leading newspaper, repeats many of the carping criticisms in Lindee's book, adding some egregious misquotations in the process. It is unfortunate that the authors of these critiques are quite unable or unwilling to grasp and publicize the worldwide significance for humanity—and especially for the Japanese—of these studies.

11

Radiation, Signal Transduction, and Modulation of Intercellular Communication

JAMES E. TROSKO

SUMMARY

The survival of multicellular organisms through evolution in an aerobic environment came about by the adaptive responses to both the endogenous oxidative metabolism in the cells of the organism and all the chemicals and low-level radiation to which they had to be exposed. Included in the defense repertoire were preventive mechanisms against excessive oxidative damage to membranes, proteins, and DNA; built-in redundancies for damaged molecules; tight homeostatic and physically coupled redox systems, pools of reductants, antioxidants, DNA repair mechanisms and sensitive "sensor" molecules such as NF-kB; and signal transduction mechanisms affecting both transcription and post-translational modification of proteins needed to cope with the oxidative stress. The biological consequences of the response of a multicellular organism to the low-level radiation that might exceed the background level of oxidative damage to a cell in a tissue could be apoptosis, cell proliferation, or cell differentiation. If these biological endpoints are not detected at rates above non-irradiated control levels in an organism, it is highly unlikely that low-level radiation would play much of a role in the multistep process of carcinogenesis. Finally, gap-junctional intercellular communication, known to be required for homeostatic regulation of cell proliferation and adaptive functions in a multicellular organism, could, by its ability to suppress cell division in coupled cells of a tissue, provide protection to any one cell receiving track-radiation through the sharing of reductants and by triggering apoptosis.

*Once you screw up how cells talk to each other and their environment,
bad things happen.*

— D. Coffey (1995)

UNDERSTANDING LOW-LEVEL RADIATION EFFECTS ON HUMAN HEALTH: FROM MOLECULAR BIOLOGY TO HUMAN EPIDEMIOLOGY

Solving the long-standing, yet extremely germane problem of predicting the risks to human health after exposure to low-level acute and chronic ionizing radiation has had to rely on the extrapolation and integration of theoretical, experimental molecular, in vitro and in vivo cellular and animal studies, in addition to epidemiological examination of therapeutic, accidental, and atomic bomb exposures to ionizing radiation. A rather narrow analysis will be made because of the complexities arising from the initial quality and quantity of the radiation exposure (e.g., high or low LET; acute, chronic, dose, etc.), as well as from the genetic, developmental, sex, and pre- and post-exogenous factors influencing the radiation effects. Specifically, the question of this brief review will be whether one can determine what potential roles ionizing radiation can play in the induction of human cancers at low-level acute or chronic exposures. Therefore, the assumptions to be made explicit are that (1) all cancers arise from a single stem cell ["the stem cell theory" or "oncogeny as partially blocked ontogeny theory" (Nowell, 1976; Potter, 1978; Fialkow, 1979)]; (2) the carcinogenic process involves both multiple steps and multiple mechanisms ["initiation," "promotion" and "progression" phases (Pitot et al., 1981)]; and (3) the ultimate appearance of a cancer is the result of the breakdown of multiple homeostatic mechanisms at the molecular, cellular, tissue, organ, and organ-system levels in a multicellular organism such as a human being (Potter, 1974; Potter, 1983). While this analysis will focus on cancer, many of the implications will relate to other potential long-term health effects related to low-level radiation exposure, such as birth defects, cataracts, atherogenesis, and other chronic diseases.

The fundamental assumption will be that, while the primary injury to a multicellular organism is the energy deposition (tracks or hits to critical molecules) in terms of ionizations and excitations in localized clusters at the atomic and molecular levels in a cell, and while the cell is considered the unit of life (Feinendegen et al., 1995), it is the intercellular coupling or syncytium of cells that is the *functional unit of life*. It is this understanding of how a "society of cells" within an organism responds to the immediate and long-term effects of radiation that will give us some insight to understand how cancers, attributed to radiation exposure by epidemiological studies, come about and how molecular effects of radiation at the cell level ultimately contribute to the disruption of all the homeostatic mechanisms at each hierarchical level of a multicellular organism.

The concept of molecular epidemiology was developed in this era of new concepts and molecular findings related to cancer—namely, that oncogenes and tumor-suppressor genes exist, and that in most tumors, physical and chemical carcinogens could induce specific lesions in DNA that could form mutations having a unique "fingerprint" (Jones et al., 1991; Shields and Harris, 1991). While this concept of "molecular epidemiology" has been somewhat supported by observations that sunlight-induced skin tumors had mutations in their p53 genes indicative of the pyrimidine dimers induced in skin cells by the ultraviolet light of the sun, associations with other carcinogenic chemicals have not been as successful in supporting the concept (Brash et al., 1991; Dumaz et al., 1993). If, in fact, ionizing radiation of any suspected exposure could induce unique DNA lesions that might cause unique mutations in oncogenes and tumor-suppressor genes that are distinct from spontaneous or background radiation, then possibly molecular biology can contribute to epidemiology in resolving the specific role that ionizing radiation plays in carcinogenesis. However, as will be discussed below, carcinogenesis is more than mutagenesis, and ionizing radiation can do more than just induce DNA damage.

RADIATION CARCINOGENESIS: NO ONE THING "CAUSES" CANCER

While there is no dispute about neoplastic transformants or tumors appearing in cells, animals, or humans exposed to radiation, in vitro and in vivo, there is no consensus as to the molecular or biological mechanism by which radiation contributes to the carcinogenic process. In addition, great uncertainty exists as to the risks to humans after either low-dose, acute, or chronic, exposure to ionizing radiation (Upton, 1989; Gerber et al., 1991; Sugahara et al., 1992; Hoel, 1993; Kondo, 1993; Gilbert, 1994; Pollycove, 1994; Sinclair, 1994; Thompson et al., 1994; Fliedner et al., 1995).

The carcinogenic process can be conceptualized as consisting of at least three operational phases: "initiation," "promotion," and "progression" (Pitot et al., 1981). The **initiation** phase has been defined as an irreversible event occurring in a stem cell that prevents it from terminal differentiation but not from proliferation (Figure 11.1). As long as that single initiated cell is surrounded by and communicates with its normal neighbor, it poses no problem to the organism; a problem only arises if that mitogenic suppressing influence is inhibited. **Promotion** is the process that allows the clonal amplification of the initiated cells (Trosko et al., 1990) and the prevention of the induced programmed cell death or apoptosis of the initiated cell (Burch et al., 1992). Finally, when during the clonal expansion of the initiated cells additional genetic alterations occur (e.g., mutation, amplification, loss of various oncogenes or tumor-suppressor genes), the initiated cell acquires the phenotype by which the cell becomes autonomous or independent of exogenous growth factors or tumor promoters, then the cell has entered the **progression** phase.

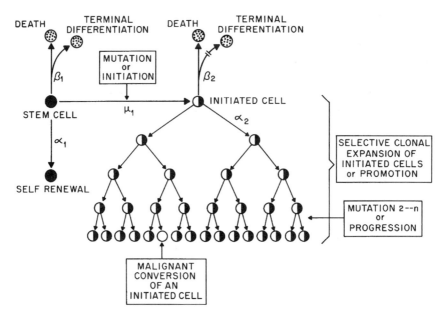

FIGURE 11.1 The initiation/promotion/progression model of carcinogenesis. β_1 = rate of terminal differentiation and death of stem cell; β_2 = rate of death, but not of terminal differentiation of the initiated cell; α_1 = rate of cell division of stem cells; α_2 = rate of cell division of initiated cells; μ_1 = rate of the molecular event leading to initiation (i.e., possibly mutation); μ_2 = rate at which second event occurs within an initiated cell.

The scientific question to be asked is, Can ionizing radiation induce each phase of carcinogenesis? That is, can ionizing radiation convert a normal stem cell to an initiated cell? Can ionizing radiation contribute to the clonal expansion of a pre-existing initiated cell? Or, can ionizing radiation bring about the neoplastic conversion of an initiated clone of promoted cells? It should be obvious in the context of an acute exposure to low-level radiation, by definition, ionizing radiation cannot be a tumor promoter, since a single exposure would not be sufficient to allow sustained proliferation of an initiated cell or sustained prevention of apoptosis of the initiated cell. It is unknown whether sustained proliferation or prevention of apoptosis can occur with low-level chronic exposure to ionizing radiation.

While in vitro experiments with animal cells suggest answers to these questions, because of limitations in the interpretations and extrapolation to human cells and the human organism, it is not possible to conclude rigorously that ionizing radiation is an initiator and a progressor. The same could also be said of the few animal experiments designed to test if ionizing radiation could affect these three phases of carcinogenesis (Trosko, 1994). On the other hand, on a theoretical level, based on the molecular concepts of oncogenes and tumor-suppressor genes (Bishop, 1995) and current understanding of the molecular nature of how ionizing radiation can damage DNA and induce various kinds of gene and chromosomal mutations (Thacker, 1986; Lucke-Huhle et al., 1990; Renan, 1992; Felber et al., 1992; Ito et al., 1993; Cox, 1994; Goodhead, 1994), one could speculate where one might expect ionizing radiation to play a role in the multistep, multiple mechanism process of carcinogenesis (Trosko, 1992).

IONIZING RADIATION AS A MUTAGEN: INITIATOR OR PROGRESSOR?

It is well established that ionizing radiation can induce DNA single- and double-strand breaks, directly or indirectly, as well as various DNA base modifications (Cox, 1994; Ward, 1995). In addition, point mutations, gene deletions, and a variety of chromosome aberrations can be induced by exposure to ionizing radiation (Thacker, 1986; Renan, 1992; Lucke-Huhle et al., 1990; Felber et al., 1992; Ho et al., 1993; Goodhead, 1994; Nelson et al., 1994). Molecular analyses of a number of genes used as targets for ionizing radiation appear to indicate that the majority of the mutant cells that survived in order to be analyzed contained deletion mutations with gross chromosomal aberrations (Thacker, 1986). Assuming for the sake of this analysis that ionizing radiation is a better deletion-mutagen/clastogen than a point mutagen, ionizing radiation would be better at deactivating a tumor-suppressor gene than activating an oncogene. While chromosome rearrangements (i.e., possibly the transposition of the *bcr/abl* gene) or small gene deletions (i.e., the truncation of the ErB2/Neu) by ionizing radiation could activate oncogenes without inducing point mutations within the genes, the likelihood of deleting a tumor suppressor gene by ionizing radiation would seem more probable. If ionizing radiation does leave a unique molecular lesion leading to unique types of mutations in either oncogenes or tumor-suppressor genes found in tumors of those known to have been exposed to ionizing radiation (i.e., "molecular epidemiology"), then one might be able to pinpoint the specific role ionizing radiation plays as a mutagen in the different stages of carcinogenesis.

IONIZING RADIATION AS A CYTOTOXICANT

Until recently, it was well established that one of the consequences of ionizing radiation was cell death caused either by DNA/chromosomal lesions or by some non-DNA-related events within the cell. With the explosion of research on the phenomenon of programmed cell death or apoptosis, it now seems clear that much of what previously was thought to be ionizing-radiation-induced cell necrosis was, in fact, apoptosis (Meyn et al., 1993).

In addition, the fact that apoptosis can be induced by targets in the cell other than DNA implies that at low-dose levels, oxidative stress induction of epigenetic events might be important (Fritch et al., 1994). This new insight is now forcing the integration of knowledge on apoptosis, stem cells, various oncogenes (Bcl-2, ras, etc.), tumor-suppressor genes (i.e., p53), non-DNA damage to cells, activation of various transcription factors (AP-1; NF-kB), and tumor promotion.

Clearly, cell removal via surgery or cell death in tissues containing initiated cells has been shown to act to promote tumors in animals (Argyris, 1985). Recently, it has been observed that many chemical tumor promoters, such as phorbol esters, phenobarbital, and DDT can inhibit apoptosis (Burch et al., 1992). The speculation is that tumor promotion is a process that requires not only the prevention of the loss of initiated cells by the blockage of apoptosis, but also the clonal amplification of these initiated cells (Trosko et al., 1983).

In the context of trying to determine if ionizing radiation can promote tumors by killing cells, several diametrically opposite effects of radiation on cell killing must be examined. On a single-cell level (only relevant in vivo for free-existing cells of the "soft tissues," other single migrating cells, and stem cells), the probability of killing the cells after low-level exposure by DNA or non-DNA mechanisms must be considered (1) for various kinds of cells [stem versus progenitor versus terminally differentiated cells (Potten et al., 1994; Vergouwen et al., 1994)]; (2) at various times in the cell cycle (Little, 1994); and (3) for cells that have effective DNA repair (for DNA lesions causing cell death) or effective quenching of non-DNA lesions [e.g., glutathione levels (Miura and Sasaki, 1991; Saunders et al., 1991; Bergelson et al., 1994; Meister, 1994; Sierra-Rivera et al., 1994)]. This consideration will relate to radiation-induced leukemias. For cells in solid tissues, the phenomenon of gap-junctional intercellular communication, which effectively couples many cells of a tissue into a syncytium, the probability of killing a cell in this matrix will be affected by additional factors to be discussed later.

At a low-dose exposure given either acutely or chronically, if radiation does not induce significant cell killing by necrosis (DNA lesions causing lethal mutations or non-DNA mechanisms not triggering apoptosis), the likelihood of ionizing radiation acting as a tumor promoter is small if not nil. Recent evidence regarding ionizing-radiation-induced apoptosis indicates that, depending on the type of cell (Vergouwen et al., 1994; Potten et al., 1994), p53 tumor-suppressor gene can be induced (Potten et al., 1994). The induction of p53, a molecule which has multiple functions (Cox and Lane 1995), would allow for more DNA repair, thus removing DNA lesions and reducing the chance of certain classes of mutations (Gotz and Montenarh, 1995). If a cell with DNA damage is not removed by either necrosis or apoptosis, then this surviving cell could produce mutations which could then contribute to the carcinogenic process (Gotz and Montenarh, 1995). Ionizing-radiation-induced p53-dependent apoptosis is then considered a mechanism by which potentially initiated cells are removed from tissue rather than promoted.

In fact, one of the current ideas for radiation therapy is to increase apoptosis in existing tumor cells (Kerr and Searle, 1979).

If radiation can induce cell killing by necrosis and apoptosis, the question now is, At low-dose exposures, does ionizing radiation primarily induce apoptosis? Also, as will be discussed later, are all cells equally susceptible to radiation-induced apoptosis? Until experiments are designed to answer these questions, it seems impossible to predict if ionizing radiation is a potential tumor promoter at low doses, given acutely or chronically.

IONIZING RADIATION AS EPIGENETIC AGENT VIA INDUCTION OF OXIDATIVE STRESS

Only in the last few years has the idea been advanced that radiation might bring about some of its biological effects by non-DNA damage-related mechanisms. Clearly, ionizing radiation can damage DNA in cells, and this damage can be the substrate for mutations and/or cell death. Under those conditions where the probability is too low that a track or hit occurs within a given cell of a tissue or at or near DNA (Goodhead, 1994; Feinendegen et al., 1995), it is known that the cell does react epigenetically. That is, the cell responds by induction of gene transcription, modified translation of gene messages, and altered post-transitional modification of existing proteins (definition of an epigenetic response) (Trosko et al., 1992).

Specific gene induction by ionizing radiation and activation of preexisting proteins have been reported and reviewed several times (Woloschak et al., 1990; Boothman and Lee, 1991; Weichselbaum et al., 1991; Uckum et al., 1992; Herrlick and Rahmsdorf, 1994). While the prevailing paradigm has been that DNA damages or arrested DNA forks were the signal for the initiation of new gene transitional activity (Devary et al., 1992), there is evidence that there exists a non-DNA damage source of a radiation-induced signal transduction pathway, including that which leads to apoptosis (Ramakrishan et al., 1993; Buttke and Sandstrom, 1995).

The single isolated cell or the syncytium of cells within a tissue functions because of the use of oxygen as the terminal electron acceptor in order to produce energy. In order to survive potential cellular damage because of the constant generation of reactive oxygen species (ROS), a delicate homeostatic regulatory system had to evolve with the evolution of living organisms in an aerobic world, which would allow a cell (1) to recognize any change in its external environment; (2) to adapt to this signal by proliferating, differentiating, or if differentiated, adaptively responding to the signal (Gotz and Montenarh, 1995). A balance of pro-oxidants and antioxidants provides elements for a homeostatic mechanism (Cerutti et al., 1994). Tipping that balance within limits allows the cell to perform its necessary survival functions. While the cell and tissue have developed mechanisms to minimize and avoid lethal disruptions of the redox homeostatic state (e.g., various repair systems and redundant targets of oxidative damage), in some cases the fact that oxidative stress is not handled adaptively by a cell may, in fact, be adaptive to the tissue and organism (Buttke and Sandstrom, 1995). In other words, non-adaptive cells would be eliminated by apoptosis.

Ionizing radiation, as well as ultraviolet light and many chemical toxicants, can induce free radicals, directly or indirectly. In those cases, especially at low doses, where there is little or no genomic DNA damage (Feinendegen et al., 1995), the ROS produced can act as signaling molecules, per se, or by oxidatively modifying cellular components. In order to contribute to the alteration of the redox state of the cell and thereby trigger some epigenetic event, the amount of ROS and/or its intracellular localization must exceed the normal background level of oxidative metabolism-generated ROS (Hill and Treisman, 1995).

While the detailed mechanisms for ionizing, UV, or chemical induction of oxidative stress are not yet known, it has been postulated that plasma- and intracellular unsaturated fatty acid-containing membranes are targets of the ROS (Devary et al., 1992; Buttke and Sandstrom, 1995). Cellular and organelle membranes are probably the sensors enabling the cell to gauge the change in the redox state of the cell. In addition, specific protein molecules (e.g., p53, NF-kB, AP-1) are important cellular monitors of the altered redox state (Angel and Karin, 1991; Devary et al., 1992; Hainaut and Milner, 1993; Baeuerle and Henkel, 1994; Hill and Treisman, 1995; Ueda et al., 1995). Their alteration by oxidated/reductive changes would dramatically affect their numerous and critical functions within a cell. The formation of reactive oxygen species by ionizing radiation, as well as by phorbol esters and H_2O_2, can stimulate mitogen-activated protein kinase activity (Stevenson et al., 1994). Free radicals at high doses can cause toxicity; however, at low doses they are associated with cell growth of certain cells (Nicotera et al., 1994; Stevenson et al., 1994; Timblin et al., 1995).

Before anything could happen after the cell experiences an ionizing track, the cellular oxidative protective mechanisms must be breached. That is, if the intracellular pool of reduced glutathione is insufficient and antioxidant deficiencies exist, then lipid peroxidation could occur, since the track is more likely to be in the vicinity of the plasma or other membranous target than in the nucleus (Feinendegen et al., 1995). Any resulting oxidative stress could trigger activation of protein kinases [i.e., tyrosine kinase (Devary et al., 1992)], inactivation of p53 (Hainaut and Milner, 1993; Baeuerle and Henkel, 1994), or activation of a transcription factor NF-kB (Baeuerle and Henkel, 1994). It has been shown that the regulation of the activity of NF-kB might be responsible for the ionizing radiation sensitivity of the ataxia telangiectasia phenotype (Jung et al., 1995).

To help stabilize the redox state against endogenous and exogenously generated oxidative stress, antioxidants—especially glutathione levels—have been shown to influence radioresistance of certain cell lines under certain conditions (Miura and Sasaki, 1991; Saunders et al., 1991). However, the endpoints measured usually included DNA damage and cell survival, and the various studies used a variety of cell types (mostly cancer cell lines) irradiated under a number of conditions.

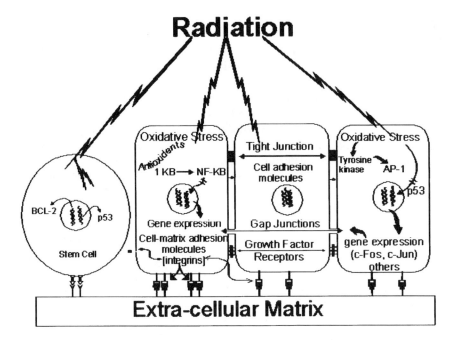

FIGURE 11.2 Schematic diagram the complex factors (stem cells; coupled progenitor/differentiated cells; intracellular antioxidant protective mechanism; redox or oxidative-stress-sensing signal transducing elements) which could influence the initial radiation exposure in tissues.

Recent studies have implied the role of glutathione depletion in activation of gene expression (Bergelson et al., 1994; Hu and Cotgreave, 1995) in accordance with the prediction that oxidative stress-sensing-molecules such as NF-kB and AP-1 would be triggered under these conditions. Moreover, since in solid tissues, most cells are coupled by gap junctions, glutathione levels would be stabilized by transfer of glutathione which can traverse the gap junctions (Barhoumi et al., 1995; Nakamura et al., 1995). The relation of drug or radiation sensitivity and the ability of cells to couple via gap junctions have been previously shown (Tofilon et al., 1984; Frankfurt et al., 1991; Pitts, 1994) (Figure 11.2).

The implication of these studies for normal cells is that ionizing radiation-induced oxidative stress would be modified by the metabolically coupled cells. To induce either cell proliferation or apoptosis in individual cells within the coupled syncytium must involve some as yet unknown factors.

The tumor-suppressor gene, p53, has been implicated in multiple functions related to the regulation of cell division, differentiation, DNA repair, and programmed cell death or apoptosis (Cox and Lane, 1995; Gotz and Montenarh, 1995). Indeed, the association of the mutated p53 gene or loss of heterozygosity of the p53 gene has been linked to a large number of human cancers (Hollstein et al., 1991). However, in those cancers having no p53 mutations or loss of heterozygosity, functional inactivation can occur by other means, including amplification of the MDM2 gene product which binds to p53 (Fakharzadeh et al., 1991). These processes involving the mutation, deletion, or inaction by gene amplification are all considered genotoxic events. In the cell whose redox state might be altered without having a genotoxic event, the altered redox state itself could bring about the reversible, but nevertheless potentially disruptive effect of ionizing radiation. Several studies have implicated the functional conversion of the normal p53 protein to a "mutated" form in cells with an altered redox state (Hainaut and Milner, 1993; Ueda et al., 1995).

Phorbol esters, powerful skin tumor promoters, can bring about the phosphorylation of the p53 protein (Baudier et al., 1992). Depending on the cell type, phorbol esters can induce cell proliferation, block apoptosis, or modulate differentiation via their ability to activate protein kinase C (Yamasaki et al., 1984). Ionizing radiation has been shown to induce protein kinase C at rather high dose levels (2Gy) (Woloschak et al., 1990; Avila et al., 1993). TPA, by down-regulation of PK-C, has been shown to block the induction of transcription by ionizing radiation, presumably by the lack of post-translational modification of critical transcription factors needed for initiation of gene transcription (Lin et al., 1990). The stimulation of tyrosine kinase activity by ionizing radiation and other inducers of oxidative stress has also been shown to activate phospholipase D, which could then lead to signal transduction pathways needed to initiate early-response gene transcription (Avila et al., 1993).

Ionizing radiation is known to induce p53, and it is implicated in DNA damage recognition and apoptosis (Gotz and Montenarh, 1995). Potten and co-workers (1994) have shown that low levels of ionizing radiation could induce apoptosis in the stem cell pool of the small intestine of the mouse, whereas the stem cells of the large intestine were resistant to radiation induction of apoptosis. The expression of Bcl-2, or lack of it, as well as the induction of p53 by radiation in these tissues, correlated with the ability of ionizing radiation to induce cancers in these two tissues. In the p53 knockout mice, the roles of p53 in radiation-induced apoptosis and carcinogenesis have been examined (Clarke et al., 1994). It turns out that the rapidly appearing apoptosis that occurs after ionizing radiation is absent in the p53 null mouse, but that spontaneous apoptosis occurs at the level found in the normal mouse (Potten et al., 1994). While contradictory results have been obtained when testing whether the functional loss of p53 would affect radiation sensitivity, cell-type specificity might have influenced the differences (Jung et al., 1992; Slichenmyer et al., 1993).

The exact cascade of events might include the oxidation of unsaturated fatty acids in the membranes, and the preferential release of the hydroperoxides by phospholipase A2 enzyme, with the subsequent increase of intracellular calcium. Increased phosphorylation of proteins such as c-Jun due to activation of kinases such as the src kinase after UV radiation (Devary et al., 1992) could activate AP-1 transcription factor. This, in turn, could activate the genes having AP-1 elements in their DNA. Simultaneously, the phosphorylation of p53, the "guardian" of the cell (Lane, 1992) could convert it to a longer lasting molecule, as well as one with dramatically different functions (Woloschak et al., 1990; Uckum et al., 1992).

The alteration in the redox state by ionizing radiation, as well as by non-genotoxic chemicals and growth regulators, can also affect the nuclear factor-kB (NF-kB) (Baeuerle and Henkel, 1994). NF-kB is normally regulated by the 1kB-a protein in the cytoplasm of the cell. Upon phosphorylation of the 1kB-a, the un-bound NF-kB translocates to the nucleus where it activates transcription, while the 1kB-a undergoes proteolysis (Jung et al., 1995).

These examples of the ability of agents such as ionizing radiation to activate pre-existing proteins quickly by the generation of radical oxygen species in cells where there has been no genomic DNA damage should support the idea that ionizing radiation can be an epigenetic agent. The rapid effect on pre-existing proteins via either direct conformational changes due to oxidation/reduction events, or by **posttranslational** modification by phosphorylation/dephosphorylation and the subsequent **transcriptional** activation of early response genes, are within the definition of **epigenetic** events. These events could not occur, however, if the radiation-induced reactive oxygen species did not exceed the background levels due to normal oxidative metabolism, for which the cell has provided a variety of protective mechanisms. The availability of the intracellular glutathione levels and antioxidants, together with the number and **intracellular** localization of the ionization tracks, would determine if there would be a disequilibrium of the redox state. Thresholds must be exceeded for this to occur (e.g., exceeding background levels to cause a change in state and the necessity for producing sufficient redox-induced transcription factors for gene activation) (Hill and Treisman, 1995). Most importantly, in the context of the process of carcinogenesis, several factors must be kept in mind, namely (1) whether single isolated cells or cells coupled by gap junctions in a syncytium are oxidatively stressed; (2) what stages of the cell cycle the cells are in at the time of oxidative stress; (3) the state of differentiation at the time of oxidative stress; (4) whether the cells are in different stages of the carcinogenic process (e.g., initiated, promoted, or neoplastically converted); and (5) the wide background of genetic or environmental factors could modify the initial radiation effect in any given individual.

RADIATION, THE REDOX STATE, AND CELL-CELL COMMUNICATION

A multicellular organism such as a human being exists because from the single fertilized egg to the adult of 100×10^{12} cells, all the cells communicated with

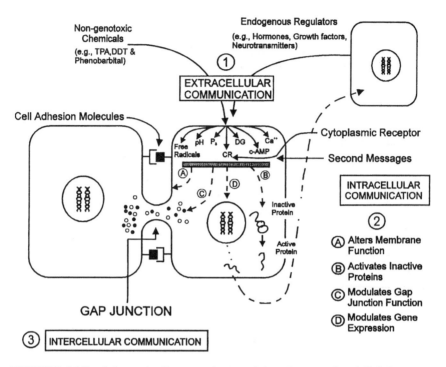

FIGURE 11.3 Schematic diagram characterizing the postulated link between extracellular and intercellular and intracellular communication through various intracellular transmembrane signaling mechanisms. Intracellular communication alters membrane function, activates inactive proteins, modulates gap-junction function, and modulates gene expression. This diagram provides an integrated view of how the neuroendocrine immune system (mind or brain/body connection) and other multisystem coordinations could occur. While not shown here, activation or altered expression of various oncogenes (antioncogenes) could also contribute to the regulation of gap-junction function.

each other via extra-, intra- and intercellular communication processes, as shown in Figure 11.3.

Although cell-matrix and cell-cell adhesion are distinct forms of communication, for the purpose of this discussion they will be viewed as a subclass of intercellular communication. Specifically, intercellular communication is mediated by the membrane-associated protein channel, the **gap junction**. The gap junction appeared very early in the evolution of the multicellular organism, where a delicate balance of the regulation of cell proliferation, cell differentiation, and adaptive functions of the differentiated cells had to be maintained (Revel, 1988).

The gap-junction structure—which consists of hemi-channels, **connexons**, in each of the two contiguous cell membranes—allows the passive diffusion of ions and molecules below 1,200 daltons between the coupled cells (Evans, 1988). By being coupled via gap junctions, cells can either equilibrate their ions and regulatory molecules or transiently send or receive signals to upset equilibria (Sheridan, 1987). Synchronization of electronic or metabolic functions occurs in various tissues by gap-junctional intercellular communication (GJIC) (Bennett et al., 1991).

The connexons are composed of a hexameric arrangement of proteins called **connexins**. These connexins are coded by a highly evolutionary conserved family of genes (Dermietzel et al., 1990). The gap-junction genes express themselves in different cells of different organs (Gimlich et al., 1990; Yancey et al., 1992). The connexin genes are regulated at the transcriptional, translational, and post-translational levels (Saez et al., 1990). Many of the second messages needed for signal transduction (e.g., Ca++, nitric oxide, diacylglycerol, c-AMP, phosphorylation by protein kinases) have been shown to either up- or down-regulate GJIC. Hormones, growth factors, neurotransmitters, as well as oncogenes (e.g., *src, neu, ras, raf, mos*), a tumor-suppressor gene, chemical tumor promoters (e.g., TPA, phenobarbital, DDT, TCDD, etc.), anti-tumor promoters and anti-carcinogens (e.g., retinoids, carotenoids, green tea components, lovastatin), have been shown to either up- or down-regulate GJIC (Trosko et al., 1993; Miyachi and Nishikawa, 1994). Most cancer cells have either defective homologous or heterologous GJIC (Kanno, 1985; Yamasaki et al., 1987).

All this evidence seems to be consistent with the original hypothesis by Loewenstein that the loss of GJIC is the causal basis of a cancer (Loewenstein, 1966). In view of the stem cell theory of cancer (Kondo, 1983) and the facts that cancer cells do not have growth control nor are they able to terminally differentiate ["oncogeny as partially blocked ontogeny" (Potter, 1978)], the recent observations that stem cells do not express gap-junction genes (Chang et al., 1987; Kao et al., 1995) might implicate gap junctions in the carcinogenic process (Holder et al., 1993), since gap junctions have been implicated in the regulation of growth control and differentiation (Loewenstein, 1979; Brissett et al., 1994). With the reports that transfection of non-communicating cancer cells with connexin genes could restore normal growth control and reduce tumorigenicity (Eghbali et al., 1991; Metha et al., 1991; Zhu, et al., 1991; Zhu et al., 1992; Jou et al., 1993), and that the antisense connexin gene could eliminate the suppression of neoplastic transformation (Goldberg et al., 1994), more evidence has been provided to support the hypothesis that GJIC is needed to prevent tumorigenesis.

The role of gap junctions in low-level radiation induction of cancer is relevant in several places in the carcinogenic process, namely, the **stem cell** (the "target" cell for the carcinogenic process), the **promotion** phase of carcinogenesis (the phase of carcinogenesis that depends on epigenetic events to trigger cell proliferation of

the single initiated cell) and **apoptosis** (the potential removal of stem cells or initiated cells). To tie these ideas together with radiation-induced redox changes and
carcinogenesis, the role of gap junctions might be seen as an integrating element.

If the stem cell is the "target" cell for carcinogenesis, then the apoptosis-related
survival of those stem cells that are sensitive to low-level radiation-induced non-
genotoxic or epigenetic changes would be affected. This appears to be the case
for the stem cells of the small intestine (Potten et al., 1994). The frequency of
small intestine cancers is very low compared to cancers of the large intestine,
where the stem cell does not seem to be affected by radiation-induced apoptosis.
If a stem cell has been initiated by some other carcinogen, a radiation-induced
disequilibrium in the redox state might not be handled as well as in stem cells that
are not initiated. Although oxidative stress has been implicated in the apoptotic
process (e.g., chemicals that induce oxidative stress appear to block apoptosis), it
is also linked to mitogenesis (Devary et al., 1992; Ramaishan et al., 1993; Herrlick
and Rahmsdorf, 1994). Therefore, the radiation-induced reactive oxygen species
might trigger very different responses, depending on the type of stem cell, which
genes are expressed in the stem cell (e.g., Bcl-2, p53), the condition of the stem
cell (e.g., whether it been initiated or not), and the availability of glutathione and
antioxidants.

One additional characteristic of stem cells is that they might not possess expressed gap-junction genes. The fertilized egg, being a totipotent stem cell, does
not have expressed or functional gap junctions at an early stage of development
(Lo and Gilula, 1979). Two presumptive human epithelial stem cells of the kidney
and breast do not express gap junctions (Chang et al., 1987; Kao et al., 1995). The
growth of these cells is regulated by extracellular communication mechanisms that
modulate intracellular communication-dependent signal transduction.

Gap juctions seem to be associated with the process of stem cell differentiation.
Cells in solid tissues coupled by these gap junctions are able to share ions and small
molecules, including $Ca++$ and glutathione, which could help stabilize the redox
state of a given cell (Kavanagh et al., 1988; Barhoumi et al, 1993). Many chemical
tumor promoters that can cause oxidative stress in cells can (1) activate NF-kB,
AP-1, and early response genes, and can stimulate protein kinases, inactivate by
phosphorylation p53 and 1kB-a, etc.; (2) block GJIC; and (3) block apoptosis
(Angel and Karin, 1991; Hainaut and Milner, 1993; Baeuerle and Henkel, 1994;
Cerutti et al., 1994; Buttke and Sandstrom, 1995; Hill and Treisman, 1995). On
the other hand, glucocorticoids and retinoids can up-regulate GJIC and induce
apoptosis, while having dramatically different effects on the redox state of the cell
and signal transduction (Hossain et al., 1989; Ren et al., 1994).

Gap junctions have been shown to be able to transfer glutathione between cells
(Barhoumi et al., 1995; Nakamura et al., 1995), and the depletion of glutathione
levels has been shown to potentiate TPA's ability to inhibit GJIC, presumably by
enhancing oxidative stress in the cells (Hu and Cotgreave, 1995). Gap-junctional
intercellular communication has been implicated in several studies on radiation

and chemical-induced toxicity. In other words, many studies have shown that cells grown in spheroids, as opposed to cells grown in sparse or even confluent two-dimensional cell cultures, are more radioresistant (Tofilon et al., 1984; Olive and Durand, 1994). This might imply the micro-environment in the spheroid could induce GJIC which is absent or reduced in non-spheroid conditions. There is evidence that the micro-environment can influence the presence and function of gap junctions in cells not normally having functional GJIC (Sasaki et al., 1984; Confort et al., 1986). The existence of contradictory literature related to radiation- and chemotherapeutic drug-resistance and their correlation with the presence or absence of GJIC might simply be related to whether the conditions in each particular experiment are conducive or not to the up-regulation of gap junctions. This is particularly true in the case where several oncogenes, which are associated with the down-regulation of GJIC (e.g., *ras, raf, neu, src*), have been associated with conferring radiation resistance to the cell (Sklar, 1988; Pirollo et al., 1993).

Since oxidative stress has been linked to the blockage of apoptosis and the block-age of GJIC, and the blockage of GJIC has been linked to tumor promotion, the question is now, Can low-level radiation exposure of solid tissues induce sufficient oxidative stress to block apoptosis in initiated cells, block GJIC, and help bring about the promotion phase of carcinogenesis? Promotion of an initiated cell can only occur if there is *sustained* application of the promoting stimulus (Siskin et al., 1982). If the low-level ionization can be sufficient to induce oxidative stress without genotoxicity, it might overcome a threshold (background oxidative metabolism) in the cell to activate early response genes, block apoptosis, and stimulate cell division. The tissue must be exposed chronically in order for this to occur. Since it has been shown that doses of 0.5 Gy can induce permeability of tight junctions and transient down-regulation of gap-junctional intercellular communication (So-mosy et al., 1993), there is a possibility that an acute exposure to ionizing radiation could cause some cells to proliferate, others to differentiate, and still others, which are terminally differentiated, to produce adaptive responses. Protracted down-regulation of the tight junctions and gap junctions in epithelial cell layers might be responsible for the lethal effects of high doses of radiation (Somosy et al., 1993). Future studies at lower acute doses must be done to study the possible correlation between radiation-induced oxidative stress and down-regulation of GJIC.

DOES LOW-DOSE IONIZING RADIATION EXPOSURE CAUSE CANCER?

Given in vitro and whole-animal experiments with ionizing radiation as well as several epidemiological studies, one would have to conclude, after extrapolation from high-dose levels to low-dose levels with the assumption of a linear no-threshold model, that radiation does indeed contribute to the carcinogenic process. Recent analyses of the incidence of solid tumors found in the survivors of the atomic bombs in Hiroshima and Nagasaki (Thompson et al., 1994) would lead some to conclude that even at very low-dose exposures, there is a finite probability of radiation-induced cancers. Caution must be taken in the interpretation of these data. As was stated by Sinclair (1994), "Clearly, the linear fit over a dose range up to 4 Gy or 5 Gy is just that—a fit. It does not imply single-hit kinetics or any other mechanism." Among the known uncertainties in trying to derive risk estimates from the Life Span Studies on the atomic bomb survivors are the epidemiological studies themselves, the dosimetry, the projection to lifetime, the generalization from conclusions about the Japanese population of that time to all populations today, and the extrapolation to low-dose and dose-rate (Trosko, 1995).

From the analysis above—together with the fact that carcinogenesis is well known to involve multiple genetic and epigenetic changes before a normal diploid stem cell becomes invasive, malignant, and metastatic—the likelihood that an acute and even chronic low-level ionizing radiation exposure can affect each and all the steps necessary is extremely remote. If it can be shown that frequencies of induced gene mutations and chromosomal aberrations from apoptotic cells versus those from cells that survive radiation are the same or different, then the role of radiation-induced apoptosis of stem-like cells in carcinogenesis can be assessed. Indeed, if radiation-induced apoptosis is a biological protective mechanism to remove genetically damaged cells from tissue, an explanation for the findings to-date on the possible genetic damage done to the survivors of the atomic bombs, which so far have not shown any measurable increase (Neel et al., 1990; Neel and Lewis, 1993), might be that the germinal stem cells of the males have been eliminated.

While this analysis has not examined the phenomenon of the "adaptive response" or "hormesis" (Wolff, 1992; Lehnert, 1995), the possibility exists that a low dose of ionizing radiation, by inducing an increment of oxidative stress above normal background levels, could induce signal transduction for the transcription of genes needed to protect the cells from additional stress-inducing agents. If this includes induction of antioxidants, reductants, and cell cycle-checkpoint inhibitors to facilitate time for DNA repair—then one could hypothesize that there might be a mechanism to explain some of the reported "adaptive responses." The role of gap-junctional intercellular communication (Trosko, 1991) would be included. In tissues where cells are coupled by gap junctions, cell division would not be occurring, thereby facilitating the process of repair before lesions in DNA are "fixed" by replication errors.

ACKNOWLEDGMENTS

These ideas were derived from research supported in part by grants from the National Cancer Institute (CA21104), the US Air Force Office of Scientific Research (No. 94-NL-196), and the Michigan Great Lakes Protection Fund. The author wishes to acknowledge the excellent word processing skills of Mrs. Robbyn Davenport.

12

Interaction Between Radiation Dose and Other Cancer Risk Factors

CHARLES E. LAND

SUMMARY

Ionizing radiation exposure is a well-established, probabilistic risk factor for cancer of various organs, based in large part on studies of the Hiroshima and Nagasaki A-bomb survivors. For many of these organ sites, other risk factors have also been identified and their associated risks quantified. Knowledge about the joint effects of radiation and other risk factors is central to understanding the role each plays in the causation of cancer, and to transferring risk coefficients to populations other than those from which they were derived. Such knowledge has been surprisingly difficult to obtain, largely because results must be fairly precise to be of much use. Investigations have been most successful for female breast cancer among A-bomb survivors in relation to reproductive history, and for lung cancer among uranium miners in relation to smoking history. In the first case, a level of synergy that is at least multiplicative has been estimated between radiation dose and age at first full-term pregnancy, whereas in the second, the relationship between radon exposure and smoking history appears to be non-additive but probably less than multiplicative. The steps by which these inferences were reached are outlined in some detail.

INTRODUCTION

Site-specific studies of radiation dose and cancer risk, in the Life Span Study (LSS) sample of the Radiation Effects Research Foundation (RERF) and in other exposed populations continually followed up over time, proceed in a series of steps beginning with the evaluation of evidence that a dose-related excess risk actually exists. Once a dose-response relationship has been strongly established, it becomes possible to study modifiers of radiation-related risk with some hope of success. Usually, the first modifiers to be considered are sex, age at exposure, age at observation (attained age), and time following exposure, since information about them is usually obtained at the same time as information on radiation exposure and disease occurrence. Modification of dose-response by factors other than those just mentioned is a more difficult problem, because it usually requires special data-gathering efforts. The discussion that follows covers interactions between radiation and smoking as causes of lung cancer and, at somewhat greater length, interactions between radiation and reproductive history as risk factors for female breast cancer.

MAIN EFFECTS: RADIATION

Radiation dose from the atomic bombings of Hiroshima and Nagasaki is a strong risk factor for female breast cancer (Tokunaga et al., 1994) and for lung cancer (Thompson et al., 1994). For breast cancer, the dose-response function is steep and linear (Figure 12.1). Excess relative risk at 1 Sv (ERR_{1Sv}) declined markedly with age at exposure, a phenomenon that can be described adequately as either a negative exponential in age ATB, or as a step function with breaks at ages 20 and 40 (Table 12.1). Within age-ATB cohorts, ERR_{1Sv} depended hardly at all on attained age (Table 12.2) except for an anomalously high value before attained age 35 among women exposed before age 20. The latter finding may reflect the existence of a highly susceptible genetic subgroup, a possibility that can be investigated in the light of recent discoveries in molecular genetics beginning with the identification of genes possibly involved in familial aggregations of breast cancer (Wooster et al., 1994), the cloning of two of them (Miki et al., 1994; Savitsky et al., 1995), and the identification of specific heritable mutations (Shattuck-Eidens et al., 1995; Struewing et al., 1995a, 1995b). With this exception, however, there is no evidence that ERR_{1Sv} depends markedly upon attained age or time following exposure. Based on individual dose estimates and a fitted dose-response model allowing for variation by age ATB, about 17% of the 591 breast cancers observed during 1950–1985, among female survivors with DS86 estimates, were caused by their exposure. By age ATB, this percentage is estimated to be 27% for women under 20 ATB, 15% for those 20–39 ATB, and 7.5% for those exposed at 40 years of age or older.

For lung cancer, the dose-response is also linear, or at least not significantly different from linearity [Figure 12.2; computed from LSS Tumor Registry data for 1958–1987 (Thompson et al., 1994) and obtained by special arrangement from the Radiation Effects Research Foundation in Hiroshima and Nagasaki]. The most remarkable finding is a substantial and highly significant ($p=0.0002$) difference in slope between males ($ERR_{1Sv}= 0.439$, 95% confidence interval 0.14–0.81) and

TABLE 12.1 Female breast cancer incidence, LSS sample, 1950–1985: estimated excess relative risk at 1 Sv (ERR_{1Sv}), by age ATB (E). Comparison of a fitted negative exponential model, $\text{ERR}_{1Sv} = 3.60 \exp(-0.0373\ E)$, and stepwise models corresponding to various intervals of age ATB.

Age ATB	Exponential model	5-year intervals	10-year intervals	20-year intervals	Combined ages ATB
0–4	3.32	4.64	3.21	2.41	1.36
5–9	2.79	2.08			
10–14	2.27	1.82	2.19		
15–19	1.92	2.49			
20–24	1.59	1.18	1.27	1.25	
25–29	1.32	1.45			
30–34	1.09	0.96	1.23		
35–40	0.91	1.60			
40–44	0.75	0.02	0.54	0.48	
45–49	0.63	1.78			
50–54	0.52	0.51	0.31		
≥55	0.38	−0.12			
Deviance (d.f.)*	7.32(1)	13.64(11)	8.99(5)	8.36(2)	0.00(0)

* Difference in deviance cf. the single interval model in the right-most column.

TABLE 12.2 Female breast cancer incidence, LSS sample, 1950–1985: Estimated excess relative risk at 1 Sv (ERR$_{1Sv}$), by age ATB and attained age.

Attained age (years)	0–19 ATB		20–39 ATB		≥ 40	
	ERR$_{1sv}$	(95% CI)	ERR$_{1sv}$	(95% CI)	ERR$_{1sv}$	(95% CI)
<35	14.40	3.85–125.00	—	—	—	—
35–44	1.45	0.51–3.03	0.96	0.05–2.86	—	—
45–54	2.64	1.29–4.81	1.49	0.59–2.96	<0	—
55–64	1.80	−0.02–8.33	1.13	0.32–2.51	<0	—
65–74	—	—	0.81	Unk–3.12	0.62	Unk–2.42
≥75	—	—	—	—	0.76	Unk–3.11
p-value for trend	0.087		0.63		0.22	

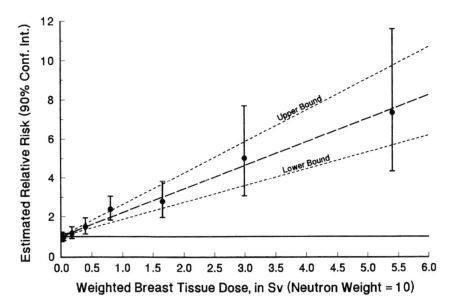

FIGURE 12.1 Breast cancer incidence among female atomic bomb survivors, Life Span Study sample, 1950–1985 (Tokunaga et al., 1994). Excess relative risk, indirectly standardized by city of exposure, age at the time of the bombings, age at observation for risk, and calendar year, by weighted radiation dose to breast tissue (neutron RBE = 10 relative to gamma rays).

females (ERR_{1Sv} = 1.85, 1.24–2.60). There is no evidence that age at exposure, attained age, or time following exposure are important modifiers of dose-response (Table 12.3). Based on individual dose estimates, a linear dose-response model, and the estimated, sex-specific values of ERR_{1Sv}, about 5.4% of the 511 lung cancers observed among exposed, known-dose LSS males during 1958–1987 and 17.7% of the 366 observed among females would not have occurred in the absence of radiation exposure.

The overall level of radiation-related lung cancer associated with exposure to inhaled radon and its decay products among various cohorts of underground miners of uranium and other minerals is considerably higher than that among A-bomb survivors. In a combined analysis of data from 11 different studies, Lubin et al. (1995) estimated a linear dose-response coefficient of 0.0049 per working level month (WLM) for excess relative risk over the range 0–400 WLM, with a less steep increase from 400 to about 2,500 WLM.

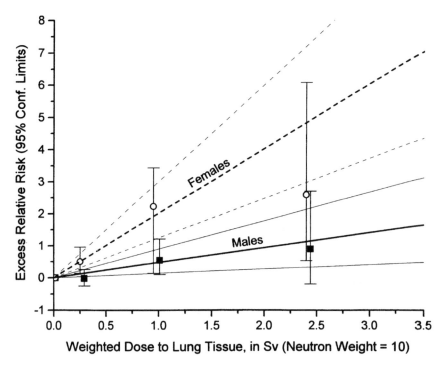

FIGURE 12.2 Lung cancer incidence among atomic bomb survivors with under 4 Gy kerma, Life Span Study sample, 1958–1987 (Thompson et al., 1994). Fitted linear dose-response curves, with 95% confidence limits and indirectly standardized by city, age ATB, and attained age, for male and female survivors: weighted radiation dose to lung tissue (neutron weight of 10 relative to gamma rays).

MAIN EFFECTS: SMOKING

Cigarette smoking is a far stronger lung cancer risk factor than the gamma rays and neutrons received from the Hiroshima and Nagasaki atomic bombs. Habitual two-pack-per-day smokers have about 24 times the lung cancer mortality of lifelong nonsmokers (Rogot and Murray, 1980) whereas, even at the near-lethal level of 5 Sv, the estimated relative risks from A-bomb survivor data, unadjusted for smoking, are only about 3 for males and 10 for females. Only very high levels of alpha radiation from inhaled radon decay products yield estimates of radiation-related lung cancer risk that approach or exceed those associated with heavy smoking habits (National Research Council, 1988).

TABLE 12.3 Lung cancer incidence, 1958–1987, LSS tumor registry data (Thompson et al., 1994). Estimated excess relative risk at 1 Sv, adjusted for sex and standardized to males, by age ATB, attained age, and calendar time.

Factor	ERR_{1Sv}	95% CI	p-value for trend
Age ATB:			
0–19	0.59	0.49–1.64	
20–39	0.41	−0.04–0.96	
≥40	0.46	−0.04–0.96	0.63
Attained age(y):			
<55	0.50	0.37–1.44	
55–74	0.44	−0.09–0.91	
≥75	0.50	0.11–1.12	0.90
Age ATB:			
1958–65	0.70	0.47–1.60	
1966–75	0.49	0.11–1.12	
1976–85	0.38	−0.19–0.83	
1986–87	0.47	0.36–1.43	0.50

The high rates of lung cancer among cigarette smokers, and the historically high prevalence of smoking in many populations with histories of substantial radiation exposure, make it difficult to study radiation as a risk factor without taking smoking history into account. This is especially true if the radiation effect is relatively small. Smoking information, available for 50% of the total LSS cohort (Shimizu et al., 1989), indicates that the proportions of smokers are 84% for males and 16% for females. An analysis restricted to mortality among the subcohort with smoking information obtained estimated excess relative risks of 0.1 and 1.2 for men and women, respectively, exposed to 1 Gy tissue kerma at age 30. The estimated relative risk (RR) of lung cancer for a person with a 40 pack-year smoking history (e.g., 20 cigarettes a day for 40 years) was about 16 compared to a nonsmoker, after adjustment for radiation dose, sex, and age ATB (Shimizu et al., 1989 Table 13). Adjustment of the mortality data for smoking increased the estimated ERR_{1Gy} for males from 0.1 to 0.5 while reducing that for females from 1.2 to 1.0.

INTERACTION: SMOKING AND RADIATION DOSE

A case-control interview study was conducted of 485 LSS lung cancer cases and 1,089 controls matched on sample status (AHS or non-AHS), city, sex, date of birth, and survival status (Blot et al., 1984); subjects (or, more often, next of kin) were interviewed about smoking histories. Analyses of these data, as reported by the BEIR IV committee (NRC, 1988), found relative risks for the highest radiation dose group (1+ Gy) of 1.6 for men and 4.0 for women; for smoking, the RR was 17.2 for men who smoked more than 20 cigarettes per day, and 3.7 for women who smoked more than 10 per day (the highest level tabulated). Separate inter-action analyses by sex, for radiation dose versus cigarettes per day (Table 12.4) and for dose versus years smoked (not shown), did not discriminate between an additive interaction model and a multiplicative model, neither of which could be rejected. The lack of information in this study probably stems from (1) the rather low radiation dose-response among men, (2) the high proportion of male smokers and the low proportion of female smokers, and (3) the relatively small number of high-dose A-bomb survivors.

The BEIR IV committee also reported an interaction analysis of data from a case-control study of New Mexico uranium miners (52 cases and 218 controls), categorized by years of mining and number of cigarettes per day. This study also was inconclusive, perhaps because years of mining is an inadequate surrogate for cumulative WLM. The committee also reported a cohort analysis of Colorado Plateau miners, based on 151 cases and 65,365 PY with exposures under 2,000 WLM, categorized by age at diagnosis, cigarette consumption, and cumulative WLM (Table 12.5). Their analysis (Table 12.6), which represented multiplicative and additive interaction models as special cases of a more general mixture model, did not reject the mixture model ($P|\chi^2_{35} > 37.8 = 0.34$), the multiplicative model ($P|\chi^2_{36} > 38.6 = 0.35$), or the additive model ($P|\chi_{36} > 44.6 = 0.15$). How-ever, in the more limited context of the mixture model and the mixture parameter λ, the estimated value $\lambda = 0.4$ was significantly different from the value $\lambda = 0$ corresponding to the additive model, but consistent with the value $\lambda = 1.0$ corre-sponding to the multiplicative model. In their report, the committee emphasized that the deviance (versus the marginally saturated model estimates tabulated in Table 12.5) was virtually the same for the multiplicative model and the best-fitting mixture model, even though 1.0 and 0.4 are quite different numbers, and that $\lambda = 0.4$ was in no sense a "halfway point" between the additive and multiplicative models. Based on their analysis, the committee concluded that the multiplica-tive interaction model was an adequate representation of the interaction between smoking and radon exposure, and that the additive model was not.

More recently, Lubin and Steindorf (1995) modeled joint relative risks for smok-ing history (ever versus never) and exposure to inhaled radon decay products among six cohorts of US uranium miners for which such information was available. They concluded that, at that level of smoking history detail, the best-fitting interaction

TABLE 12.4 Analysis of case-control data on smoking rate and radiation exposure from a case-control study of lung cancer among A-bomb survivors: relative risks by sex, number of cigarettes per day, and radiation dose. (Adapted from BEIR IV Tables VII-5 and VII-8, National Research Council, 1988; based on data from Blot et al., 1984).

Cigarettes/ day	Observed			Fitted additive models			Fitted multiplicative models		
	<0.01 Gy	0.01 – 0.99 Gy	≥1 Gy	<0.01 Gy	0.01 – 0.99 Gy	≥1 Gy	<0.01 Gy	0.01 – 0.99 Gy	≥1 Gy
Males									
0	1.0	1.3	3.3	1.0	0.9	3.5	1.0	0.8	1.6
1–10	3.7	2.4	7.2	3.0	2.9	5.5	2.7	2.2	4.3
11–20	6.9	6.6	10.6	6.0	5.9	8.5	5.5	4.4	8.8
≥20	26.5	13.2	24.8	19.4	19.3	21.9	17.2	13.8	27.5
χ^2 (d.f.) goodness-of-fit				$\chi^2(5)=2.2\ p = 0.90$			$\chi^2(5)=2.3\ p = 0.89$		
Females									
0	1.0	0.7	5.2	1.0	0.6	4.9	1.0	0.6	4.0
1–10	2.3	2.5	5.2	2.4	2.0	6.1	2.2	1.3	8.8
≥11	4.2	2.1	—	3.3	2.9	(7.2)	3.3	2.0	(14.8)
χ^2 (d.f.) goodness-of-fit				$\chi^2(4)=0.7\ p = 0.95$			$\chi^2(4)=1.7\ p = 0.79$		

TABLE 12.5 Relative risks for lung cancer among Colorado Plateau uranium miner cohort: risks relative to rate in the entire cohort, by age, number of cigarettes per day, and cumulative working level months (WLM). (Adapted from BEIR IV, Table VII-10).

Cigarettes/day	Cumulative exposure in working level months					
	0–59	60–119	120–239	240–479	480–959	960+
Age <65						
0–4	0.00	0.00	0.00	0.18	0.48	0.80
5–19	0.12	0.00	0.17	0.13	0.34	1.00
20–29	0.10	0.20	0.36	0.41	0.64	2.20
≥40	0.12	0.58	0.18	0.81	1.04	2.87
Age ≥65						
0–4	0.00	0.00	0.50	0.60	1.00	0.55
5–19	0.00	0.00	0.98	0.00	0.78	0.63
20–29	1.65	0.00	1.16	0.77	3.08	0.00
≥40	0.00	3.41	0.99	0.74	1.17	2.48

model was intermediate between the additive and multiplicative interaction models; according to a mixture model like that described in Table 12.7, the best fit was obtained for $\lambda = 0.12$, with 95% confidence interval 0.1–0.6. The multiplicative model ($\lambda = 1$) was statistically rejected but fit the data somewhat better than the additive model ($\lambda = 0$), which was rejected at a more extreme level of significance. Thus, the ratio of radon-related to non-radon-related cancers was estimated to be higher among nonsmokers than among smokers, although for a given radon exposure level more radiation-related lung cancers were estimated to occur among smokers.

MAIN EFFECTS: REPRODUCTIVE HISTORY

In all populations in which it has been studied, breast cancer risk has a strong dependence upon reproductive history (Kelsey et al., 1993). Within populations, nulliparous women have risks comparable to those of parous women whose first full-term pregnancies occurred at about age 30, while a first full-term pregnancy before age 18 is associated with a risk that is only one third as high. Studies differ on whether number of children and length of lactation history are independently related to risk or are merely strong negative correlates of age at first full-term pregnancy. Late age at menarche and early age at natural menopause are protective in many studies, and a bilateral oophorectomy prior to menopause is strongly protective.

A case-control interview study was conducted at RERF, first, to confirm that various aspects of reproductive history identified in other populations are also determinants of breast cancer risk in the LSS population (Land et al., 1994a) and, second, to investigate the interaction of such factors with radiation dose (Land et al., 1994b). At the time of case selection, 473 cases had been identified, of whom 233 were still alive and living in or near Hiroshima or Nagasaki. Six of the latter number were later determined not to be cases and were dropped from the analysis. Refusals, usually because of ill health, reduced the final number to 196. Controls (n=566), individually matched to the cases by city, age ATB, exposure status and radiation dose if exposed, were selected using a variable matching ratio. Four controls were selected if the case had a tissue dose greater than 0.5 Gy or was from Nagasaki; otherwise, 2 controls were selected. Losses of potential controls due to refusal (9 percent) or residence outside the contacting area (6 percent) were approximately the same as for the cases. Subjects were interviewed in their homes or at RERF.

Matching of cases and controls by radiation dose was the major design inno-vation. The rationale for this was that all the information that could be obtained about radiation dose-response would be available from past, concurrent, and future studies of breast cancer incidence in the entire LSS cohort, and that information about dose-response from the case-control study therefore would be superfluous. Furthermore, matching on radiation dose, and incidentally ensuring that the case-control study could provide no useful information about dose-response, would minimize any need to adjust for radiation dose when investigating other variables as main effect factors. The most important consideration, however, was that the design change would improve statistical power for investigating interactions with radiation dose (Land, 1990).

A number of reproductive history factors, many of them highly correlated, were strongly related to breast cancer risk. Among them were age at first full-term pregnancy, number of births, and cumulative lactation history. Age at first full-term pregnancy was at least marginally significantly related to risk even after adjustment for number of births and/or cumulative lactation, whereas the latter two variables were more highly correlated with each other (Table 12.7). All three of these variables were somewhat more strongly associated with risk for premenopausal than for postmenopausal cancer, a difference that was marginally significant for cumulative lactation (Land, 1994a).

TABLE 12.6 Results of BEIR IV interaction analysis of smoking and radiation exposure among Colorado Plateau uranium miners (adapted from National Research Council, 1988, Table VII-11).

Model	Deviance	d.f.	Deviance difference (1 d.f.)	Prob.
$R_{add} = 1 + \alpha(WLM) + \beta(smoking)$	44.6	36	6.8	0.014
$R_{mult} = (1 + \alpha(WLM))(1 + \beta(smoking))$	38.6	36	0.8	0.370
$R_{mix} = (R_{mult})^{\lambda}(R_{add})^{1-\lambda}*$	37.8	35	—	—

*Fitted value $\lambda = 0.4$.

TABLE 12.7 Main-effect analyses for breast cancer risk in relation to reproductive history (Land et al., 1994a). Age at first full-term pregnancy, number of births, and cumulative lactation history, in years. Estimated relative risks by factor level.

Main effect	Relative risk
Age at first full pregnancy	
Nullip.	1.00
30–39	0.95
27–29	0.85
24–26	0.61
21–23	0.50
17–20	0.39
p-value for trend	
Unadjusted	<0.0001
Adjusted[a]	—
Adjusted[b]	0.066
Adjusted[c]	0.070
Number of births	
0	1.00
1	0.76
2	0.73
3–4	0.39
5–6	0.34
≥7	0.14
p-value for trend	
Unadjusted	<0.0001
Adjusted[a]	0.030
Adjusted[b]	—
Adjusted[c]	0.16
Cumulative lactation (years)	
0	1.00
0.1–0.5	0.61
0.6–1.0	0.78
1.1–2.0	0.56
2.1–4.0	0.36
4.1–9.8	0.31
p-value for trend	
Unadjusted	<0.0001
Adjusted[a]	0.041
Adjusted[b]	0.40
Adjusted[c]	—

[a] Adjusted for age at first full-term pregnancy.
[b] Adjusted for number of births.
[c] Adjusted for cumulative lactation period.

Age at menarche and age at menopause, somewhat surprisingly, were not related to risk. Another surprise was that the data provided no useful information on family history of breast cancer, apparently because subjects often or usually lacked information about cancer diagnoses among their close relatives.

INTERACTION: RADIATION AND REPRODUCTIVE HISTORY

The question of interaction between radiation dose and other breast cancer risk factors is important for a number of reasons. One is that women deemed for various reasons to be at unusually high risk of breast cancer conceivably might be especially sensitive to the carcinogenic effects of x-rays used in monitoring for early cancer detection. More generally, we might hope through investigations of interaction to refine our estimates of radiation-related risk by taking proper account of ancillary information about populations and individuals at risk. Perhaps even more important from a long-term perspective is the possibility of gaining insights into why certain personal characteristics (e.g., age at first full-term pregnancy) are associated with increased or decreased breast cancer risk, by observing their modifying influences on the effects of an independent exposure to a known carcinogen, ionizing radiation.

Like the analyses of radiation, smoking, and lung cancer among A-bomb survivors and uranium miners mentioned above, the analysis concentrated on two simple interaction models, the additive and the multiplicative, and a parametric mixture model containing the two simple models as special cases. Relative risk was assumed to be linear in radiation dose, with age-ATB-specific coefficients as estimated from the 1950–1985 LSS cohort breast cancer incidence data (Tokunaga et al., 1994), and linear in each of the reproductive history factors tested:

$$R_{mult}(D, X; \beta) = (1 + \alpha_E D)(1 + \beta X) \qquad (12.1)$$

$$R_{add}(D, X; \beta) = 1 + \alpha_E D + \beta X. \qquad (12.2)$$

Here, α_E = 3.6 exp(-0.03735 E) as estimated by Tokunaga et al. from the 1950–1985 LSS cohort data, where E denotes age ATB. The parametric mixture model,

$$R_{mix}(D, X; \beta, \theta) = (1 + \alpha_E D)(1 + \beta X/(1 + \alpha_E D)^\theta), \qquad (12.3)$$

was chosen for computational convenience. It was somewhat different from the geometric mixture model used by the BEIR IV committee but fulfilled the same function, i.e., the analysis essentially depended on a single parameter, θ, for discriminating between the additive ($\theta = 1$) and multiplicative ($\theta = 0$) interaction models.

In all analyses involving interaction models, the reproductive history variables were translated to have zero mean in the case-control set. This was done because the radiation dose-response coefficients had been estimated from the entire cohort, without reference to reproductive history. Thus they are defined with respect to a reference set with zero dose, and reproductive history covariates equal to their

TABLE 12.8 Supplementary interaction analysis of breast cancer case-control study data (Land et al., 1994b), using the approach of Table 12.3, $\alpha_E = 3.6 \exp(-0.03735 \, E)$ as estimated by Tokunaga et al. (1994) from the 1950–1985 LSS cohort data, where E denotes age ATB, D denotes weighted breast tissue dose (DS86) in Sv (neutron RBE = 10). X denotes age at first full-term pregnancy, with nulliparous women assigned default value 30, minus the sample mean for that variable.

Model		Deviance	Chi-square (1 d.f.)[a]	p-value
Additive:	$R_{add}(D, X; \beta) = 1 + \alpha_E D + \beta X$	393.88	10.08	0.0015
Multiplicative:	$R_{mult}(D, X; \beta) = (1 + \alpha_E D)(1 + \beta X)$	387.27	3.586	0.0580
Mixture:[b]	$R_{mix}(D, X; \beta, \lambda) = (R_{mult})^{\lambda} (R_{add})^{1-\lambda}$	383.69	—	—

[a]Difference in deviance between the designated model and the mixture model.
[b]Fitted value $\lambda = 3.55$, with 95% confidence interval 0.94–47.7.

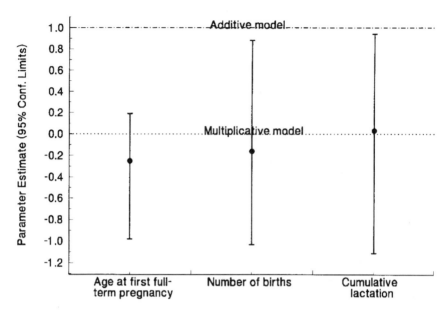

FIGURE 12.3 Summary of interaction analyses for each of three reproductive history variables as joint breast cancer risk factors with radiation dose (Land et al., 1994b). Point estimates and 95% confidence intervals for the parameter in the model, $R_{mix}(D, X; \beta, \theta) = (1 + \alpha_E D)(1 + \beta X/1 + \alpha_E D)^\theta)$, where $\alpha_e = 3.6$ exp(-0.03735 E) as estimated by Tokunaga et al. (1994) from the 1950–1985 LSS cohort breast cancer incidence data, and where E denotes age ATB, D denotes weighted breast tissue dose (DS86) in Sv (neutron RBE = 10), and X denotes age at first full-term pregnancy with a default age of 30 years for nulliparous women, number of births, or cumulative lactation history in years. In this model, $\theta = 0$ corresponds to the multiplicative model, $R_{mult}(D, X; \beta) = (1 + \alpha_E D)(1 + \beta X)$, and $\theta = 1$ to the additive model, $R_{add}(D, X; \beta) = 1 + \alpha_E D + \beta X$.

respective population means. It follows, then, that these covariates should have the same reference set, which is accomplished, approximately, by translating each covariate by its sample mean in the case-control set.

The results are as shown in Figure 12.3. The estimated value of the parameter θ was significantly less than one for all three reproductive history variables, but consistent with zero. Thus, for all three variables, the additive model was rejected, whereas the multiplicative model was consistent with the data. Thus, in this population, early age at first full-term pregnancy, multiple births, and lengthy lactation history were all protective against breast cancer risk generally, and also protective against radiation-related breast cancer risk.

When separate analyses were done by age ATB, the discrimination between models appeared to be somewhat stronger for women under 20 ATB, perhaps because the radiation dose response was stronger. There were six cases with diagnoses at ages under 35 for whom the radiation-related risk may have been substantially higher than at older attained ages (Table 12.2); however, deleting those six case-control sets did not change the results appreciably.

None of the women under age 17 ATB reported experiencing a pregnancy before age 17. It is interesting that the interaction analysis for age at first full-term pregnancy, restricted to that age cohort, yielded the same multiplicative (or super-multiplicative) relationship as that for all women under 20 ATB (estimated value of θ, -0.26 with upper confidence limit 0.31). Thus, excess risk among women exposed as young girls was reduced by a *subsequent* full-term pregnancy at a young age, and increased by nulliparity or a late first pregnancy. The finding suggests that terminal differentiation of cells for milk secretion, induced by a full-term pregnancy, may reduce the proliferative potential even of cells already initiated by radiation.

The above interpretation is consistent with experimental results obtained by Clifton et al. (1975, 1978). In their study, female rats were irradiated and injected with prolactin-secreting, transplantable pituitary tumors. One group received no further treatment; another received adrenalectomy, which precluded the production of adrenal corticoids necessary for cell differentiation for milk secretion, and another received both adrenalectomy and glucocortisol replacement therapy. High levels of radiation-induced mammary cancer were experienced by the adrenalectomy-only group compared to rats with intact adrenals or adrenalectomized rats given glucocortisol replacement therapy.

SUPPLEMENTARY ANALYSES

The following analyses may provide additional insight into the underlying interactive relationship between radiation dose and one of the reproductive history variables, age at first full-term pregnancy, as breast cancer risk factors. Figure 12.4 shows logistic model estimates of the relative risk multiplier per additional year of age at first full-term pregnancy, separately for matched case-control sets grouped according to whether the radiation-related ERR estimated from the cohort data was zero, between zero and 0.5, between 0.5 and 1.0, or over 1.0. According to the multiplicative model, the ERR associated with age at first full-term pregnancy does not depend upon the common radiation doses of the matched case-control sets (the dose terms should cancel out, as in the ratio AD / BD, where D represents the common term for dose-related risk), whereas the additive model says that the ERR should approach unity with increasing dose, as in the ratio $(D + A)/(D + B)$. If anything, the estimates diverge from one or remain constant with increasing dose; they do not approach one as predicted by the additive interaction model.

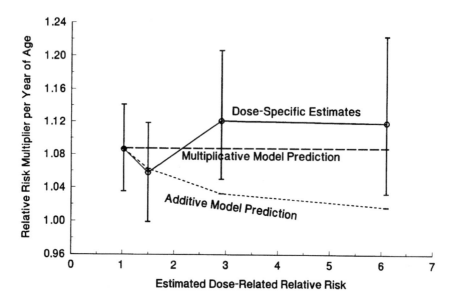

FIGURE 12.4 Logistic model estimates of the relative risk multiplier per additional year of age at first full-term pregnancy (default age 30 for nulliparous woment), by radiation dose (estimated dose-related relative risk), compared to predicted values according to the multiplicative and additive interaction models.

Table 12.8 contains the results of an analysis according to the geometric mixture model of Table 12.6, in which the radiation dose component for each subject was fixed at the function of dose and age ATB estimated from the LSS cohort data (Tokunaga et al., 1994). The deviances corresponding to the additive ($\lambda = 0$) and multiplicative ($\lambda = 1$) models were identical to those obtained in the analysis summarized in Figure 12.3. The best-fitting estimate of λ was 3.55, with 95% confidence limits 0.94–47.7. Thus, as in the analysis of Figure 12.3, the best-fitting model might be described as "super-multiplicative" in the sense that the effects on ERR of changes in reproductive history appear to increase with increasing radiation dose. Also, as in Figure 12.3, the data were consistent, in the context of a particular general model, with the multiplicative model but not with the additive model.

Figure 12.5 summarizes the results of an exploration of the sensitivity of the interaction analysis of Figure 12.3 for radiation dose and age at first full-term pregnancy, to variations in the dose-response estimated from the cohort data. There was little difference in the analysis if the age-specific dose-response coefficients were replaced by their upper or lower 90% confidence limits. On the right is a stem and leaf plot of the distribution of the point estimate based on 100 simulation

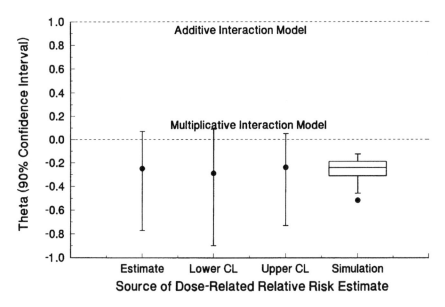

FIGURE 12.5 Sensitivity analysis of the interaction analysis of Figure 12.3 with respect to radiation dose and age at first full-term pregnancy. On the left is the analysis of Figure 12.3, but with 90% confidence limits for the estimate of θ. Next are point estimates and confidence limits for θ when α_E was replaced by its lower or upper 90% confidence limit as estimated by Tokunaga et al. (1994). On the right is a stem and leaf plot of the sample distribution of 100 point estimates of θ obtained from 100 simulation runs, in each of which individual dose estimates were allowed to vary randomly and independently according to a simulated lognormal distribution with median equal to the DS86 value and a 30% geometric standard deviation. The maximum point estimate was -0.13, the minimum -0.52.

runs, in each of which individual doses for all of the subjects in the case-control study were allowed to vary randomly and independently according to a lognormal error distribution, with median equal to the respective DS86 estimate and a 30% geometric standard deviation. Thus, both consistent and random changes in the estimated radiation-related excess relative risk estimates, within reasonable ranges of uncertainty, had only minor effects on the results of the interaction analysis.

CONCLUSIONS

The contrast between the rather clear interaction results obtained for breast cancer with respect to radiation and reproductive history, and the rather murky findings for lung cancer with respect to radiation and smoking in the LSS sample mainly reflects the relatively low level of the lung cancer dose-response, particularly among males. This conclusion is reinforced by the more conclusive results obtained for the uranium miner cohort. General uncertainty about smoking information may be another factor. There is more to be learned about radiation–reproductive history interactions in the LSS sample for cancers of other sites, like ovary, as well as breast cancer. Age at first full-term pregnancy is highly correlated with age at first delivery, and information about that variable, like number of children, is relatively accessible and free of uncertainty. At this time there are over 1,000 confirmed cases of female breast cancer among LSS sample members that could be included in a replication of the study described above. Such a study is planned for the near future, and may result in more detailed findings and, possibly, new insights into underlying relationships.

13

A Simplified Model of Radiation Carcinogenesis in the Atomic Bomb Survivors

MORTIMER L. MENDELSOHN

SUMMARY

A simplified model of carcinogenesis combines the linear dose response of solid cancer incidence in the exposed, and the age response of cancer in the controls. Using the Armitage and Doll formulation, the model assumes that cancer in the controls is the result of n time-driven somatic mutations in the evolving stem cell that gives rise to the cancer, where $n - 1$ is the slope of the logarithm of the cancer rate versus logarithm of age relationship. In radiation-induced cancer, the atomic bomb exposure, in proportion to dose, contributes one and only one of the n mutational events.

The model agrees with actual cancer rates when it uses n taken from the control data, and from reasonable estimates for mutation rate in somatic cells, for numbers of stem cells at risk, and for the number of relevant target oncogenes and cancer-suppressor genes. Also, the model correctly predicts a slope of $n - 2$ for the log of radiation-induced cancer rate versus log age. The advantages and limitations of the model are discussed.

INTRODUCTION

Fifty years after the atomic bombings of Hiroshima and Nagasaki, the cancer experience of the survivors provides remarkable detail on human radiation carcinogenicity. The foremost example is the exquisitely linear, nonthresholded, dose response for the induction of solid cancer (i.e., all cancer except leukemia) as reported by Thompson et al. (1994). Such linearity of response is elegantly simple but is difficult to reconcile mechanistically with two-hit chromosomal induction, a mutation-growth-mutation model, or any other multi-mutational mechanism of cancer induction. It was in thinking about possible explanations for this situation that I happened upon a novel way to frame the carcinogenesis problem (Mendelsohn, 1995, 1996). Donald Pierce and I are currently in the final stages of a formal mathematical description of this evolving formulation, and the present paper will be a qualitative review of the model.

HYPOTHESIS

The simplest mechanism for explaining linear dose responses in radiation biology is by somatic gene mutation. Thus I hypothesize, based on the linear cancer process in the atomic bomb survivors, that the only significant difference between the background and radiation-induced solid cancers is the occurrence of one and only one radiation-induced somatic mutation in the oncogenes and suppressor genes of the evolving stem cell that eventually gives rise to the induced cancer. Based largely on the age distribution of cancer rates and the now-extensive information on multiple genetic defects in human solid cancers, I consider most human solid cancers to arise through a multi-gene mutational mechanism involving a half-a-dozen or so loci. The supporting arguments for these positions follow.

SOMATIC MUTATION IN SURVIVORS

The only somatic specific-locus gene-mutation dosimeter that shows a consistent radiation response in the survivors is the flow-cytometric measurement of glycophorin A mutants in erythrocytes (Akiyama et al., 1995). Glycophorin A is a major constituent of the erythrocyte membrane. It has no known function and comes in two equally probable alleles named M and N, following the precedence for the two corresponding and related blood types. In heterozygotes of glycophorin A, every normal erthrocyte simultaneously expresses M and N glycoproteins. Using fluorescent antibodies and flow-cytometric high-speed analysis of erythrocytes, two deviant patterns are detectable. The $N\emptyset$ or $M\emptyset$ outcome in a single erythrocyte is a gene loss (or hemizygosity) of one of the alleles of glycophorin A, with the other allele expressing normally. The NN outcome is a gene duplication (or reduction to homozygosity) in which M glycoprotein is no longer detectable and N is present in double quantity. These outcomes occur about 29 and 20 times, respectively, per million erythrocytes in normal adult subjects. In survivors grouped by dose, the two dose responses are linear and nonthresholded, and differ from each other significantly in slope and y-intercept (Akiyama et al., 1995).

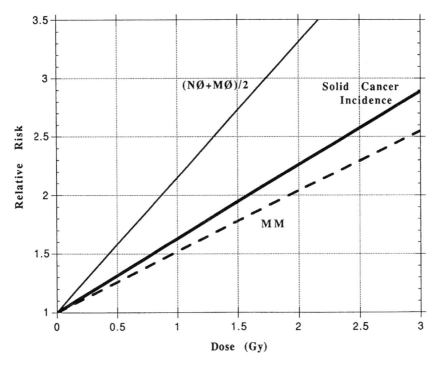

FIGURE 13.1   The dose response for relative risk of three endpoints in the atomic bomb survivors: the incidence of solid cancers, the combined frequencies of the glycophorin A gene loss mutations M∅ and N∅, and the frequency of the glycophorin A reduction to homozygosity mutation MM. The mutation data are replotted from Akiyama et al. (1995), and the cancer data from Thompson et al. (1994).

The dose responses for cancer and glycophorin A mutation can be compared by expressing them on a common scale of dose and relative risk, as shown in Figure 13.1.

The cancer dose response lies midway between the two genetic dose responses, and the three have common shape and roughly common magnitudes of response. Based on this, and in keeping with my hypothesis, one can imagine the cancer induction mechanism to be a combination of gene loss and reduction to homozygosity of a gene with approximately the size and vulnerability of glycophorin A.

Three other somatic genetic dosimeters show inconsistent or absent responses in the survivors (Akiyama et al., 1995). All three are tested in the lymphocyte and are known to respond well to recent radiation exposures both in vitro and in vivo. After radiotherapy, such human lymphocytic mutants have a half-life of a few years

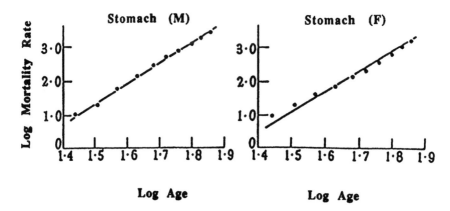

FIGURE 13.2 Log death rate for stomach cancer versus log age, in males (M) and females (F), adapted from Armitage and Doll (1954). In each case, the lines are drawn with a slope of 6, indicating the reasonableness of fit of stomach cancer mortality to age to the 6th power.

in vivo, thus explaining the lack of response in the atomic bomb survivors at four to five decades after exposure.

Cytogenetic analyses of the survivors' lymphocytes show stable increases in balanced chromosomal translocations (Awa, 1991). The dose response is curvilinear upwards and an order of magnitude steeper than the mutation and cancer responses. I will return to this later.

THE ROLE OF MUTATION IN CARCINOGENESIS

In 1954, Armitage and Doll published a landmark paper in which they confirmed that the time course of many human cancers is a simple power function of age. It is the property of such functions to be linear in log-log plots of mortality rate versus age. An example, reproduced from their paper, is shown in Figure 13.2. Their accompanying mathematical analysis interprets the power function as a reflection of a cancer process with n successive stages, each independent, irreversible, and roughly similar in likelihood. The biological representation of a stage was unclear to them at the time, but they included the possibility that stages might be due to successive somatic mutation.

Many cancer analysts have dealt with the Armitage and Doll formulation, further confirming the frequent but not universal pattern of the age response and exploring its validity. The prevailing notion has been that somatic mutation is too rare a process to be the sole mechanism for the stages. In recent years, following the lead of Moolgavkar and Knudson (1981), the trend in cancer modeling has been to focus instead on two mutational stages with an intervening accelerated growth phase that increases the number and sensitivity of target cells and hence the likelihood of the second mutation (see, for example, Little in this volume).

A parallel line of argument in support of multiple mutation stems from the molecular oncologists who, in the past decade, have assembled an impressive body of evidence pointing to oncogene activation and suppressor-gene loss based on gene mutation, chromosome aberration, and gene amplification in human cancer. This wave of discovery culminated in the finding by Fearon and Vogelstein (1990) that in the colon the molecular genetic changes correlate with an evolving succession of precancerous and cancerous lesions, thus emphasizing the link between genomic damage and stages of this disease. While it may not be possible to identify similar staging in other organs, the general pattern has been that each type of adult human cancer has its own constellation of oncogenic change. The patterning, while not exact, fits the notion that each type of solid cancer in adults has ~ 6 changes out of a somewhat larger set of candidate genetic sites. I believe that such changes may well represent the stages of Armitage and Doll (1954).

Recently, Vogelstein's group (Papadopoulos et al., 1994) and Kolodner's group (Bronner et al., 1994) simultaneously published the evidence for two mutator genes in human hereditary nonpolyposis colon cancer. Mutations in the two mismatched repair genes occur in approximately 90% of these cancer-prone patients. Since such a mutator mechanism increases the probability of subsequent mutations in the evolving cancer clone, it and other mutator processes make it easier to fit multistage models such as the one proposed here using current somatic mutation and cancer rates.

THE FIT OF A MUTATIONAL MODEL IN THE SURVIVORS

Analysis of the cancer rates in the controls for the atom bomb survivors show linearity of log cancer rate versus log age for solid cancer overall, as well as for the few most common types of cancer which have sufficient data to be analyzed separately. The overall slope is in the range of four to five, suggesting a five- to six-stage mechanism. Breast cancer is a clear exception, showing a two-phased response, with a change of slope around the time of menopause.

With the help of Don Pierce, the feasibility of this modified Armitage and Doll model was tested by using representative values for cancer rate, number of stages, somatic mutation rate, and numbers of stem cells. As an example—assuming six stages, a somatic mutation rate of 5×10^{-5}/gene/year, 30 candidate onco- and suppressor genes, and 10^9 stem cells—the model matches the known cancer rate at age 60. Such a somatic mutation rate is roughly in accord with the mutation rate of human cells in vivo and in vitro, but is likely to be highly dependent on a number of poorly known factors such as the kinetic activity of the stem cells. Similarly, the number of stem cells is an elusive concept, depending heavily on definition and stage of life. This simplified calculation assumed zero mutations at birth, and a constant-sized stem cell compartment. Embryonic life is very likely a time of high mutation rate, and we know that newborns have significant numbers of several somatic mutations (Manchester et al., 1995, McGinniss et al., 1989). Also, the size of the stem cell compartment probably changes with age, growing in the early years and declining late in life.

Returning to the radiation-induced cancers, a one-mutation dynamic for radiation induction would lead to the magnitude and linearity of dose response shown in Figure 13.1. It would also predict that the radiation-induced cancers by organ would be almost proportional to the control cancers by organ. This result is consistent with organ distributions of control and induced cancers in the atomic bomb survivors (Thompson et al., 1994).

The age pattern of induced cancer should have a close and well-defined relationship to the age pattern of the controls. That is, where n is the number of stages, $n - 1$ is the slope of the log cancer rate versus log age of the controls, and $n - 2$ should be the slope of the log cancer rate versus log age of the induced solid cancers. Calculations by Pierce using the atomic bomb data have confirmed this relationship.

In its present state, this model puts no constraint on the ordering of mutations. However, the biology of cancer development may require that certain mutations precede others. Such restrictions would reduce the effective target size for mutation, thereby decreasing the integrated yield of relevant mutations and the corresponding yield of cancers. Severe restrictions could invalidate the model by making background cancer a too-unlikely event.

The model clearly predicts cancer rates that progressively rise with age for most solid cancers, including those that are radiation induced. Biology may again intervene in the form of age-related changes that close off or make less likely a particular channel for cancer development. Two broad examples come to mind. Hormone changes associated with menarche, menopause, or aging could dramatically change the sexually responsive tissues. Diminution of growth in the aged, associated with loss of stem cells, should reduce cancer induction and could be the cause of the frequently seen downward bend in log cancer rate versus log age in the elderly. I predict that the radiation-induced cancers will show parallel changes, regardless of the age at which the subjects are irradiated.

SOME CAVEATS

It is almost certainly a gross oversimplification to reduce the bulk of human carcinogenesis to a sequence of time-driven mutational events. We know that there are profound metabolic and kinetic differences, organ by organ, and that these are confounded by complex secondary factors based on growth, promotion, and environmentally induced mutational injury. One suspects that there are contributions as well from many other aspects of lifestyle and biology. However, regardless of these complexities in background carcinogenesis, the essence of this model is that the added effect of radiation induction may be quite simple.

These results for background and radiation-induced solid cancer are not expected to apply to childhood leukemia. The types of leukemia that arose steeply in the first decade after the bombing and then declined afterward must involve fewer stages and some kind of age-related turn-off of susceptibility. Also, these leukemias have dose responses similar in magnitude and shape to those of chromosomal translocation (Preston et al., 1994).

Similar exceptions may apply to the solid cancers of childhood, genetically based and otherwise. Retinoblastoma needs only one additional mutation in its heritable form, and this occurs so frequently in the involved kindreds that independent bilateral disease is common (Knudson, 1985). The sporadic form of the disease involves two mutations and is unilateral. In both types, the disease is limited to early life and most likely involves a turn-off of susceptibility as the retina matures. Other embryonic tumors share the same kind of early onset and low numbers of required mutation.

A corollary of cancer being caused by time-based mutational processes is that somatic mutation itself should be cumulative with age. While the frequency of glycophorin A mutants in the atomic bomb survivors does increase with age, the response is too flat to fit this model. There are several explanations for why this may be happening. One is that the observed mutations are a composite of (1) long-lived and integrating stem cell mutations, (2) short-lived and non-integrating mutations in the differentiating cells of the committed erythrocyte chain, and (3) false positive, non-integrating technical errors of the method. In overestimating the background, the two non-integrating components of the measurement also lead to underestimation of the relative risk of mutation as given in Figure 13.1.

Yet to be understood with this model is the likely effect of fractionated or low-dose-rate irradiation. At one extreme, the case of a high radiation background, this should appear to the model as an increase in the underlying somatic mutation rate rather than as any effect on the slope of log cancer rate versus log age. At the other extreme, of two acute doses well divided in time, the model predicts essentially no difference from the same total dose delivered in a single fraction. This is largely because of the linearity of the process, or to put it another way, because of the enormous unlikelihood that the same cell lineage would be affected by both fractions.

Finally, I acknowledge that my use of Occam's razor may have carried this model too far on the scale of simplicity. This leaves me in good company, since many others in cancer research have taken the same path in their search for insight into this seemingly impenetrable mystery. I, like they, have embraced simplicity in the hopes that even a brief flash of light may illuminate the truth; the complexity can always be added later. I am grateful to Donald A. Pierce for his contribution to the evolving mathematical formulation and analysis of this material. This work was performed under the auspices of the US Department of Energy by Lawrence Livermore National Laboratory under contract No. W-7405-ENG-48.

14

Mechanistic Modeling of Radiation-Induced Cancer

MARK P. LITTLE

SUMMARY

A variety of models of carcinogenesis are reviewed, and in particular, the multistage model of Armitage and Doll and the two-mutation model of Moolgavkar, Venzon, and Knudson. Both the latter models, and various generalizations of them also, are capable of describing at least qualitatively many of the observed patterns of excess cancer risk following ionizing radiation exposure. However, there are certain inconsistencies with the biological and epidemiological data for both the multistage model and the two-mutation model. In particular, there are indications that the two-mutation model is not totally suitable for describing the pattern of excess risk for solid cancers that is often seen after exposure to radiation, although leukemia may be better fitted by this type of model. Generalizations of the model of Moolgavkar, Venzon, and Knudson that require three or more mutations are easier to reconcile with the epidemiological and biological data relating to solid cancers.

INTRODUCTION

One of the principal uncertainties that surround the calculation of population cancer risks from epidemiological data results from the fact that few radiation-exposed cohorts have been followed up to extinction. For example, 40 years after the atomic bombings of Hiroshima and Nagasaki, two-thirds of the survivors were still alive (Shimizu et al., 1990). In attempting to calculate lifetime population cancer risks it is therefore important to predict how risks might vary as a function of time after radiation exposure, in particular for that group for whom the uncertainties in projection of risk to the end of life are most uncertain, namely those who were exposed in childhood.

One way to model the variation in risk is to use empirical models incorporating adjustments for a number of variables (e.g., age at exposure, time since exposure, sex) and indeed this approach has been used in the Fifth Report of the Biological Effects of Ionizing Radiations (BEIR V) Committee (NRC, 1990) in its analyses of data on the Japanese atomic bomb survivors and various other irradiated groups. Recent analyses of solid cancers for these groups have found that the radiation-induced excess risk can be described fairly well by a relative risk model (ICRP, 1991). The *time-constant relative risk model* assumes that if a dose of radiation is administered to a population, then, after some latent period, there is an increase in the cancer rate, the excess rate being proportional to the underlying cancer rate in an unirradiated population. For leukemia, this model provides an unsatisfactory fit, consequently a number of other models have been used for this group of malignancies, including one in which the excess cancer rate resulting from exposure is assumed to be constant, i.e., the *time-constant additive risk model* (UNSCEAR, 1988).

It is well known that for all cancer subtypes (including leukemia) the excess relative risk (ERR) diminishes with increasing age at exposure (UNSCEAR, 1994). For those irradiated in childhood there is evidence of a reduction in the ERR of solid cancer 25 or more years after exposure (Little et al., 1991; Little, 1993; Thompson et al., 1994). For solid cancers in adulthood the ERR is more nearly constant, or perhaps even increasing over time (Little and Charles, 1991; Little, 1993), although there are some indications to the contrary (Weiss et al., 1994). Clearly then, even in the case of solid cancers, various factors have to be employed to modify the ERR.

Associated with the issue of projection of cancer risk over time is that of projection of cancer risk between two populations with differing underlying susceptibilities to cancer. Analogous to the relative risk time projection model one can employ a *multiplicative transfer* of risks, in which the ratio of the radiation-induced excess cancer rates to the underlying cancer rates in the two populations might be assumed to be identical. Similarly, akin to the additive risk time projection model one can use an *additive transfer* of risks, in which the radiation-induced excess cancer rates in the two populations might be assumed to be identical. The data that are available suggest that there is no simple solution to the problem (UNSCEAR, 1994). For example, there are weak indications that the relative risks of stomach cancer following radiation exposure may be more comparable than the absolute excess risks in populations with different background stomach cancer rates (UNSCEAR, 1994). The breast cancer relative risks observed in the most recent analysis of the Japanese atomic bomb survivor incidence data (Thompson et al., 1994) are rather higher than those seen in various other datasets, particularly for older ages at exposure (Shore et al., 1986; Miller et al., 1989; Boice et al., 1991).

The observation that sex differences in solid tumor relative excess risk are generally offset by differences in sex-specific background cancer rates (UNSCEAR, 1994) might suggest that absolute excess risks are more alike than relative excess risks. Taken together, these considerations suggest that either a relative or absolute transfer may be warranted; or indeed, the use a hybrid transfer approach, such as that which has been employed by Muirhead and Darby (1987).

Arguably, models that take into account the biological processes leading to the development of cancer can provide insight into these related issues of projection of cancer risk over time and transfer of risk across populations. For example, Little and Charles (1991) have demonstrated that a variety of mechanistic models of carcinogenesis predict an ERR that reduces with increasing time after exposure for those exposed in childhood, while for those exposed in adulthood the ERR might be approximately constant over time. Mechanistic considerations also imply that the interactions between radiation and the various other factors that modulate the process of carcinogenesis may be complex (Leenhouts and Chadwick, 1994), so that in general one would not expect either relative or absolute risks to be invariant across populations.

ARMITAGE-DOLL MULTISTAGE MODEL

Mechanistic models of carcinogenesis were originally developed to explain phenomena other than the effects of ionizing radiation. One of the more commonly observed patterns in the age-incidence curves for epithelial cancers is that the cancer incidence rate varies approximately as $C \cdot [age]^{\beta}$ for some constants C and β. At least for most epithelial cancers in adulthood, the exponent β of age seems to lie between four and six (Doll, 1971). The so-called multistage model of carcinogenesis of Armitage and Doll (1954) was developed in part as a way of accounting for this approximately log-log variation of cancer incidence with age. The model supposes that at age t an individual has a population of $X(t)$ completely normal (stem) cells and that these cells acquire one mutation at a rate $M(0)(t)$. The cells with one mutation acquire a second mutation at a rate $M(1)(t)$, and so on until at the $(k-1)$th stage the cells with $(k-1)$ mutations proceed at a rate $M(k-1)(t)$ to become fully malignant. The model is illustrated schematically in Figure 14.1. It can be shown that when $X(t)$ and the $M(i)(t)$ are constant, a model with k stages predicts a cancer incidence rate which is approximately given by the expression $C \cdot [age]^{k-1}$ with $C = M(0) \cdot M(1) \cdot ... \cdot M(k-1)/(1 \cdot 2 \cdot ... \cdot k - 1)$ (Armitage and Doll, 1954; Moolgavkar, 1978).

In developing their model Armitage and Doll (1954) were driven largely by epidemiological findings, and in particular by the age distribution of epithelial cancers. In the intervening thirty years, there has accumulated substantial biological evidence that cancer is a multistep process involving the accumulation of a number of genetic and epigenetic changes in a clonal population of cells. This evidence is reviewed by the United Nations Scientific Committee on the Effects of

FIGURE 14.1 The Armitage-Doll multistage model.

Atomic Radiation (UNSCEAR, 1993). However, there are certain problems with the model proposed by Armitage and Doll (1954) associated with the fact that to account for the observed age incidence curve $C \cdot [age]^\beta$ with β between four and six, between five and seven stages are needed. For colon cancer there is evidence that six stages might be required (Fearon and Vogelstein, 1990). However, for other cancers there is little evidence that there are as many rate-limiting stages as this. BEIR V (NRC, 1990) surveyed evidence for all cancers and found that two or three stages might be justifiable, but not a much larger number. To this extent the large number of stages predicted by the Armitage-Doll model appears to be verging on the biologically unlikely. Related to the large number of stages required by the Armitage-Doll multistage model are the high mutation rates predicted by the model. Moolgavkar and Luebeck (1992) fitted the Armitage-Doll multistage model to datasets describing the incidence of colon cancer in a general population and in patients with familial adenomatous polyposis. Moolgavkar and Luebeck (1992) found that Armitage-Doll models with five or six stages gave good fits to these datasets, but that both of these models implied mutation rates that were too high by at least two orders of magnitude. The discrepancy between the predicted and experimentally measured mutation rates might be eliminated, or at least significantly reduced, if account were taken of the fact that the experimental mutation rates are locus-specific. A "mutation" in the sense in which it is defined in this model might result from the "failure" of any one of a number of independent loci, so that the "mutation" rate would be the sum of the failure rates at each individual locus.

Notwithstanding these problems, much use has been made of the Armitage-Doll multistage model as a framework for understanding the time course of carcinogenesis, particularly for the interaction of different carcinogens (Peto, 1977). Thomas (1990) has fitted the Armitage-Doll model with one and two radiation-affected stages to the solid cancer data in the Japanese Life Span Study (LSS) Report 11 cohort of A-bomb survivors. Thomas (1990) found that a model with a total of five stages, of which either stages one and three or stages two and four were radiation-affected, fitted significantly better than models with a single radiation-affected stage. Little et al. (1992, 1994) also fitted the Armitage-Doll model with up to two radiation-affected stages to the Japanese LSS Report 11 dataset and also to data on various medically exposed groups, using a slightly different technique from that of Thomas (1990). Little et al. (1992, 1994) found that the optimal solid cancer model for the Japanese data had three stages, the first of which was radiation affected, while for the Japanese leukemia data the best fitting model had three stages, the first and second of which were radiation affected.

Both Thomas (1990) and Little et al. (1992, 1994) assumed the ith and the jth stages or mutation rates $(M(i-1), M(j-1), j > i)$ in a model with k stages to be (linearly) affected by radiation, and the transfer coefficients [other than $M(i-1)$ and $M(j-1)$] to be constant [as is the stem cell population $X(t)$]. In these circumstances it can be shown (Little et al., 1992) that if an instantaneously administered dose of radiation d is given at age a, then at age t $(> a)$ the cancer rate is approximately:

$$\mu \cdot t^{k-1} + \alpha \cdot d \cdot a^{i-1} \cdot [t-a]^{k-i-1} + \beta \cdot d \cdot a^{j-1} \cdot [t-a]^{k-j-1} + \gamma \cdot d^2 \cdot a^{i-1} \cdot [t-a]^{k-j-1}$$

$$(14.1)$$

for some positive constants μ, α, and β, and where γ is given by:

$$\gamma = \begin{cases} \frac{\alpha \cdot \beta \cdot \Gamma(k-i) \cdot \Gamma(j)}{2 \cdot \mu \cdot \Gamma(k)} & \text{if } j = i+1 \\ 0 & \text{if } j > i+1 \end{cases}$$

$$(14.2)$$

and $\Gamma(\cdot)$ is the gamma function (Abramowitz and Stegun, 1964).

The first term $(\mu \cdot t^{k-1})$ in expression 14.1 corresponds to the cancer rate that would be observed in the absence of radiation, while the second term $(\alpha \cdot d \cdot a^{i-1} \cdot [t-a]^{k-i-1})$ and the third term $(\beta \cdot d \cdot a^{j-1} \cdot [t-a]^{k-j-1})$ represent the separate effects of radiation on the ith and jth stages, respectively. The fourth term $(\gamma \cdot d^2 \cdot a^{i-1} \cdot [t-a]^{k-j-1})$, which is quadratic in dose d, represents the consequences of interaction between the effects of radiation on the ith and the jth stages and is only non-zero when the two radiation-affected stages are adjacent $(j = i+1)$. Thus if the two affected stages are adjacent, a quadratic (dose plus dose-squared) relationship will occur, whereas the relationship will be approximately linear if the two affected stages have at least one intervening stage. Another way of considering the joint effects of radiation on two stages is that for a brief exposure, unless the two radiation-affected stages are adjacent, there will be insignificant interaction between the cells affected by radiation in the earlier and later of the two radiation-affected cell compartments. This is simply because very few cells will move between the two compartments in the course of the radiation exposure. If the ith and the jth stages are radiation-affected the result of a brief dose of radiation will be to cause some of the cells which have already accumulated $(i-1)$ mutations to acquire an extra mutation and move from the $(i-1)$th to the ith compartment.

Similarly, it will cause some of the cells which have already acquired $(j - 1)$ mutations to acquire an extra mutation and so move from the $(j - 1)$th to the jth compartment. It should be noted that the model does not require that the same cells be hit by the radiation at the ith and jth stages, and in practice for low total doses, or whenever the two radiation-affected stages are separated by an additional unaffected stage or stages, an insignificant proportion of the same cells will be hit (and mutated) by the radiation at both the ith and the jth stages. The result is that, unless the radiation-affected stages are adjacent, for a brief exposure the total effect on cancer rate is approximately the sum of the effects, assuming radiation acts on each of the radiation-affected stages alone. One interesting implication of models with two or more radiation-affected stages is that as a result of interaction between the effects of radiation at the various stages, protraction of dose in general results in an increase in cancer rate, i.e., an inverse dose-rate effect (Little et al., 1992). However, it can be shown that in practice the resulting increase in cancer risk is likely to be small (Little et al., 1992)

The optimal leukemia model found by Little et al. (1992, 1994), having adjacent radiation-affected stages, predicts a linear-quadratic dose-response, in accordance with the significant upward curvature which has been observed in the Japanese LSS 11 dataset (Shimizu et al., 1990; Pierce and Vaeth, 1991). This leukemia model, and also that for solid cancer, predicts the pronounced reduction of ERR with increasing age at exposure (see Figure 14.2) which has been seen in the Japanese atomic bomb survivors and other datasets (UNSCEAR, 1994). The optimal Armitage-Doll leukemia model predicts a reduction of ERR with increasing time after exposure for leukemia. At least for those exposed in childhood, the optimal Armitage-Doll solid cancer model also predicts a reduction in ERR with time for solid cancers. These observations are consistent with the observed pattern of risk in the Japanese and other datasets (Little, 1993; UNSCEAR, 1994). Nevertheless, there are indications that the Armitage-Doll model may not provide an adequate fit to the Japanese leukemia data (Little et al., 1995). For this reason, and because of the other problems with the Armitage-Doll model discussed above, one needs to consider a slightly different class of models.

TWO-MUTATION MODEL

In order to reduce the biologically implausible number of stages required by their first model, Armitage and Doll (1957) developed a further model of carcinogenesis, which postulated a two-stage probabilistic process whereby a cell following an initial transformation into a pre-neoplastic state (initiation) was subject to a period of accelerated (exponential) growth. At some point in this exponential growth a cell from this expanding population might undergo a second transformation (promotion) leading quickly and directly to the development of a neoplasm. Like their previous model, it satisfactorily explained the incidence of cancer in adults, but was less successful in describing the pattern of certain childhood cancers.

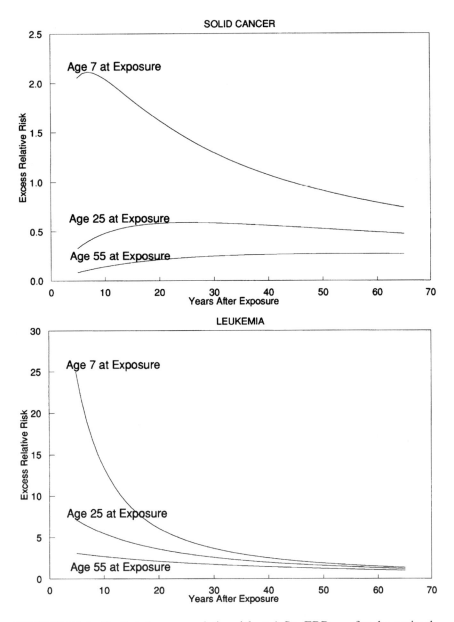

FIGURE 14.2 Predicted excess relative risk at 1 Sv, ERR$_{1Sv}$, for the optimal Armitage-Doll models fitted to the Japanese A-bomb survivors for cancers other than leukemia (with three stages, the first of which is radiation-affected) and for leukemia (with three stages, the first and second of which are radiation-affected).

Normal   Intermediate   Malignant

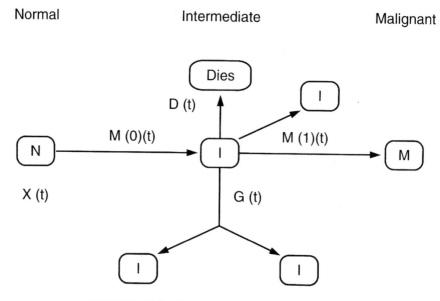

FIGURE 14.3   The two-mutation (MVK) model.

The two-mutation model developed by Knudson (1971) to explain the incidence of retinoblastoma in children took account of the process of growth and differentiation in normal tissues. Subsequently, the stochastic two-mutation model of Moolgavkar and Venzon (1979) generalized Knudson's model by taking account of cell mortality at all stages, as well as allowing for differential growth of intermediate cells. The two-stage model developed by Tucker (1967) is very similar to the model of Moolgavkar and Venzon but does not take account of the differential growth of intermediate cells. The two-mutation model of Moolgavkar, Venzon, and Knudson (MVK) supposes that at age t there are $X(t)$ susceptible stem cells, each subject to mutation to an intermediate type of cell at a rate $M(0)(t)$. The intermediate cells divide at a rate $G(1)(t)$; at a rate $D(1)(t)$ they die or differentiate; at a rate $M(1)(t)$ they are transformed into malignant cells. The model is illustrated schematically in Figure 14.3. In contrast with the case of the (first) Armitage-Doll model, there is a considerable body of experimental biological data supporting this initiation-promotion type of model (see, e.g., Moolgavkar and Knudson, 1981; Tan, 1991). The model has recently been developed to allow for time-varying parameters at the first stage of mutation (Moolgavkar et al., 1988). A further slight generalization of this model to account for time varying parameters at the second stage of mutation was presented by Little and Charles (1991), who also demonstrated that the ERR predicted by the model, when the first mutation rate was subject to instantaneous perturbation, diminished at least exponentially for a sufficiently long time after the perturbation. Moolgavkar et al. (1990) and Luebeck et al. (1996) have used the two-mutation model to describe the incidence of lung cancer in rats exposed to radon, and in particular to model the inverse dose-rate effect that has been observed in these data. Moolgavkar et al. (1993) have also applied the model to describe the interaction of smoking and radiation as causes of lung cancer in the Colorado Plateau uranium miner cohort.

Recently Moolgavkar and Luebeck (1992) have used models with two or three mutations to describe the incidence of colon cancer in a general population and in patients with familial adenomatous polyposis. They found that both models gave good fits to both datasets, but that the model with two mutations implied biologically implausibly low mutation rates. The three-mutation model, which predicted mutation rates more in line with biological data, was therefore somewhat preferable. The problem of implausibly low mutation rates implied by the two-mutation model is not specific to the case of colon cancer, and is discussed at greater length by Den Otter et al. (1990) and Derkinderen et al. (1990), who argue that for most cancer sites a model with more than two stages is required.

GENERALIZED MVK AND MULTISTAGE MODELS

More recently, a number of generalizations of the Armitage-Doll and two- and three-mutation models have been developed (Tan, 1991; Little, 1995). In particular two closely related models have been developed, whose properties have been described in the paper of Little (1995). The first model is a generalization of the two-mutation model of Moolgavkar, Venzon, and Knudson and so will be termed the *generalized MVK model*. The second model generalizes the multistage model of Armitage and Doll and will be referred to as the *generalized multistage model*. For the generalized MVK model it may be supposed that at age t there are $X(t)$ susceptible stem cells, each subject to mutation to a type of cell carrying an irreversible mutation at a rate of $M(0)(t)$. The cells with one mutation divide at a rate $G(1)(t)$; at a rate $D(1)(t)$ they die or differentiate. Each cell with one mutation can also divide into an equivalent daughter cell and another cell with a second irreversible mutation at a rate $M(1)(t)$. For the cells with two mutations there are also assumed to be competing processes of cell growth, differentiation, and mutation taking place at rates $G(2)(t)$, $D(2)(t)$, and $M(2)(t)$ respectively, and so on until at the $(k-1)$th stage the cells that have accumulated $(k-1)$ mutations proceed at a rate $M(k-1)(t)$ to acquire another mutation and become malignant. The model is illustrated schematically in Figure 14.4. The two-mutation model of Moolgavkar, Venzon, and Knudson corresponds to the case $k = 2$. The generalized multistage model differs from the generalized MVK model only in that the process whereby a cell is assumed to split into an identical daughter cell and a cell carrying an additional mutation is replaced by the process in which only the cell with an additional mutation results, i.e., an identical daughter cell is not produced. The classical Armitage-Doll multistage model corresponds to the case in which the intermediate cell proliferation rates $G(i)$ and the cell differentiation rates $D(i)$ are

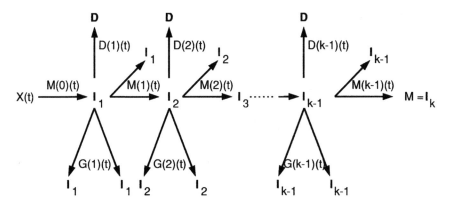

FIGURE 14.4 The generalized MVK model.

all zero. It can be shown (Little, 1995) that the ERR for either model following a
perturbation of the parameters will tend to zero as the attained age tends to infinity.
One can also demonstrate that perturbation of the parameters $M(k-2)$, $M(k-1)$,
$G(k-1)$, and $D(k-1)$ will result in an almost instantaneous change in the cancer
rate (Little, 1995).

Figure 14.5 shows the calculated ERR for the generalized MVK model in the
case where the underlying process has $X(t) \equiv 1$, $G(1) = G(2) = G(3) = 1\ y^{-1}$,
$D(1) = D(2) = D(3) = 0.8\ y^{-1}$, $M(0) = M(1) = M(2) = M(3) = 10^{-4}\ y^{-1}$
and the various parameters are individually altered. In each case the perturbation
lasts for a period of a year, after which each parameter returns to its original value.

The first thing to note from Figure 14.5 is that the curves obtained by augment-
ing the intermediate cell growth rate $G(1)$ for the two-mutation model are almost
identical to those obtained by decreasing the intermediate cell death and differentiation rate $D(1)$ by the equivalent amount. Examinations of the perturbed hazard
function when $G(i)$ is augmented by $0.4\ y^{-1}$ and when $D(i)$ is decreased by 0.4
y^{-1} suggest only very slight differences between these cases for the three-mutation
model (or for models with a greater number of mutations). Consequently, only the
case in which $G(i)$ is augmented is shown in the lower two panels of Figure 14.5.
It can be shown that the hazard function for the generalized multistage model be-
haves very similarly to that of the corresponding generalized MVK model (Little,
1995). Little (1995) gives additional details of the consequences of perturbation of
the model parameters at older ages; the most significant difference is made to the
two-mutation model, for which the ERR following perturbation of $M(0)$ is less
pronounced shortly after exposure.

Figure 14.5 illustrates the theoretical results of Little (1995) and shows that
immediately after perturbing any of the parameters in either of the two-mutation
models the ERR quickly increases. Figure 14.5 also shows that perturbing the

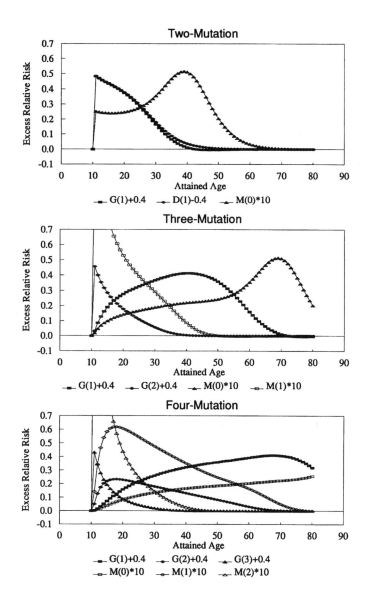

FIGURE 14.5 ERR for the generalized MVK model when the number of mutations is between two and four when $X \equiv 1$, $G(1) = .. = G(3) = 1\ y^{-1}$, $D(1) = .. = D(3) = 0.8\ y^{-1}$, $M(0) = .. = M(3) = 10^{-4}\ y^{-1}$ normally, subject to perturbations at age 10: $G(1), .., G(3)$ increased by $0.4\ y^{-1}$; or $D(1)$ decreased by $0.4\ y^{-1}$ (two-mutation model only); or $M(0), .., M(2)$ multiplied by 10 for 1 year, after which each parameter returns to its original value.

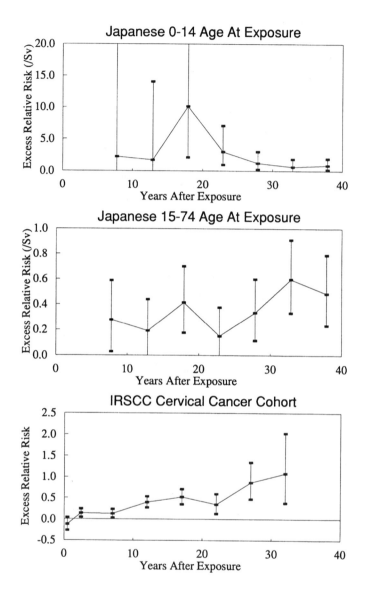

FIGURE 14.6 ERR$_{1Sv}$ and 90% CI for cancers other than leukemia in the Japanese LSS Report 11 data using a neutron relative biological effectiveness of 20, attained age less than 75 years, for ages 0–14 and 15–74 at exposure (using for dose the average dose to breast, large intestine, lung, and stomach); ERR and 90% CI for cancers other than leukemia, lung, breast, ovary, and uterus in the IRSCC Cohort Study (Boice et al., 1985).

parameters $G(2)$ or $M(1)$ produces a similar effect in the three-mutation model, and that an analogous change in the hazard function is produced when $G(3)$ or $M(2)$ are perturbed in the four-mutation model. Although they are not shown in Figure 14.5, the results of perturbing the final mutation rate $M(k-1)$ are even more striking, leading in all cases to an ERR which increases for the duration of the increase of $M(k-1)$, and which then immediately decreases to near zero; again this result is in accord with the theoretical predictions of Little (1995).

To what extent is the behavior of the excess risk resulting from perturbation of the various parameters in agreement with what has been observed a short time after irradiation in various groups exposed to ionizing radiation? Data on the first five years after exposure are missing for the follow up of the Japanese Λ bomb survivor LSS cohort (Shimizu et al., 1990), but the available data suggest that the ERR for solid cancers increases progressively in the first few years after exposure. As the top two panels of Figure 14.6 indicate, there is only weak evidence for a sudden increase in risk for cancers other than leukemia immediately after the bombings. (The two leftmost datapoints in the top panel correspond to follow-up in the periods approximately 6–10 and 11–15 years after the bombings.) However, apart from the problems of the missing first few years of follow-up, at least in the youngest age-at-exposure group (those under 15 at the time of the bombings) the excess risk is estimated with low precision. For example, the leftmost two ERRs for this age group in the bomb survivors are 2.23 Sv^{-1} (90% CI <0, 20.48) and 1.73 Sv^{-1} (90% CI <0, 14.06) (as shown in Figure 14.6) and so are not significantly elevated. There are no strong indications of an elevation in risk in the first five years after radiotherapy for cancers other than leukemia and of the reproductive organs in a study of women followed up for second cancer after radiotherapy for cervical cancer (Boice et al., 1985). This corresponds to the first two datapoints in the bottom panel of Figure 14.6. [Lung cancers are also excluded from the International Radiation Study of Cervical Cancer (IRSCC) data shown in Figure 14.6 because of indications of above-average smoking rates in this cohort (Boice et al., 1985)]. In various other irradiated groups in the first few years after exposure a pattern of risk is observed that is similar to the pattern seen in the atomic bomb survivors and the IRSCC dataset. In particular, for very few of those groups irradiated in childhood are there indications of a sudden increase in risk shortly after irradiation (Little, 1993). To this extent, there are indications of inconsistency between the predictions of the two-mutation model, concerning variation in risks a short time after exposure, and this body of epidemiological evidence relating to solid cancers. When generalized MVK models with two, three, and four mutations are fitted to the Japanese A-bomb survivor solid cancer data, only the two-mutation and three-mutation models provide a satisfactory fit, and in each case only when the first mutation rate $M(0)$ is assumed to be affected by radiation (Little, 1996). Despite the fact that both the two-mutation and three-mutation models give equivalent fits, the predictions of the two models of the pattern of excess risk in the first five years

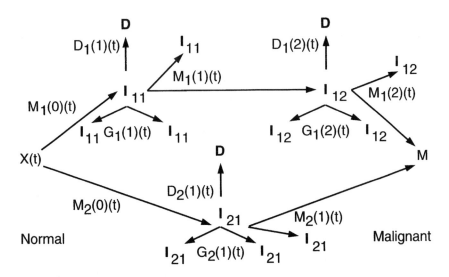

FIGURE 14.7 A multiple pathway (MVK-type) carcinogenesis model.

after exposure are very different (Little, 1996), as might be expected given the results of Little (1995).

It must be pointed out that Moolgavkar et al. have to some extent circumvented the problem posed by this instantaneous rise in the hazard function after perturbation of the two-mutation model parameters. In their analysis of the Colorado Plateau uranium miners data (1993), Moolgavkar et al. assumed a fixed period of 3.5 years between the appearance of the first malignant cell and the clinical detection of malignancy. However, the use of such a fixed period of latency would only translate a few years into the future the sudden step-change in the hazard function. To achieve the observed gradual increase in excess risk shortly after exposure, a stochastic process must be used to model the transition from the first malignant cell to detectable cancer, such as is provided by the final stage in the three-mutation models described here. In particular, an exponentially growing population of malignant cells could be modeled by a penultimate stage in which $G(k-1)$ is non-zero and $D(k-1)$ is zero, the probability of detection of the malignant clone being determined by $M(k-1)$.

The evidence with respect to the time distribution of excess risk induced by a more or less instantaneously administered dose of radiation for leukemias is rather different. In the IRSCC data (Boice et al., 1985) there is a significant excess risk for acute non-lymphocytic leukemia (ANLL) in the period between one and four years after first treatment. This rapid rise in risk shortly after exposure is paralleled in many other irradiated groups (Little, 1993). Although the data on early years of follow-up are lacking, the indications are that the same is true in the Japanese

A-bomb survivor data (Shimizu et al., 1990; Preston et al., 1994). Comparison of Figure 14.5 with the observed pattern of variation of ERR for leukemia in the Japanese LSS cohort and in other exposed groups indicates that the two-mutation model might well describe the time course of radiation-induced leukemia (Little, 1995). The results of fitting the generalized MVK model, with various numbers of mutations, to the Japanese A-bomb survivor leukemia data indicate that the three-mutation model provides the most satisfactory fit, and then only when the first mutation rate $M(0)$ and the second mutation rate $M(1)$ are assumed to be affected by radiation (Little, 1996).

MULTIPLE PATHWAY MODELS

Little et al. (1995) fitted a generalization of the Armitage Doll model to the Japanese atomic bomb survivor leukemia data which allowed for two cell populations at birth, one consisting of normal stem cells carrying no mutations, the second a population of cells each of which has been subject to a single mutation. The leukemia risk predicted by such a model is equivalent to that resulting from a model with two pathways between the normal stem cell compartment and the final compartment of malignant cells, the second pathway having one fewer stage than the first. This model fitted the Japanese leukemia dataset significantly better, albeit with biologically implausible parameters, than a model that assumed just a single pathway (Little et al., 1995). The findings of Kadhim et al. (1992, 1994), that exposure of mammalian hematopoietic stem cells to alpha particles could result in a general elevation of mutation rates to very much higher than normal levels, implies, if these findings are at all relevant to carcinogenesis, that there might be multiple pathways in the progression from normal stem cells to malignant cells. The mutation rates and indeed the number of rate-limiting stages might be substantially different in these two or more pathways, as Figure 14.7 illustrates. A number of such models are described by Tan (1991), who also discusses at some length the biological and epidemiological evidence for such models of carcinogenesis.

CONCLUSIONS

The classical multistage model of Armitage and Doll and the two-mutation model of Moolgavkar, Venzon, and Knudson, and various generalizations of them also, are capable of describing, at least qualitatively, many of the observed patterns of excess cancer risk following ionizing radiation exposure. However, there are certain inconsistencies with the biological and epidemiological data for both the multistage and two-mutation models. In particular, there are indications that the two-mutation model is not totally suitable for describing the pattern of excess risk for solid cancers that is often seen after exposure to ionizing radiation, although leukemia may be better fitted by this type of model. Generalized MVK models that require three or more mutations are easier to reconcile with biological and epidemiological data relating to solid cancers.

ACKNOWLEDGMENTS

This paper makes use of data obtained from the Radiation Effects Research Foundation (RERF) in Hiroshima, Japan. RERF is a private foundation funded equally by the Japanese Ministry of Health and Welfare and the US Department of Energy through the US National Academy of Sciences. The conclusions in this paper are those of the author and do not necessarily reflect the scientific judgment of RERF or its funding agencies.

15

The Distinction in Radiobiology Between Medical and Public Health Functions

VICTOR P. BOND AND LUCIAN WIELOPOLSKI

SUMMARY

Starting with the classical threshold-sigmoid medical-toxicological plot, reasons are advanced for why the coordinates of this function are not appropriate for the analysis of public health–epidemiological data. Misunderstandings with respect to both the level of biological organization and the word "dose" are pointed out, which explain why public health–epidemiological data, anomalously, yield linear functions on medical-toxicological coordinates. It is then shown why substantially different coordinates must be used to obtain a function that describes properly and completely the cancer data obtained from epidemiological studies on the atomic bomb survivors. Arguments are put forth that contradict the linear, no-threshold hypothesis. Reasons are advanced for why, if the amount of radiation energy is expressed in the proper terms, the numerical value for the cancer "risk coefficient" becomes substantially smaller than it now is.

INTRODUCTION

The basic elements of present radiation quantities and units were developed in the era when the dogma was still firmly entrenched that radiation-attributable cancers could be induced only with repeated large exposures, i.e., a threshold was thought to pertain. Thus, most studies at the time were oriented around early effects on, and early responses of organs and organisms to moderate and large exposures.

The model used was that of classical medicine-toxicology, using pharmaceuticals or other chemicals, and characteristic threshold-sigmoid dose-response functions were observed. As will be shown below, such studies with pharmaceuticals require that the mass of biological subjects be used in order to determine the proper amount of chemical to administer.

However, when presumably the same model was employed for the early acute effects of the agent radiation energy, the practice of obtaining the mass, and thus the proper amount of the agent energy, was not followed. Furthermore, the reason for not doing so was neither mentioned nor explained. As a result, when late effects, particularly cancer, became an issue, the crucial step of including the mass and thus the imparted energy, first to both the individual elements and then to the subject population system of interest, was completely ignored.

The two related terms "medical-toxicological" (MT) and "public health–epidemiological" (PH), must be distinguished. With the MT biomedical discipline, the subject of interest is an individual; with the PH discipline, the subject of interest is a well-defined population.

A principal objective of this communication is to explore the reasons for and the severe consequences of not recognizing the proper role of the mass of the subject and thus the total energy imparted. To accomplish this objective, it is highly informative first to examine how the mass of the subject, and thus the total amount of agent transferred, enters into the formulation of classical pharmaceutical dose-response functions.

CLASSICAL MT FUNCTIONS

For Pharmaceuticals

Figure 15.1 shows schematically a classical MT "dose response" function for pharmaceuticals or other chemicals, which typically has a threshold and is sigmoid in shape. The abscissa is here labeled the "prescription dose," D_P, in units of the amount of agent per unit mass of biological subject, e.g., mg/kg. It is so labeled because it constitutes an instruction, once a value of D_P is selected, to prepare the amount of agent that must actually be administered. This is the product of the prescription dose and the mass of the subject, m_s. That product is here termed the true dose, T_D, in units of mg. This use of the word dose conforms to actual practice in medicine, and to the definition of the word "dose" in both medical and general dictionaries (D_P is not mentioned).

Although neither m_s nor T_D appear at all in the D_P-response curve, it is obvious that, once D_P is determined, it is T_D and not D_P that is actually administered, and that m_s is used explicitly when determining the proper value for T_D.

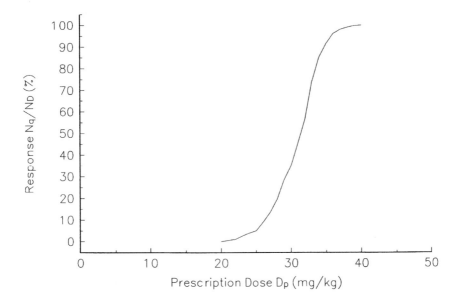

FIGURE 15.1 A classical MT dose-response function, used in medicine and toxicology for pharmaceuticals or other chemicals.

For Acute Radiation Responses

Figure 15.2 shows a typical MT curve for acute mortality in mice, which is a frequently used endpoint in radiobiology. The coordinates are identical to those in Figure 15.1, except that the units of D_P, mg/kg for the chemical agent, are replaced by Gy (J/kg) for the radiation agent, energy. These quantities express the concentration and not the amount of agent. With radiation only, T_D is also termed "imparted energy," with the symbol ε (ICRU, 1993). Similarly, D_P is ε/m.

With radiation exposure the animals are immersed in a radiation field and it is assumed that the resulting irradiation is to the whole body and is nominally uniform. Thus agent transfer, rather than indirectly via the bloodstream as with chemicals, is direct, i.e., the radiation field penetrates throughout the body so that energy is uniformly deposited and the agent concentration, ε/m, is equivalued throughout.

However, ε/m on the abscissa of Figure 15.2 is an intensive quantity, i.e., it does not depend on the mass of the system, nor is it additive (i.e., D_P can be the same throughout a relatively small or large subject). Accordingly, when an animal is uniformly irradiated, the subject perforce receives the value of ε appropriate for

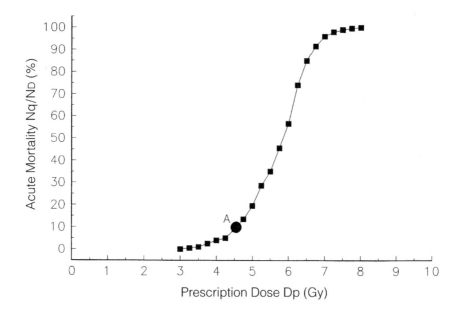

FIGURE 15.2 The same plot shown in Figure 15.1, but for acute mouse mortality in mice exposed to 250 kVp x-rays.

the stated values of ε/m and mass, whatever those precise values may be. Thus, there is usually no requirement to weigh or use the mass of the animal (although, of course, the weight and thus ε could be obtained easily). Nevertheless, exactly as with chemicals, it is not D_P that is transferred. Rather, it is $(D_P \cdot m_s)$, or ε, in units of joules and in an amount sufficient to ensure a constant value of D_P throughout the subject, that actually is administered.

Moreover, the radiotherapists early coined, and still use the concept of "gram-rad," equal to ε, applied to the tumor bed. Every effort is made to minimize this value. However, the lack of a need to obtain the weight of the subject was seriously misinterpreted to mean that with radiation, unlike with the chemical MT model, there is no need even to be aware that the response could not be obtained unless a value of ε appropriate for the entire subject were delivered. The misunderstanding was reinforced and compounded by the early decision to define absorbed dose purely in terms of a concentration, energy per unit mass (ICRU, 1968). This confusingly gave a new and parochial meaning to a word, dose, that has been used for decades to define a quite different quantity, the amount of agent to an entire subject. As a result of this untenable dual definition of "dose," there is little reason for anyone, including radiobiologists, to be familiar with or appreciate the importance of the terms ε or T_D, at least in this specific context. Of equal significance, this definition enhances the misconception that absorbed dose (i.e., D_P) is a measure of the amount of radiation when it is a concentration. The seriousness of these misunderstandings will become even more apparent below.

LINEAR FUNCTIONS

Although with MT the focus is always on an individual subject seeking or under medical care, when the emphasis is shifted to public health and epidemiology (PE), the biological subject of interest is a well defined population, N_E, of "normal" subjects. These persons are frequently or continuously exposed in a "field" of potentially harmful objects or substances, which results in uncontrolled ("stochastic") interactions (accidents), in which some N_E subjects become hit and dosed. The hit and dosed (N_D) subjects receive values of T_D of varying magnitudes, and thus sustain injury of different degrees of severity.

With this change in biomedical discipline (from MT to PE), different quantities assume importance, and the nature of the graphs changes, i.e., a completely different analysis of the same set of data, which includes all values of T_D, D_P, m_s, N_E, N_D and N_q, is required. The principal question of interest in PH then becomes, What is the absolute number of quantally responding subjects, N_q, that resulted from individuals in a population being "hit" and dosed? rather than, What are the chances that an individual will respond quantally? Consequently T_D, which is an extensive quantity that can be summed over the entire exposed population, is required because the total number responding quantally depends on this total amount of agent.

However, T_D or ε can also be expressed as $\varepsilon = N_D \bullet D_p \bullet m_s$. With m_s being a constant, and if D_P is assumed to be constant, then $\varepsilon = kN_D$. Thus ε is proportional to N_D. With uncontrolled delivery of an agent to a population, inevitably one encounters a distribution of values of D_P. However, for ease of presentation, the entire distribution is here replaced by a single value (mean) of D_P.

With the above simplification, it is easily possible to state succinctly the differences between the classical MT curve and the function for PE. If N_D is held constant (e.g., on the ordinate of Figure 15.2) and only the agent concentration, D_P, is permitted to vary, one obtains the threshold-sigmoid classical MT curve shown in Figure 15.2. However, if the agent concentration, D_P is held constant (e.g., at point A in Figure 15.2) and only N_D (or ε) is permitted to vary, one obtains, as is indicated schematically in Figure 15.3, a linear PH function.

With the current interpretation of linear radioepidemiological functions, one might hypothesize from Figure 15.3 that, "any amount of radiation, however small," can cause some acute mortality in an exposed population. However, the linear function in Figure 15.3 represents only an approximation of a step function, each step representing an integer number of subjects (Figure 15.4). Because no fewer

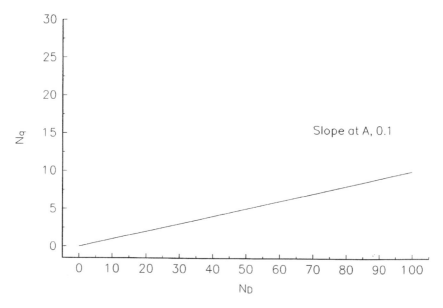

FIGURE 15.3 A PE function. The slope of the line is the value of point A in the linear function obtained when D_P is held constant (point A in Figure 15.2), and N_D is permitted to vary. The slope of the line equals the value of point A.

than one subject can die, the range of the function can extend only from one to the level at which all subjects are dead. Similarly for the integer N_D in the abscissa in Figure 15.3, because no fewer than one subject can be exposed, the lower limit of the domain is one.

When integer steps of the otherwise continuous variable ε are used, instead of N_D, they represent the mean value of a perhaps normal distribution that is not related to the linear regression line. It is the variance in this distribution that determines, as one moves away from the mean, the probability of a quantal response. In any case, the strict nature of the domain and range of the function precludes linear (or any other kind of) extrapolation of the function to 0,0 coordinates. Thus the linear, no-threshold hypothesis is seriously challenged.

PH PRINCIPLES APPLIED TO ATOMIC BOMB SURVIVORS

Selected for this examination is a sample of about 40,000 atomic bomb survivors (N_E subjects, all of whom became N_D subjects), in which, at the time of the study, a total of 245 radiation-induced lethal solid cancers, N_{ca}, had appeared (Shimizu et al., 1989; Pierce, 1989, as combined by Bond et al., 1991). The survivors were

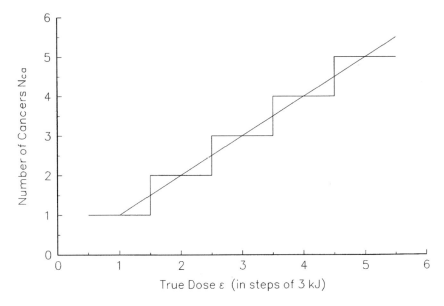

FIGURE 15.4   The linear function shown in reality represents a continuous approximation to a step function, integer numbers of observed cancers. The number of quantally responding subjects cannot be less than one.

divided into eight dose intervals of arbitrary width, containing widely differing numbers of survivors. It is assumed that all 40,000 survivors received essentially uniform whole body irradiation, from penetrating photons only (i.e., the data are for a standard, "low LET" radiation).

Figure 15.5 shows the linear dose-response function that has been used widely to represent such PH data.[1]   Although clearly the graph appears to conform to that for a PH function, the coordinates correspond to those of an MT function. Furthermore, no information on the absolute number of cancers induced can be obtained from the plot. Thus, the function alone cannot be of value in PH studies.

In order to obtain a function that is suitable for PH studies, one must go back to the original data for the atomic bomb survivors and plot the absolute number of persons with a lethal cancer, as a function of either ε or its equivalent N_D. The

---

[1] It is often stated that with Figure 15.5 one needs only to "multiply the 'risk' by the number dosed," to obtain an absolute value of N_{ca}. This is unsatisfactory for PH purposes because this applies only to each D_P group separately, when what is needed is the accumulated increments of dose and response for the entire dosed population. Thus, data external to the function are required.

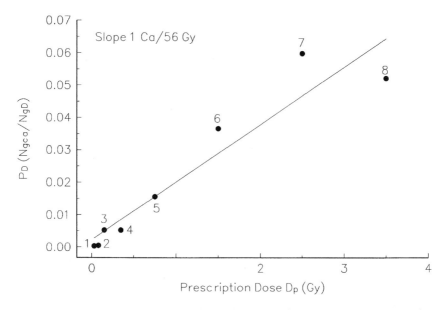

FIGURE 15.5 A plot frequently employed to display cancer data obtained from the atomic bomb survivors. Values for the absolute number of (persons with) an excess cancer are not obtainable from this graph.

result would be N_{ca} alone on the ordinate, and ε (or N_D) on the abscissa, both proper coordinates for PH functions.

However, as shown in Figure 15.6, a first step in this direction has already been taken with the introduction of the hybrid quantity person-Gy, i.e., $N_D D_P$. Thus, although this function may be regarded as representing some improvement over Figure 15.1, in that N_D now appears on the abscissa, it alone is still inadequate for PH purposes because person-Gy retains the non-additive unit of Gy. Also, the ordinate cannot yield more than about 80 cancers, about one third of the total of 245 induced.

To remedy the problems with Figure 15.6, it is necessary to recognize that each member of the population, a person who has received a determinable amount of energy, is an element of the defined population system of $\sim$40,000 dosed persons, and that the total amount of imparted energy transferred to the system, on which the total cancer yield depends, is determinable only by summing over all increments of agent delivered to the elements. Such a suitable quantity can be obtained from the person-Gy in Figure 15.6 by multiplying this hybrid quantity by the mass of

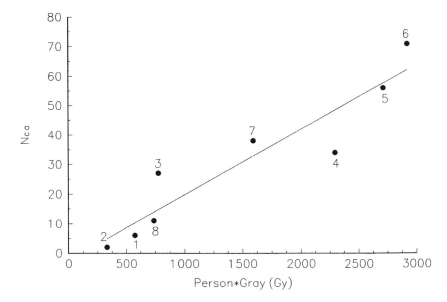

FIGURE 15.6 Number of solid cancers in the atomic bomb survivors, plotted as a function of person-Gy. Note that the actual yield, 245 cancers, is not reachable using this plot.

a person (mass of the average atomic bomb survivor, taken to be 55 kg). The resulting quantity, ε, being extensive, is additive and dependent on the mass of the system.

Figure 15.7 shows the results of a conversion, dose group by dose group, of person-Gy to ε, but without summation over the groups. Although this does yield coordinates in terms suitable for PE, it still fails to yield the correct number of induced cancers. Figure 15.8 shows the results of summing ε over all groups (this is not extrapolation but rather calculation of the energy imparted to the entire system). It is only this function that can represent the PE data—that is, 245 induced solid cancers—completely and accurately, with no requirement for additional external information.

The fact that only the summed increments yield the correct number of cancers reinforces the major point made earlier: that is, an additive quantity or number (ε or N_D) that is dependent on the mass of the system must appear as the independent variable. This confirms the fact that the function in Figure 15.5 cannot be appropriate for the PH situation.

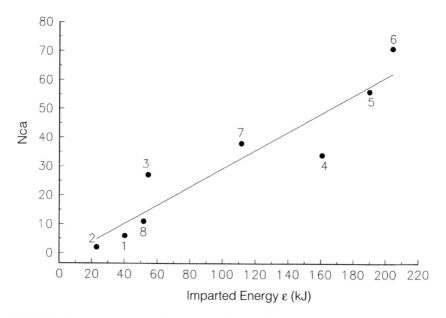

FIGURE 15.7 Absolute solid tumor yield as a function of partial summations of imparted energies in the constituent groups. The total of 245 cancers does not appear on this plot.

It is often claimed, incorrectly, that "doses" (ε/m) are additive if delivered to the same mass, or different masses of nominally the same value, e.g., the weights of radiation workers. However, additivity in such situations is illusory. The summation performed under these circumstances is not $\sum_i (\varepsilon/m_s)$ but rather $\frac{1}{m_s}\sum_i (\varepsilon)$. Thus it is actually ε that is being summed: ε/m is not additive under any circumstances.

Cellular Analysis

We return now to the anomaly alluded to above: namely, why, as shown in Figure 15.5, do the cancer data yield a linear function, even though plotted on MT coordinates, are expected to yield a threshold-sigmoid function? The answer begins with the fact that, although overt cancers obviously appear only at the organ-organism level, essentially all human cancers originate at the cell level. Furthermore, there appears to be a one-to-one relationship, that is, one overtly expressed cancer represents one completely transformed cell. This means that the relevant question is not precisely that asked in the first sentence above, but rather,

Why do cell quantal responses yield a linear function when plotted on MT axes that traditionally have yielded a threshold-sigmoid function?

First, the root of the problem is a disparity in the levels of biological organization: the response is at the cellular level, while the physical insult D_P applies only at the tissue level. But D_P at the higher organizational level cannot provide any information on what happens at the cell level. The solution lies in removing the disparity and recognizing that one can obtain either a linear or a threshold-sigmoid function entirely at the cellular level, just as was done above for the organ-organism level. This places both response and physical insult entirely at the level where the causative interaction between cells and a radiation field transpires. With respect to a threshold and curvilinearity at the cell level, such a function, termed a "hit size effectiveness function" (HSEF), has been described (Bond et al., 1985; Morstin et al., 1989; Sondhaus et al., 1990). Linear functions associated with the HSEF, with either N_H (the number of hits on cells), or ε as the independent variable, have been demonstrated (Bond et al., 1995; Sondhaus et al., 1990).

With respect to linear functions at the cellular level, if the radiation quality is held constant at any region of the HSEF, a linear function, N_q as a function of N_H (or ε) can be obtained. Therefore, the answer to the original question, or as modified for single cell effects generally, is that, because D_P at the organ-organism level is proportional to both N_H and ε at the cell level, it follows that either a cell response in terms of the number of incipient cancers, or an organ-organism response in terms of N_{ca}, would be linear if plotted against either D_P or ε, with the latter accumulated at either the cell or organ-organism level.

Returning now to the linear functions shown in Figure 15.7 and 15.8, and recalling that N_D can be used on the abscissa in place of ε and that one cancer means one fully transformed cell, it might be attractive to hypothesize that only one hit on (dose to) a susceptible cell may cause a cancer. However, for this to actually happen, and provided that a number of other factors are favorable, the cell single hit dose z_1 must be greater than the threshold for the HSEF.

The function for mixed solid tumors, obtained from the atomic bomb survivor data, is also unquestionably compatible with linearity (although the data would easily accommodate quadratic, linear-quadratic or other functions) clearly the function is not the threshold-sigmoid type of relationship shown in Figure 15.2. Earlier work with mixed type breast tumors in rats showed a strictly linear dose-response function up to about 4 Gy, where a peaking occurred (Bond et al., 1960). It is tempting to adopt for this phenomenon the same cellular explanation given above, i.e., that, with the microdosimetrically evaluated radiation quality being identical for all values of D, it is only quantity (N_H or ε) that could vary.

However, this explanation for linearity was developed in the context of "low level" radiation, where the microdosimetric cell distribution (z_1 spectrum) and its mean, $\bar{z}_1$, are constant, and the mean number of hits per exposed cell is in the range of 1.0 or smaller. With the linearity continuing to quite high doses, a possible explanation is that multiple hits on cells, for some endpoints, can act independently, with no interaction up to quite high doses (e.g., 4 Gy).

Another factor that enters with mixed solid tumors is the relatively recent demonstration that, at least for some tumor types, a number of separate mutations, presumably within a stem cell, may be required for the induction and expression of a neoplasm. Supposedly, a large number of environmental agents, including ionizing radiation, can cause any one of these mutations. All such events are presumably independent, and the affected cell does not reveal its carcinogenic potential until the full complement of genetic changes (e.g., five) is attained (Trosko, 1990).

With a short radiation exposure as from the bomb, it is virtually certain that radiation could not have caused more than one of the five or so necessary mutations in any one cell. Also, presumably, that mutation could be number one, five or any number in between. Only in those cells in which the necessary five mutations are induced could the process of expression also be activated; all others would have to await the completion of the five, from other sources of radiation, or other mutagens. Because a simple hit on a cell with $(n - 1)$ hits could be effective, linearity would be expected.

In any event, all excess solid cancers are attributable to a necessary mutation caused by the bomb radiation. Each mutation would have to be regarded as an independent event, especially because the individual mutations are for the most part well separated in time. It is attractive to consider that this independence of event action may play a key role in determining the linearity of response.

ADDITIONAL RAMIFICATIONS: RISK COEFFICIENTS

In showing that the abscissa of PH functions must be in terms of T_D, in units of joules (Figure 15.8) rather than in units of J/kg (Figure 15.6), the numerical value of the *risk coefficient* (slope of the function) is reduced by a factor equal to the mass of the average person (55 kg for the average atomic bomb survivor). However, because the risk coefficient clearly must be in terms of the amount of the agent energy, the coefficient can be only in terms of T_D. Thus, the numerical value of the present risk coefficient (but not the absolute number of tumors) is reduced by the ratio of slopes of the functions in Figure 15.6 and Figure 15.8, i.e., 0.0215/0.00039=55. for the ICRU standard man, the factor would be 70.

The Linear Hypothesis

In Figure 15.8 the PH yield, N_{ca}, as a function of the total amount of the agent energy, has significant consequences with respect to the linear, non-threshold hypothesis. First, the fact that the abscissa is T_D, the actual amount of energy and not its concentration, indicates that this, and not the absorbed dose plot in Figure 15.5, is the appropriate quantity for expressing and evaluating the validity of the hypothesis. Also, as indicated by the inverse of the slope of the function (Bond

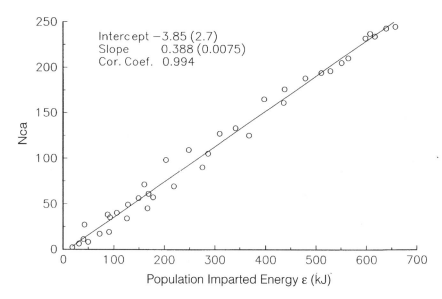

FIGURE 15.8   Absolute solid tumor yield as a function of the total amount of imparted energy. This complete function does show the total of 245 cancers.

et al., 1991), the average amount of radiation energy required to cause a cancer is about 3 kJ (about ten times the acute lethal dose, if delivered to one person). This is not compatible with a conclusion that any amount, however small, may cause a cancer.

To explore the linear hypothesis question further, in Figure 15.9 the group average values of T_D required to cause one cancer were plotted as a function of the concentration dose D_P (the errors for the first two points are quite large). This makes clear the fact that, however small the energy concentration (absorbed dose) to the average person may be, the quite large (3 kJ) average requirement for one induced cancer remains the same. Also, for this function (Figure 15.9) to be supportive of the linear hypothesis, one would expect the amount of energy required for cancer induction to decrease with decreasing D_P. Thus, these findings are incompatible with the linear hypothesis.

A key point is that, with one complete set (table) of relevant data, either type of function can be obtained at will. This shows that it is highly misleading, and the source of enormous confusion, to emphasize that a linear function is "without threshold." Thus, one agent cannot be described simultaneously as being so extremely toxic that there is no threshold for a lethal response, on the basis of one function derived from a given complete set of data, and then described as being

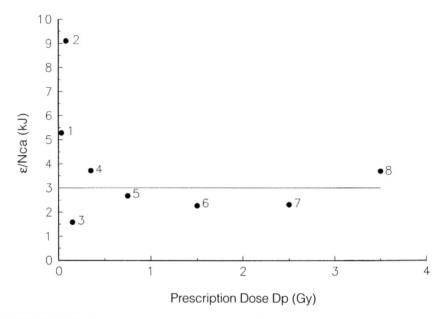

FIGURE 15.9 Average amount of energy required per induced solid tumor (3 kJ), as a function of absorbed dose. This value remains the same, however small the energy concentration (absorbed dose) may be.

so benign as to have a substantial threshold, on the basis of another very different function derived from the identical set of data. Because threshold-sigmoid and linear relationships are completely different functions, it is nonsensical to state or imply that any of the characteristic features of one should also apply to the other.

Lastly, apparently quite convincing to defenders of the *linear, nonthreshold* hypothesis is the fact that, using cellular systems, a loss of statistical significance of the response as D_P is reduced can be overcome by including a suitably larger number of cells. This is done to "improve the statistical significance." However, this argument is fallacious because the independent variable, D_P, being intensive and thus not dependent on mass, will remain precisely at the same location, even though the increase in the number of exposed cells may have produced many more quantally responding calls. This gives the impression that one can defy thermodynamic laws and produce more responding cells with no additional expenditure of energy. However, if the extensive, and thus mass-dependent and additive quantity, T_D, is used as the independent variable, any addition of mass (e.g., cells) and thus energy at a statistically weak point, will force that point upward on the function (Bond et al., 1995). This fact removes any defensible basis for *extrapolation* of the function downward from the region in which statistically reliable actual response data are obtainable.

DISCUSSION AND CONCLUSION

Strong arguments can be made from basic MT considerations that D_P (ε/m) is a measure of agent quality, and T_D (ε) a measure of radiation quantity (amount). Accordingly, the plots with these two independent variables would be termed *quality* and *quantity* functions, respectively. Because substantial space would be required to develop this theme, and because it would further complicate an already difficult subject, it was not included in this paper.

Had the idea of separate quality and quantity functions been developed, a family of linear functions, rather than only one, could have been provided in Figure 15.3. In historic and classical MT functions (Figure 15.1), agent concentration has been used extensively as the most relevant physical measure of agent quality (but always recognizing that the biological response must be the final judge of agent quality). By this measure, the increase in effectiveness with agent concentration (absorbed dose) in Figure 15.1 represents an increase in agent quality.

A strong emphasis on "risk" diverts one from the primary practical responsibility of both the public health officer and the epidemiologist. This is the proper collection of data on the actual number of quantal responses in a given population, for example, the 245 observed excess cancers in a sample of the atomic bomb survivors (a measure of the severity of effect on the dosed *population*, but not on the individual). Although these data may later be used to formulate probabilities, only observable excess tumors can play a role in determining the primary PH outcome, the absolute number of excess cancers observed. In other words, the overall exercise is clearly a PH problem. Extensive emphasis on average values of "risk" or probability, presumably intended for the individual, projects the incorrect image of the problem being in the MT realm.

The conclusions from the above are: (1) T_D or N_D, to the total exclusion of D_P, are the proper imparted-energy–related independent variables for describing PH data adequately; (2) it is only this approach that can provide the correct total yield of cancers; (3) the PH data do not provide sufficient evidence to even formulate, let alone prove, the linear hypothesis (i.e., linear functions alone cannot provide a basis for it); (4) the numerical value for the risk coefficient for cancers is about 70 times smaller if T_D is used, correctly, as the amount of agent energy, than it is when D_P is used, incorrectly, as the amount of energy; and (5) the linear, no-threshold hypothesis is flawed.

16

Molecular Analyses of In Vivo *hprt* Mutations in Humans

RICHARD J. ALBERTINI, J. PATRICK O'NEILL,
AND JANICE A. NICKLAS

SUMMARY

Somatic mutations in critical genes are initiating events in cancer. Although studies of oncogenes and tumor-supressor genes are crucial for an understanding of pathogenesis, mutations in non-disease-related reporter genes may be more informative regarding the mutation process itself. One such reporter is the human hypoxanthine-guanine phosphoribosyltransferase (*hprt*) gene. In vivo *hprt* mutations in T lymphocytes can be assessed by either a short-term or a cloning assay. The latter permits isolation of mutant cells for molecular analyses. A worldwide database has now established the normal background mutant frequency (MF) for *hprt*. This background, some modifying factors, and the results of illustrative studies of the mutagenic effects of chemical and radiation exposures are summarized. A synopsis of the molecular spectra as currently defined for both background and radiation-induced in vivo *hprt* mutations is also presented. These spectra show that carcinogenic mutagenic mechanisms are captured in the *hprt* reporter gene, especially those that are operative in several of the lymphoid malignancies. The possibility of assessing in vivo *hprt* mutations in humans in cell types other than lymphocytes is considered.

INTRODUCTION

Somatic mutations arise in normal individuals. Some are spontaneous mistakes due to replication errors, while others result from exposures to endogenous or exogenous mutagens. The vast majority of these mutations are probably harmless to the

253

individual. However, rarely, critical genes or genetic regions are involved, leading to disease. This is perhaps best illustrated in cancer, where somatic mutations are the initiating pathogenic events.

The study of genes of oncological significance holds the central position in current cancer research. While molecular analyses of mutations in cancer-relevant genes define the relationship between their malfunction and pathogenesis, studies of mutations in the harmless or reporter regions of the genome may be more useful for defining the mutation process itself. Part of the reason is that cancer-related genetic events are usually non-selectable, making it difficult to recognize generalized classes of rare events. More importantly, mutations in oncogenes or tumor-supressor genes often alter cell proliferation, making it impossible to measure primary frequencies. It is these primary frequencies that are the important measurements both for defining mutagenic mechanisms and for public health.

Assays for in vivo arising mutations are of considerable value for monitoring human populations for potential health hazards due to environmental mutagens and carcinogens. This value is enhanced if the mutant cells can be recovered for molecular analyses. In this case, the in vivo mutants become material for basic research into mutational mechanisms.

The gene for hypoxanthine-guanine phosphoribosyltransferase (*hprt*), located at position Xq26, is constitutively expressed but dispensable for cell viability in virtually all tissues. The encoded HPRT enzyme phosphoribosylates hypoxanthine and guanine for reutilization through the purine salvage pathway. Without this activity, the de novo pathway of purine biosynthesis is used by the cell so that viability is maintained. HPRT phosphoribosylates the purine analogues 8-azaguanine (AG) and 6-thioguanine (TG) as well as its normal substrates. In the absence of this metabolic conversion, these analogues are not cytotoxic. Cells that lack HPRT activity are therefore resistant to killing by these agents, while HPRT-proficient cells are not.

Mutations of the *hprt* gene provide a selectable reporter event for in vivo somatic mutations in humans (Albertini et al., 1982, 1990; Morley et al., 1983). Two methods are available for assessing these mutations in peripheral blood T lymphocytes. One is a short-term phenotypic assay, which has the advantage of speed and potential for automation but the disadvantage of consuming the mutant cells, which are then not available for study (Albertini et al., 1981, 1988; Ammenheuser et al., 1988, 1991, 1994; Ostrosky-Wegman et al., 1988, 1990; Morley et al., 1983; Montero et al., 1991; Ward et al., 1994). The other is a cloning assay which allows for molecular analyses (Albertini et al., 1982, 1990; O'Neill et al., 1987, 1989). Issues of historical importance for in vivo mutation research such as demonstrating unequivocally the genetic basis of the measured endpoints (Albertini et al., 1985; Turner et al., 1985) and the relationship between mutational events and their mutant progeny (Nicklas et al., 1986, 1989) have been resolved for the *hprt* system using the cloning assay.

This chapter describes both methods for assessing *hprt* mutations arising in vivo in human T cells. Quantitative results from our laboratory and elsewhere are presented, with emphasis on results using the cloning assay. The bulk of the chapter considers the molecular bases of *hprt* mutations using results from several laboratories but primarily from our own. The importance of characterizing T-cell receptor (TCR) genes among mutant isolates, allowing mutations to be analyzed in the context of in vivo clonality, is detailed. Finally, the degree to which carcinogenic mutagenic mechanisms are captured by the reporter *hprt* mutations is examined. This capture determines the validity of using these mutations as surrogates for cancer-causing genetic events in humans.

THE ASSAYS FOR hprt MUTATIONS ARISING IN VIVO IN HUMAN T LYMPOCYTES

Short-Term Assays

The T lymphocytes of human peripheral blood are isolated by density centrifugation, cryopreserved, and then stored in liquid nitrogen. Cryopreservation is required for the short-term assays so that cycling T cells in the peripheral blood do not distort results (see below) (Albertini et al., 1981). The measured endpoint in the short-term assays is T cells' in vitro resistance to TG inhibition during DNA synthesis when stimulated by phytohemagglutinin (PHA). Culture intervals are short—42 hours; that is, 24 hours before and 18 hours after addition of label in the absence (control) or presence (test) of TG. The label is either ^{3}H-thymidine for enumeration of resistant cells by autoradiography (Albertini et al., 1981; Ammenheuser et al., 1988) or 5-bromodeoxyuridine (Montero et al., 1991) for the scoring of TG-resistant cells by fluorescence differences. The labeling index of test cultures divided by the total number of enumerable cells in these cultures (labeling index of controls × total cells) gives the variant frequency (VF). VF is assumed to be the frequency of mutants in the peripheral blood. The term "variant" is used because, strictly speaking, mutation cannot be demonstrated because resistant cells are consumed in the assay. The short-term assays are thus phenotypic assays for mutation.

Cloning Assay

Peripheral blood T lymphocytes are obtained in the same way as for the short-term assays. Cryopreservation in this case is optional but is often convenient so that groups of assays can be performed at the same time. The measured endpoint in the cloning assay is clonal growth in the presence of TG. The T cells are cultured in limiting dilutions in the wells of microtiter dishes in the absence (control) or presence (test) of TG and scored microscopically in 8 to 14 days (O'Neill et al., 1987, 1989). Control plates contain few cells per well, typically 1, 2 or 5. Test wells contain larger numbers per well, typically 10^4 or 2×10^4, as only rare cells will be capable of growth in TG. The wells also contain a source of growth factor (interleukin-2), PHA, and irradiated accessory cells. Cloning efficiencies are determined from the Poisson relationship $P_0 = e^{-x}$, where x is the average number of clonable cells per well. The ratio of the cloning efficiency in the presence of TG to the cloning efficiency in its absence gives the mutant frequency (MF). The term "mutant" is used because the genetic basis of the resistant cells can be demonstrated. Growing colonies can be isolated and propagated in vitro for molecular and other studies.

Quantitative Results: Estimations of In Vivo hprt Mutations in Humans

Initial estimates of frequencies of *hprt* mutations in humans came from the autoradiographic short-term assay (Strauss and Albertini, 1977, 1979). These reports, mostly from our laboratory, gave VF values in the range of 10^{-4}, which were clearly in error. The early studies did not recognize the fact that infrequent cycling T cells in the peripheral blood may continue DNA synthesis in vitro in the presence of TG, even if non-mutant. Although these "phenocopies" are eventually killed by TG, they do incorporate label during the early hours of culture in the short-term assays and thus are scored as variants. As outlined elsewhere, cryopreservation removes this effect (Albertini et al., 1981). Reports of the last several years are based on methods that remove the phenocopy effect and do reflect the numbers of true variants (Albertini et al., 1981, 1988; Ammenheuser et al., 1988, 1991, 1994; Ostrosky-Wegman et al., 1988, 1990; Montero et al., 1991; Ward et al., 1994). However, the initial reports demonstrate the potential for error inherent in phenotypic assays unless special precautions are taken.

Table 16.1 summarizes typical quantitative VF and MF values obtained in studies in several laboratories over the last decade. Mean frequencies are remarkably similar for young adults whether determined by short-term or by cloning assay. This does not necessarily mean that these two kinds of measurement detect exactly the same thing. It is quite possible that different subpopulations of T cells are being assayed by each method, that the different endpoints (i.e., single round of DNA synthesis versus continual clonal growth) reflect different kinds of mutants, or that sibling mutants (see below) are represented differently in the two kinds of assay. Despite this, however, both the short-term and the cloning assays show a reproducible age effect in that VF and MF both increase approximately 2–5% per year of life, at least after the teen-age years (Albertini et al., 1988; Robinson et al., 1994). Both kinds of assays have been used to determine MF values during fetal life as reflected in placental cord blood samples (McGinniss et al., 1990; Ammenheuser et al., 1994; Manchester et al., 1995). These are about tenfold lower than what is observed for young adults. The cloning assay also has been used to carefully determine *hprt* MF values in normal children (Finette et al., 1994). These frequencies rise rather steeply until approximately age 16, after which the 2–5% per year rise with age obtains. The several factors that affect background *hprt* mutant frequencies using the cloning assay have recently been reviewed and analyzed in detail with results from four laboratories (Cole and Skopek, 1994; Robinson et al., 1994). As reflected in Table 16.1, the various inherited human DNA repair deficiency states show the expected elevations in in vivo VF or MF values that most likely reflect an increased rate of mutation.

A major stimulus for developing assays of in vivo somatic mutations in humans is the desirability of monitoring for environmental mutagenesis. Several studies of populations exposed to ionizing radiation or to chemical mutagens have been conducted by laboratories worldwide (Seifert et al., 1987; Ammenheuser et al., 1988, 1991; Messing et al., 1989; Nicklas et al., 1990, 1991a; Branda et al., 1991; Bridges et al., 1991; Natarajan et al., 1991; Tates et al., 1991; Perera et al., 1992, 1993, 1994; Ward et al., 1994). Although some have been negative, reflecting either a true lack of mutations or insensitivity of the assays, numerous studies have been positive, indicating that these measurements do detect exogenously induced gene mutations. The issue of sensitivity is now being addressed by field studies that measure *hprt* mutations, which are clearly biomarkers of effect, in the same individuals in whom other biomarkers, such as those reflecting exposure, are being assessed. It is only through such studies that the sensitivities and ultimate use of these assays for human genotoxicity monitoring can be established.

Three examples of monitoring humans for increases in *hprt* mutation frequency as a result of exposure to a mutagenic agent will be summarized here. The first is a study of patients receiving ^{131}I conjugated antibodies for cancer therapy (Nicklas et al., 1990, 1991a). A single treatment or multiple treatments resulted in significant increases in *hprt* MF and these elevated MF persisted for at least 24 months after treatment. In addition, the increases in MF were proportional to the amount of radiation received as a single dose. The second example is a study of foundry workers exposed primarily to polyaromatic hydrocarbons (Perera et al., 1993, 1994). There was a significant correlation between *hprt* MF and DNA adducts. This was the first report linking in vivo exposure to a mutagenic agent with the level of DNA adducts and the frequency of somatic cell mutation. The third example is a study of ethylene-oxide-exposed workers (Tates et al., 1991). Two populations that differed in exposure level were studied. While increases in sister chromatid exchanges (SCE) were found in both groups, increases in *hprt* MF were found only in the higher exposure group. These results demonstrate the difference in sensitivity between a biomarker of exposure (SCE) and a biomarker of effect (*hprt*). The level of detection of *hprt* mutations is being actively studied in a variety of human populations exposed to mutagenic agents.

Although sensitivity is not considered in this paper, the biology of the *hprt* mutational response in T cells is relevant to the interpretation of recent attempts to detect radiation-induced somatic mutations in atomic bomb survivors (Hakoda et al., 1988a, b, 1989; Hirai et al., 1996). Interpretation must include considerations of the survival of mutants and the difference between *mutants* and *mutations*. Most in vivo *hprt* mutations in T cells in adults arise in the peripheral lymphoid tissue rather than in the true stem cell compartment. In fact, without thymic function which

TABLE 16.1 Summary of in vivo variant frequency (VF) and mutant frequency (MF) values for *hprt* mutation in human T lymphocytes.

Subjects	Short-term assays (VF)	Cloning assay (MF)
Background mean newborn	$0.8\text{-}2.1 \times 10^{-6}$	$0.7\text{-}1.2 \times 10^{-6}$
Background mean young adult	$1.5\text{-}9.0 \times 10^{-6}$	$5\text{-}10 \times 10^{-6}$
Age effects (beyond 16 years)	5.0%/year	1.6-5.0%/year
Smoking effect	Increase (100–200%)	30–50% increase (not statistically significant in all studies)
DNA repair defects	Increase in Bloom syndrome, Fanconi anemia, and Werner syndrome	Increase in ataxia telangiectasia, xeroderma pigmentosum, ± increase in Fanconi anemia
Chemical exposures	Multiple reports of increases	Multiple reports of increases
Radiation exposures	Multiple reports of increases	Multiple reports of increases

is largely absent in adults, T-cell maturation cannot occur. This plus empirical evidence indicate that the lifespan of T-cell mutants will be less than the life of the individual and, for some, may be a period of months to years (Ammenheuser et al., 1988). This can be advantageous, depending on the purpose of a human monitoring study. However, somatic mutations in other than the stem cell compartments are not useful for detecting a mutational response that occurred more than half of a human lifetime in the past. It is in this light that the reports of *hprt* mutations in the T cells of atomic bomb survivors should be interpreted. The slight increase in MFs found in the cohort of individuals with high radiation exposures compared to non-exposed controls has been interpreted as relative insensitivity of the *hprt* reporter gene to ionizing radiation. In contrast, the response to acute ionizing radiation in the form of radiotherapy of cancer is quite robust (Nicklas et al., 1990, 1991a; Sala-Trepat et al., 1990; Ammenheuser et al., 1991). These superficially discordant observations are easily reconciled when it is recognized that *mutants* are not the same as *mutations*. The infrequent elevations in MF in individual high-exposure survivors probably result from clonal amplifications of the *mutants* that descend from stem cell *mutations* in persons who were young at the time of the bomb blast. Traces of most of the *mutations* induced by the bombs which occurred in the peripheral lymphocyte tissues of adults have long disappeared in the 40+ years between exposure and assay. Therefore, there probably is no simple causal relationship between the slope of the MF values in the atomic bomb survivors and their radiation exposures. More will be said below of *mutants* versus *mutations*.

QUALITATIVE RESULTS: MOLECULAR MUTATION SPECTRA

The ability to isolate and propagate mutant colonies from cloning assays allows characterization of the underlying *hprt* mutations. A substantial armamentarium of methods is now available for these molecular analyses (Table 16.2). Mutation events ranging from single-base changes to deletions and translocations of the *hprt* gene can be identified and mutational spectra under a variety of circumstances described. However, a precise distinction between *mutants* and *mutations*, briefly considered above for its quantitative implications, is even more important for defining mutational spectra, which describe specific kinds of *mutations* among all *mutations*. Molecular characterization of another set of genes in mutant isolates [i.e., the T-cell receptor (TCR) genes] is used to define in vivo clonality, sibling mutants, and primary independent *hprt* mutational events.

TCR Gene Rearrangements in hprt Mutants

Cells of the immune system recognize the universe of antigens via their surface receptors—that is, surface immunoglobulins (Ig) on B lymphocytes and TCRs on T lymphocytes. There is an enormous diversity of these antigen-specific receptors which is generated at the somatic level by rearrangements of germ-line encoded

TABLE 16.2 Armamentarium of molecular methods.

Method	Types of mutations detected	Reference
Southern blot	Large deletions, translocations, inversions	Albertini et al., 1985, Morley et al., 1983, Nicklas et al., 1987, 1989
Multiplex PCR	Large deletions, small deletions	Gibbs et al., 1990
RT-PCR and DNA sequencing	Base substitutions, (missense, nonsense, splice), frameshifts, small deletions	Yang et al., 1989
Linker marker loss	Size of deletion	Nicklas et al., 1991b
Pulsed field gel electrophoresis	Size of deletion	Lippert et al., 1995a,b
Long PCR	Pinpoint breakpoint sites, translocations	B. Van Houten, personal communication
Genomic breakpoint PCR and sequencing	Large internal deletions	Rainville et al., 1995
Inverse PCR	Large external deletions, translocations	Williams et al., 1996
Cytogenetics & X chromosome paint	Translocations	B. Hirsch, in preparation
T-Cell receptor analyses (Southern blot, RT-PCR restriction)	Clonality determinations	Nicklas et al., 1986, Clark and Nicklas, 1996

variable (V), diversity (D), junctional (J), and constant (C) regions of the Ig and TCR genes. This process of rearrangement proceeds in an orderly fashion during lymphocyte ontogeny and is mediated by an enzyme activity system termed V(D)J recombinase. DNA cleavage by V(D)J recombinase is directed by highly conserved consensus sequences consisting of a heptamer and nonamer separated

by 12 or 23 bases. V(D)J recombinase–mediated rearrangements are characterized by certain hallmarks in the junctional regions—namely, nibbling back from
the point of incision, the presence of palindromic or P nucleotides, and the inser-
tion of non-templated bases, presumably by terminal deoxynucleotidyl transferase
(TdT) activity. These all contribute to the hypervariability of the junctional regions
which confer the requisite diversity for antigen binding. At the molecular level,
the Ig or TCR gene rearrangement patterns in B or T cells, respectively, constitute
molecular signatures of clonality. For the T cells, there are four TCR genes: the
α and β genes, located at chromosome positions 14q11 and 7q35, respectively,
which are co-expressed on the majority of cells; and the γ and δ genes, located at
chromosome positions 7p15 and 14q11, respectively, which are co-expressed on
the remainder.

The diversity of TCR gene rearrangements is so vast that identity for all rear-
ranged genes in two or more cells or isolates from an individual indicates their in
vivo clonal lineage dating from the time of these rearrangements. Thus, two or
more *hprt* mutant isolates with identical TCR gene rearrangements are clonal descendants of the same T-cell progenitor. If the *hprt* changes are also identical at the
molecular level, which is by far the usual case, these are *sibling* mutants deriving
from the same in vivo *hprt* mutation. The decidedly unusual occurrence of different *hprt* changes in isolates with identical TCR gene rearrangements may indicate
an in vivo clone with a mutator phenotype. In any event, this observation indicates
that the *hprt* mutational events arose in vivo after the TCR gene rearrangements,
that is, in a post-thymic cell in the periphery. The much more common finding
of one or more *hprt* mutant isolates with identical *hprt* changes plus one or more
non-mutant or wild-type isolates from the same individual, all with the same TCR
gene rearrangements, indicates the same post-thymic peripheral origin of the in
vivo *hprt* mutation. It is from such observations that we know that most in vivo
hprt T-cell mutations arise in the peripheral lymphoid tissue rather than in the stem
cell compartment in adults.

It should be obvious that analyses of TCR gene rearrangements among *hprt*
mutant isolates, by identifying sibling mutants, provides a basis for correcting
the in vivo mutant frequencies determined by cloning assays to in vivo mutation
frequencies. When this was analyzed for 413 wild-type and 1,736 *hprt* mutant
isolates from 58 individuals, in vivo clonality occurred among mutants but not
wild-type isolates (O'Neill et al., 1994). Thirty-five (60%) individuals had at least
one instance of clonality among their mutant isolates. Despite this, adjusting the
measured MF values by clonality to obtain the mutation frequency only had a major
effect on nine samples, all of which had MF values of $> 40 \times 10^{-6}$. Table 16.3
gives illustrative results from this study. These differences, in vivo clonality among
individuals may explain the large inter-individual variability in MF values, even
after age corrections, compared to the relatively low level Poisson variability seen
for intra-individual MF values determined by repeat assays (Robinson et al., 1994).

TABLE 16.3 Effect of *hprt* mutant clonality on the in vivo mutation frequency.

Sample #	Measured mutant frequency ($\times 10^{-6}$)	Fraction with different TCR patterns	Clonality	Corrected mutation Frequency ($\times 10^{-6}$)
2(E)6	11.6	65/73	2-2mers 1-3mer 1-5mer	10.3
a(J40+40-2)	2.6	30/30	-	2.6
33(1372)	41.9	18/32	1-2mer 1-14mer	23.6
51(MA4)	69.5	17/32	1-2mer 1-3mer 1-13mer	36.9
54(MA7)	69.5	18/18	-	69.5
5(28-4)	620	2/15	1-14mer	82.5

Molecular Characterization of In Vivo hprt Mutations

Molecular analyses at differing levels of resolution have now given results on thousands of background ("spontaneous") human in vivo arising *hprt* T-cell mutants (reviewed in Albertini et al., 1990; Cariello and Skopek, 1993; and Cole and Skopek, 1994). Our laboratory alone has analyzed approximately 1,600 isolates by Southern blot and, more recently, several hundred more by multiplex PCR of genomic DNA and by RT-PCR and sequencing of the cDNA product. Many laboratories are analyzing *hprt* T-cell mutants, and a computerized database of published results is available to all investigators in this field (Cariello et al., 1992; Cariello, 1994). The following is based on these sources.

The molecular spectrum for background in vivo *hprt* mutations in human T lymphocytes includes approximately 15% with gross structural alterations such as deletions, insertions or other rearrangements, and 85% with "point mutations" (Nicklas et al., 1989). The gross changes were defined primarily by Southern blot analyses and therefore do not include most involving less than 300 basepairs. The proportion of gross alterations when defined by multiplex PCR is somewhat smaller, probably because generally only deletion events are identified. "Point mutations" are all other kinds of events, not simply base changes. It must be emphasized that this overview applies only to the background mutants in adults. Studies of background mutants in placental cord blood, which reflect mutations arising in the fetus, show approximately 75-85% with large structural alterations, often of one kind, with the remainder being the smaller mutations. This reversal of the adult pattern persists in young children until approximately age five (Finette et al., 1996). The discussion of carcinogenic mutagenic mechanisms below will consider these fetal and childhood mutations in detail.

Another way to characterize the human background in vivo mutation spectrum is to concentrate on the "point mutations." Although data are available from several sources, those in the computerized database essentially reflect the current findings. Table 16.4 gives the results on 254 mutations. Of these, 81 or 32% are base substitutions in coding regions, with somewhat more transversions than transitions, and 19 or 7% are base substitutions in introns that affect splicing. The most frequent class of change is base substitution (81/165 for sequence-defined mutations) followed by frameshifts and small genomic deletions (23/165 each). The remaining 89 mutations have genomically undefined alterations in cDNA ("splice" mutations). Several investigators are describing sites of increased mutation ("hot spots"), but these results are still preliminary. However, a high frequency of both transitions and transversions at G_{197} has been reported by several laboratories (Cariello and Skopek, 1993).

A major reason for characterizing the background mutational spectrum at *hprt* is for comparison with mutational spectra determined in individuals who have been exposed to environmental mutagens. The rationale is that different mutagens or classes of mutagens should induce characteristic mutational changes which then can be used to define the nature of the exposure. In this sense, *hprt* mutations

TABLE 16.4 In vivo human *hprt* point mutations (n=254).

Type	Number	%
Genomic DNA alterations		
Single base substitution (coding region) Transitions (number=34, 42%) Transversions (number=47, 58%)	81	32
Single base substitution (intron, splice) Transitions (number=12, 63%) Transversions (number=7, 37%)	19	7
Frameshifts	23	9
Small genomic deletions	23	9
Insertions	4	2
Tandem mutations	2	1
Complex	13	5
cDNA ALTERATIONS (Genomic alteration unknown)		
Single exon loss	61	24
Multiple exon loss	17	7
Partial exon loss	9	4
Partial intron addition	2	1

Note: Based on *hprt* database of N. Cariello.

TABLE 16.5 Large alteration mutations at *hprt* in vivo and in vitro.

Group	Number of mutations studied	Percentage with alteration (n)
Normal adults	619	79 (13)
Pre-radioimmunotherapy patients	118	20 (17)
Post-radioimmunotherapy patients	235	90 (38)
In vitro radiation (3 Gy)	25	86 (54)

serve as restricted biomarkers of exposure, providing specificity by defining the offending mutagen. Clearly, exposures of human fibroblasts to agents such as benz[a]pyrene in vitro result in an *hprt* mutational spectrum characterized by G→T transversions (R.H. Chen et al., 1990). It is significant that mutations in the p53 gene in lung cancers from smokers but not from non-smokers are also characterized by such changes (Suzuki et al., 1992; Hollstein et al., 1994). As regards *hprt*, the early results from studies of chemical exposures in humans have been mixed, probably because insufficient numbers of mutants have been studied (Vrieling et al., 1992; Burkhardt-Schultz et al., 1993; Cole and Skopek, 1994). By contrast, *hprt* mutations in humans exposed to ionizing radiation clearly show a spectrum that becomes increasingly dominated by large alterations such as deletions as a function of radiation dose (Nicklas et al., 1990, 1991a). Table 16.5 gives the overall frequency of large-alteration *hprt* mutations in cancer patients receiving irradiation from an internal ^{131}I emitter, as compared to the background frequencies of such events in normals and in pre-treatment patients, and to the frequency of such mutations in human T lymphocytes exposed in vitro to 300 cGy external-beam acute γ-irradiation. Clearly, an induced mutational spectrum can be differentiated from the background spectrum, indicating that at least some degree of specificity regarding mutagen exposures can be achieved.

Mutagenic Mechanisms with Carcinogenic Potential are Captured in hprt

Although the original reason for defining in vivo *hprt* mutational spectra was to provide specificity for human monitoring, another, perhaps more important reason, is emerging. The original reason viewed *hprt* mutations as measures of exposure. However, these mutations really are biomarkers of effect, and changes in their frequencies and kinds may have a human health significance beyond simply reflecting mutagen exposures. *Hprt* or other specific reporter gene mutations may identify

similar events occurring elsewhere in the genome, some with pathogenic consequences. If reporter gene mutations are surrogates for mutations in cancer-relevant
genes, increases in frequencies may indicate increases in cancer risk. For this to
be true, mutagenic mechanisms with carcinogenic potential must be recorded in
the reporter gene. The emerging reason for defining in vivo *hprt* mutational spectra therefore is to determine if mutations here, measured in a gene and tissue of
convenience, are valid surrogates for cancer-causing mutations arising in target
tissues.

Hprt cannot undergo the homologous somatic recombination as often seen in
cancers because of its X-chromosomal location. It does, however, capture a variety
of other carcinogenic mutagenic mechanisms. Large deletions are seen as well as
translocations (Nicklas et al., 1989). These are common in tumors. Almost 25%
of internal *hprt* deletions have breakpoints in DNA sequences with high homology
to topoisomerase II consensus cleavage sequences (Rainville et al., 1995). Similar
chromosome breakpoints are often seen in leukemia. *Hprt* even undergoes fusions
with downstream elements resulting in fusion transcripts (Lippert et al., 1997)—
an occurrence frequently encountered in cancer. All of the above are general
mutagenic mechanisms.

There is rather striking evidence that *hprt* also captures a much more specific carcinogenic mutagenic mechanism. Human lymphoid malignancies frequently show
non-random chromosome rearrangements with one of the breakpoints near one of
the Ig genes in B-cell malignancies or one of the TCR genes in T-cell malignancies
(Finger et al., 1986; Boehm and Rabbits, 1989; Tycko and Sklar, 1990; Breit et
al., 1993). The precise site of this breakpoint is often the heptamer-nonamer consensus recognition signal sequence (RSS) that directs canonical rearrangements of
these genes. The other breakpoint required to produce the characteristic chromosome translocation, which is usually in the region of an oncogene, is often in a
cryptic consensus heptamer. Junctional regions of the new, translocated chromosome also frequently bear the hallmarks of a V(D)J-recombinase-mediated event,
so named because of the enzyme activity that directs the rearrangement of these
characteristic regions of the Ig and TCR genes, as described above. Therefore,
this specific maturational Ig and TCR gene rearrangement mechanism frequently
proceeds illegitimately with carcinogenic consequences.

A characteristic translocation t(1;14)(p32;q11) occurs in approximately 3% of
all childhood T-cell acute lymphoblastic leukemias (T-ALL) (Q. Chen et al., 1990;
Bernard et al., 1991). In the leukemic cells, most of the breakpoints on chromosome
14 cluster in the D_δ-J_δ region of the TCR δ locus, while those on chromosome 1 are
in a 1 kb 5' region of the so called *tal-1* (for T-cell acute leukemia) gene. The *tal-1*
gene codes for a protein thought to be important in early hematopoietic cells. Its
disregulation by the t(1;14) translocation probably produces an oncogene function
in the genesis of T-ALL.

In an additional 20-30% of childhood T-ALL, a submicroscopic deletion of
90kb of the *tal-1* 5' region juxtaposes its coding sequences to the first non-coding

exon of an upstream *sil* (for stem cell leukemia interrupting locus) gene (Aplan et al., 1990, 1991; Brown et al., 1990; Breit et al., 1993). This puts the fused *sil-tal-1* gene under the transcriptional control of the *sil* gene promoter (Bernard et al., 1991; Aplan et al., 1991, 1992). Again, disregulation of *tal-1* with leukemogenic consequences occurs. Importantly, the 5' breakpoint of this deletion (in the *sil* gene) occurs in only one V(D)J recombinase heptamer recognition sequence, while the 3' breakpoint occurs in one of four locations, each however in a V(D)J heptamer-nonamer recognition sequence (Bernard et al., 1991; Aplan et al., 1992; Breit et al., 1993). The junctional sequences of these deletions show the hallmarks of V(D)J recombinase activity (Aplan et al., 1991; Bernard et al., 1991; Breit et al., 1993). Therefore, in both the submicroscopic deletion and the t(1;14) translocation, illegitimate V(D)J recombinase activity produces a leukemogenic somatic mutation.

As implied above, the background *hprt* mutational spectrum in the human fetus, as determined in placental cord blood T cells, differs markedly from the background spectrum found in adults. In the fetus, up to 85% of the *hprt* mutations show gross structural alterations of the gene, with up to 50% of these being a seemingly identical exon 2-3 deletion (McGinniss et al., 1989). Because these mutations arise in the fetus at a time when T lymphocytes are undergoing thymic differentiation, it seemed reasonable that some aberration of the TCR gene rearrangement mechanism was involved. This, in fact, was what was found by sequence analysis of the breakpoint and junctional regions of 18 *hprt* deletion mutants isolated from 13 newborns. All mutants have a single 5' breakpoint in intron 1 at base 21971. However, three different 3' breakpoints in intron 3 are observed in the group of 18 mutants, occurring either at base 22250 (15 instances), base 20156 (two instances), or base 22570 (one instance) (Fuscoe et al., 1991). The breakpoint site in intron 1 is in a V(D)J recombinase heptamer, while each of the intron 3 breakpoint sites is in a V(D)J recombinase heptamer-nonamer RSS. Therefore, there is complete analogy between the breakpoint sites for the *hprt* mutations and those for the *sil-tal* fusion gene in childhood T-ALL. All of the *hprt* mutants showed the characteristics of a V(D)J-mediated recombination, i.e., heptamer or heptamer-nonamer RSS, nibbling back, P-nucleotides, and non-templated additions.

More recent studies have indicated that the V(D)J recombinase–mediated *hprt* mutations remain frequent during childhood, being the most common single kind of spontaneous mutational event in normal children up to age five (Finette, 1996). The age-frequency distribution of V(D)J recombinase–mediated *hprt* mutations shows a striking similarity to the age-frequency distribution of childhood acute T-ALL. Therefore, this pathogenic, mutagenic mechanism is captured by *hprt* as a harmless mistake, although it can give rise to serious disease when it occurs in other genetic regions. The *hprt* gene clearly is a valid surrogate for at least those mutational events that give rise to the lymphoid malignancies in childhood.

CONCLUSIONS AND FUTURE DIRECTIONS

Somatic mutations in the *hprt* gene arise as harmless errors in humans and can be measured in peripheral blood T lymphocytes. As the mutant cells are not consumed in the cloning assay for these events, the underlying mutations can be analyzed at the molecular level. Several investigations have now shown that exposures to both chemical and physical mutagens increase *hprt* MF values; the sensitivity of this biomarker for detecting such exposures is being defined. The observation that the T-cell *hprt* mutations arise in adults in the peripheral lymphoid tissue rather than in the stem cell compartment indicates that the mutant cells will have a lifetime less than the lifetime of the individual, probably best measured in months or years. In an attempt to provide specificity for assessing mutagen exposures, molecular mutational spectra are being described for both the spontaneous mutations and for those associated with a variety of genotoxic agents. It clearly has been shown that ionizing radiation shows a characteristic mutational spectrum, inducing a high frequency of *hprt* mutations with gross structural alterations. Definition of mutational spectra mandates identification of in vivo clonality of mutants. Characterization of TCR gene rearrangement patterns allows this identification, and permits calculation of mutation frequencies from measured MF values. *Hprt* molecular mutational spectra are being defined to provide specificity for human monitoring and, more recently, for determining if mutagenic mechanisms with carcinogenic potential are reported by this gene. Studies in newborns and children have shown that the most frequently arising spontaneous *hprt* mutation in this age group is a large deletion mediated by the V(D)J recombinase system. This is also the mutational mechanism responsible for a wide variety of lymphoid malignancies, including T-ALL of childhood. *Hprt*, therefore, captures precisely this mutagenic mechanism. It is perhaps significant that this particular mutation is reported in T lymphocytes that are the target cells for the malignancies of concern. The question regarding the relevance of reporter mutations in tissues of convenience for predicting pathological events in tissues of concern remains open.

This latter point too is being addressed by studies of in vivo *hprt* mutations in humans. Myeloid stem cells (CD34+) from human bone marrow or from peripheral blood, especially from placental cord blood, may be cloned in methylcellulose (Fauser and Messner, 1978). Differentiated myeloid and erythroid cell colonies are recognized after several days in the presence of critical cytokines. It is possible to incorporate TG in the methylcellulose and select for TG-resistant colony growth and differentiation. Such colonies are present and *hprt* mutant frequencies can be determined. Analyses of DNA extracted from these colonies by either RT-PCR or multiplex PCR followed by sequencing have demonstrated that they are true *hprt* mutants. As for the T-cell system, a molecular mutational spectrum is being defined. There will soon be a myeloid stem cell system analogous to the T-cell system for identifying, quantifying, and characterizing *hprt* mutations in this target tissue. The validity of *hprt* as a surrogate gene for detecting matched mechanisms that give rise to the more common adult hematopoietic malignancy can be assessed.

This chapter has reviewed the experience to date with *hprt* mutations in humans. The different assay systems are being exploited practically for mutagenicity monitoring. There is every expectation that such monitoring can identify human environmental health hazards before they cause disease. The cells recovered from the mutagenicity studies are providing material for basic studies, which are beginning to define mechanisms in humans that influence the mutational process itself.

We are grateful for the technical support of Linda M. Sullivan and Timothy C. Hunter and the secretarial assistance of Inge Gobel. This work was supported by the US Department of Energy (FG0287ER60502), the National Institutes of Health (5PO1-ES05294-05), and the American Cancer Society (CB-45 and CN-141). DOE and NIH support does not constitute endorsement of the views expressed.

17

Evolution of the Glycophorin A Assay for Measuring Biological Effects of Radiation on Humans

RONALD H. JENSEN AND DANIEL H. MOORE II

SUMMARY

The glycophorin A–based somatic cell mutation assay was originally created to measure the amount of damage that occurred in each individual as a result of exposure to mutagenic phenomena. Early in its existence, this assay was applied to blood samples obtained from A-bomb survivors and found to display a correlation with the amount of ionizing radiation exposure. Over the years, the glycophorin A assay has been improved, and the quality of the improvements have been determined by performing analyses of samples from A-bomb survivors to define dose response, variance, and persistence of effects. As it has now developed, the glycophorin A assay is rapid and easy to perform, with variance better than any other biodosimetry assay, and requirement of only very small blood samples. It serves as a lifetime biodosimeter that can be used to monitor large human populations for biological effects of radiation exposure and should be useful for predicting increased cancer risk in high-risk populations.

INTRODUCTION

This review begins only ten years ago, since the capability to perform the glycophorin A–based somatic cell mutation assay was first developed in 1986. Nevertheless, even though the time frame is small, the studies of samples from A-bomb survivors have had a large impact on the development of our analytical procedures.

Basis of the Assay

Glycophorin A (GPA) is a human blood type antigen that appears on the surface of red blood cells (erythrocytes) in two common forms, called the N and M allelic forms. Since the gene for this blood type antigen is inherited in a conventional Mendelian mode, 50% of the population expresses the heterozygous blood type NM. Essentially all the erythrocytes in individuals who are NM blood type contain both GPA^N and GPA^M on their surface. However, if a cell in the bone marrow is impacted with ionizing radiation at the glycophorin A gene (Figure 17.1), one of the two allelic forms of the GPA gene could be inactivated. As a result, all progeny of this cell would express only one of the two forms. In the example shown, the GPA^M allele has been inactivated, and, since the sister chromosome has not been affected by the ionizing radiation, GPA^N is expressed normally on all the progeny of this mutant cell. When these progeny cells mature and are secreted into the peripheral blood, they appear as variant erythrocytes, called N∅ or N-null cells. In order to detect the presence of such cells, a fluorescence immunolabeling technique was developed. The ideal form of this technique is to isolate two different monoclonal antibodies, each of which is specific for one of the two allelic forms of GPA. Then these antibodies are labeled with fluorescent conjugates so that immunostaining of erythrocytes from a heterozygous (GPA^{NM}) individual would yield primarily double-stained cells. Those cells which are variants (N∅ in the figure) would be single-stained, and the somatic cell mutation assay would be performed by carefully enumerating the frequency of single-stained erythrocytes in each individual. This requirement for expression of one of the two allelic forms of GPA provides an internal control which prevents mis-identification of degraded erythrocytes, which may have damaged membranes or cell surfaces, as being progeny of mutant bone marrow cells.

EVOLUTION OF THE ASSAY

1W1: The First Glycophorin A Assay

Description

The original version of the glycophorin A assay suffered from precision difficulties in enumerating variant erythrocytes, since they are very rare events. During the early period of developing this assay, the techniques for isolating specific antibodies for particular antigens were only beginning to be devised. Since the two allelic forms of GPA differ only by two amino acids in a sialoglycoprotein of molecular weight 36,000 daltons, we were unable to isolate an antibody that would label cells which contain GPA^N only. Thus, the first assay utilized one monoclonal antibody that was specific for GPA^M (this antibody was named 6A7), but the second antibody (named 10F7) recognized both allelic forms of glycophorin A. Thus, the assay could determine the loss of GPA^M but not GPA^N. For this reason

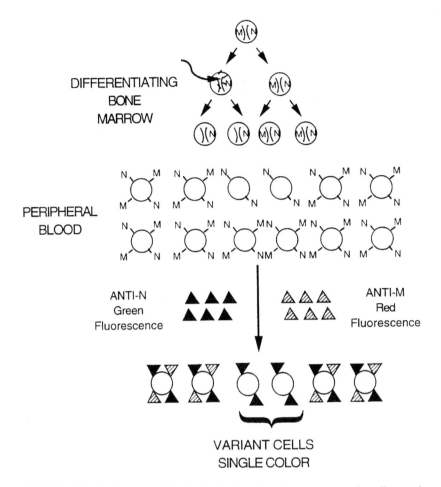

FIGURE 17.1 Schematic of the basis for the glycophorin A somatic cell mutation assay.

it was named the one-way-one (1W1) assay (Langlois et al., 1986). This was not a major defect, in that one would expect to obtain an equal frequency of both variant cell types, and enumeration of only one type should reflect the frequency of bone marrow mutant events.

A second technical problem of the times was a more difficult defect in the analysis procedure. At the time there existed only a small cadre of well-characterized fluorescent labels for immunofluorescence, so that the only two labels which were available for this application were fluorescein and Texas red. These gave well-resolved color separation of the signals, but required very different excitation wavelengths to obtain bright signals. Since the assay needed to enumerate hundreds of thousands of erythrocytes, the technique of choice was to perform high-speed flow analysis. In order to perform such analysis with two very different excitation sources, we used dual-laser-beam cell sorting, a very sophisticated and technically difficult analytical procedure. To obtain reproducible results, the single-labeled variant cells were sorted onto microscope slides and enumerated visually under a fluorescence microscope. The primary reason for this manual counting was that fluorescent artifacts were detectable in the stained erythrocyte suspension at a high enough frequency to introduce significant errors in the variant cell frequency if the flow sorter was used as a counting device. Using fluorescence microscopy, one could eliminate these false positive signals by enumerating morphologically identifiable erythrocytes.

Application to A-bomb Survivors

In collaboration with scientists from the Radiation Effects Research Foundation (RERF) in Hiroshima, we were able to demonstrate at Lawrence Livermore National Laboratory (LLNL) the first somatic cell mutation dose response of radiation exposure to humans on a population of Hiroshima A-bomb survivors (Langlois et al., 1987). This population was composed of 43 proximally exposed individuals and 20 survivors who were outside the exposure zone at the time of the A-bomb. The dose response appeared to be a linear function but with large variance among the data points, such that any of several response functions (including logarithmic or linear-quadratic) could fit the data equally well. At the time (1986–1987) the estimated dose was based on T65DR, but shortly thereafter DS86 was calculated for the individuals in our study. The resultant dose response was determined to be:

$$N\emptyset = 10.9 \times 10^{-6} + 40 \times 10^{-6} \, \text{Dose (Gy)} \quad (R^2 = 0.21).$$

SBS Assay

Description

Shortly after this study was completed, our collaborators at RERF developed a new glycophorin A flow analysis so as to perform the assay on a large number of A-bomb survivors in Hiroshima (Kyoizumi et al., 1989). The major change in this second of the GPA assays was the use of a newly available fluorophor, phycoerythrin. This fluorophor can be excited at the same wavelength as can fluorescein (488 nm) and emits fluorescence at a long enough wavelength (580–590 nm) to be separated from fluorescein emission (520–530 nm) using dichroic mirrors, band pass filters, and electronic signal compensation. As a result a single-beam laser sorter could be used for the analysis. As in original assay, the SBS assay used 6A7, which was conjugated with biotin, and a secondary avidin reaction for attaching the fluorophor, but this time with avidin-phycoerthrin.

Application to A-bomb Survivors

At RERF our colleagues used this single-beam sorter (SBS) assay to perform analysis on 62 exposed Hiroshima A-bomb survivors to obtain a dose response of

$$N\emptyset = 27 \times 10^{-6} + 34 \times 10^{-6} \text{ Dose (Gy)} \quad (R^2 = 0.33).$$

This is very similar to what was obtained for the 1W1 assay in our original analysis, and was further substantiated by running the 1W1 assay in parallel on 39 of these exposed donors plus 15 unexposed controls, with a correlation between the two assays of R^2 equalling 0.91.

BR6 Assay

Description

The next step in the evolution of the GPA assay was an improvement in eliminating false positive events by using a new monoclonal antibody which specifically recognized GPA^N. This antibody (named BRIC 157) was isolated and characterized by Anstee et al. (1988). At LLNL we conjugated this antibody with fluorescein and used it in combination with the GPA^M-specific antibody, 6A7, to perform an analysis on a FACScan flow cytometer (Langlois et al., 1990). The new assay was termed BR6 as an acronym for the two antibodies (BRIC 157 & 6A7). Parallel measurements were performed using this assay and the 1W1, dual-beam flow sorter assay on blood samples from unexposed donors to show that the assays gave comparable variant cell frequencies, and that the new assay showed improved measurement precision. For N$\emptyset$ variant cells, the group coefficient of variation for the 1W1 assay was 64%, while for BR6 the CV was 30%, a significant improvement. Another advantage of the new assay was the fact that it could be performed without cell sorting. This resulted in two improvements: (1) the new assay was less labor-intensive, since sorting is a long and tedious procedure, and microscopic enumeration is very labor-intensive; and (2) a higher precision in counting statistics could be obtained. The FACScan can perform analysis at 4,000 cells per second, while the cell sorter can be operated at a maximum of 1,000 cells per second. Using the BR6 assay, we analyzed 5×10^6 cells per sample, whereas the 1W1 and SBS assay could analyze only 5×10^5 cells per sample.

Application to A-bomb Survivors

To confirm that the BR6 and SBS assays were in agreement, a collaborative study between RERF and LLNL was again performed on identical blood samples from Hiroshima A-bomb survivors (Langlois et al., 1993). In this study, samples were obtained from 33 proximally exposed subjects (estimated DS86 doses ranged from 0.53 to 4.95 Gray). The SBS assay was performed at RERF and the BR6 assay at LLNL. With this population study, the dose responses were:

$$SBS \quad N\emptyset = 18 \times 10^{-6} + 21 \times 10^{-6} \, \text{Dose (Gy)} \quad (R^2 = 0.29).$$

$$BR6 \quad N\emptyset = 16 \times 10^{-6} + 27 \times 10^{-6} \, \text{Dose (Gy)} \quad (R^2 = 0.36).$$

As can be seen by comparing these equations, the assays again agreed nicely within their variance limits.

Comparison to Cytogenetics

At the same time that the glycophorin A assay was being improved, cytogenetic analysis was evolving in a collaborative fashion between RERF and LLNL. Historically, scientists at RERF had been performing chromosomal aberration analysis on metaphase chromosomes of lymphocytes obtained from A-bomb survivors and had acquired a large dataset indicating a significant dose response for this biological endpoint (Stram et al., 1993). At LLNL the human genome project had been developing a chromosome-specific human DNA library carried by a bacterial vector (Collins et al., 1991). Using this library, cytogeneticists were able to apply fluorescence in situ hybridization (FISH) of chromosome-specific DNA to human metaphase chromosomes and perform chromosomal aberration analysis in a rapid and quantitative fashion with fluorescence microscopy (Lucas et al., 1989). In a collaborative effort to compare this new technique with the already established cytogenetic analysis, the scientists at RERF performed conventional translocation frequency measurements, and those at LLNL performed FISH translocation frequency measurements on the same samples. The results of this comparison showed that the two cytogenetic analyses agreed with each other, with R^2 equalling 0.92, indicating that they are indeed measuring similar phenomena (Lucas et al., 1992).

Since the collaborative cytogenetic study was occurring at the same time as was the collaborative glycophorin A comparisons, the same blood samples were used for all four assays. Thus, the results for somatic mutation assays can be compared in precision with those obtained with aberration assays. Here we perform a comparison between the results from FISH translocation analysis and those of the BR6 assay. In order to compare these results, a best-fit function for the two datasets was derived and found to be a logarithmic translocation of the response data. After performing a log-linear regression analysis on each of the datasets, we performed inverse regression (Draper and Smith, 1981) to determine how well each assay could be used to predict dose. The results (Table 17.1) show that the two assays displayed very similar precision.

All the data that have been gathered on GPA analysis of blood samples from A-bomb survivors indicate that this assay shows effects from radiation long periods of time after exposure. The time since exposure for these blood donors has ranged

TABLE 17.1 Comparison of 95% confidence intervals (CI) for dose estimated by inverse regression of log-response on dose.

Dose (Gy)	TL[a]	N/O[b]	95% CI (Gy)	
			TL-based	N∅-based
0.00	0.007	4.8	9.62	12.19
0.00	0.014	6.6	9.07	10.74
0.64	0.062	12.0	8.22	8.56
0.78	0.099	19.9	7.72	8.26
0.86	0.075	27.4	7.53	8.41
1.13	0.109	23.1	7.61	8.22
1.15	0.189	25.3	7.56	8.23
1.28	0.140	192.0	8.63	8.18
1.30	0.431	39.7	7.44	8.83
1.41	0.121	29.1	7.50	8.20
1.53	0.168	26.1	7.55	8.20
1.62	0.247	103.6	7.79	8.35
2.01	0.179	89.0	7.75	8.21
2.02	0.061	58.7	7.51	8.57
2.07	0.756	98.8	7.84	9.59
2.28	0.248	48.9	7.45	8.35
2.51	0.413	173.8	8.49	8.78
2.79	0.582	132.5	8.15	9.20
3.15	0.302	36.5	7.44	8.49
4.95	0.301	121.5	8.05	8.48
		Avg.	7.95	8.80

[a]Frequency of translocations per cell (Lucas et al., 1992).
[b]N∅ variant cell frequency per million cells (Langlois et al., 1993).

from 40 to 50 years. An important question which remains is the effect on frequency of GPA variant cells at short times after exposure.

Chernobyl

The Chernobyl accident occurred in 1986 and has allowed us to study the GPA response under rather different conditions from our previous studies. We have performed a study on individuals who were on site during the emergency conditions, and other individuals who were sent to that site to assist in the cleanup of the radioactive contamination (Jensen et al., 1995). The exposure of these individuals was rather different from that experienced by the A-bomb survivors. First, the levels of exposure of a majority of the individuals involved in the Chernobyl cleanup was lower than 1 Gy. Only those who were on site during the emergency or shortly afterwards were exposed to greater than this level. In addition, the majority also were exposed over longer periods of time than the very rapid A-bomb blast. Of the 102 exposed Chernobyl accident victims who were analyzed by the GPA assay, only 10 were on site at the time of the explosion and received exposure that was rapid and extreme enough to suffer acute radiation sickness. The other 92 were cleanup workers who received their radiation dose over a period of minutes to hours. Those who received the lowest doses were exposed very slowly during the cleanup operation, at which some worked for several months. In addition, the time since exposure was much shorter for this population. The GPA assay was performed on all these individuals during the first decade after their exposure.

The results from the Chernobyl study show that the overall dose response was very similar to that seen for the A-bomb survivors:

$$N\emptyset = 6.1 \times 10^{-6} + 21 \times 10^{-6} \text{ Dose (Gy)} \quad (R^2 = 0.22).$$

At first this seems very surprising in that the extent and rate of exposure seem very different from those at Hiroshima, and Russell and Kelly (1982) have shown that rate of exposure greatly affects the specific-locus mutation frequencies in mice. However, a more careful analysis of the Chernobyl victims' data gives a plausible explanation for this result. If we calculate the dose-response for individuals who received less that 1 Gy, then only 1 of the 80 victims showed symptoms of acute radiation sickness, and the dose response was:

$$N\emptyset = 8.7 \times 10^{-6} + 4.1 \times 10^{-6} \text{ Dose (Gy)} \quad (R^2 = 0.008).$$

This shows a much lower dose response, which is not significantly different from zero. Thus, the low dose rate appears to induce mutation effects that are much lower than in people exposed at high dose rate.

Another important observation from the GPA analysis on Chernobyl accident victims is that during the first decade after exposure, the frequency of variant erythrocytes appears to be unchanged. A group of 10 individuals donated more than one blood sample over a period of seven years, with a mean interval of 16 months between donations. The $N\emptyset$ variant cell frequency for these individuals did not change systematically over time. This is very similar to the results we obtained when we performed a similar study on A-bomb survivors between 45 and 50 years since exposure (Langlois et al., 1993). Thus, it appears that the GPA assay provides lifetime biodosimetry for ionizing radiation exposure. An individual's $N\emptyset$ variant cell frequency will register at about 6×10^{-6} at age 10 and increase slowly to about 16×10^{-6} by age 65 (Grant et al., 1991) unless that person is exposed to a significant level of ionizing radiation. At that time, his $N\emptyset$ variant cell frequency will increase proportional to his exposure and remain near that level with a slow

TABLE 17.2 Overview of the changes in the glycophorin A assay and the dose response measured for A-bomb survivors.

Date	Assay	Instrument	Antibodies	N/∅ dose response (per 10^6 cells)
1987	1W1	Sorter	6A7 + 10F7	11 + 40 Dose(Gy)
1989	SBS	Sorter	6A7 + 10F7	18 + 21 Dose(Gy)
1990	BR6	Cytometer	6A7 + BRIC 157F	16 + 27 Dose(Gy)
1995	BR6-Chernobyl	Cytometer	6A7 + BRIC 157F	6.2 + 21 Dose(Gy)
1996	DB6	Cytometer	6A7-PE + BRIC 157 F	Not Done

increase continuing with aging unless another radiation exposure is experienced. Exactly how sensitive the assay is to low levels of exposure has been difficult to define. Monitoring a large population of individuals who are at high risk of exposure would be the best approach to determining the assay's feasibility.

DB6 Assay

Evolution of the GPA assay is continuing. Recently we found that the previous fluorescent labeling approach for antibody Direct BRIC 157 6A7 ("6A7") would occasionally give high frequencies of variant cells in samples for which such frequencies might not be expected. Careful monitoring of such events indicated that blood samples shipped or stored under inappropriate conditions gave these variable results.

In a study just completed, we showed that 6A7 labeled with the secondary avidin fluorescent labeling procedure was not binding to slightly damaged cells in a way that put their fluorescence intensities in the N∅ variant cell window of the bivariate histogram obtained by FACScan analysis (Jensen and Bigbee, 1996). If the same antibody was directly conjugated with phycoerythrin, the fluorophor of choice for this assay, these same samples gave cleaner histograms and lower variant cell frequencies for a number of such samples. A parallel cell sorting, fluorescence microscopic, and re-immunolabeling procedure showed that cells that would not bind the secondary complex would successfully bind the direct conjugate. We now have adopted this assay as a standard protocol for the GPA assay and recommend that others do the same.

CONCLUSIONS

Evolution of the GPA assay as a biodosimeter for ionizing radiation is tabulated in Table 17.2. The assay has undergone four significant changes in design that have improved the ease of analysis and the precision of measurement. Nevertheless, it can be seen from column 5 in Table 17.2 that the dose response has changed only a modest amount. It is clear that results from the earliest of assays can be compared reasonably with results of the most recently adopted method.

The progression of this analytical procedure over the last decade has been strongly dependent on the cooperation between US scientists at Lawrence Livermore National Laboratory, and Japanese scientists at the Radiation Effects Research Foundation, Hiroshima. The fact that the final entry in column 5 of Table 17.2 is not completed indicates that cooperation and collaboration between these groups will continue to aid GPA evolutionary advancement.

PART 4
PSYCHOSOCIAL FACTORS

18

Psychological Effects of
Radiation Catastrophes

EVELYN J. BROMET

SUMMARY

The psychological aftermath of three radiation catastrophes is reviewed, namely, the bombings of Hiroshima and Nagasaki, the accident at the Three Mile Island nuclear power plant, and the accident at the Chernobyl nuclear power plant. The major psychiatric and psychological symptoms found in the affected populations were a preoccupation with somatic symptoms, clinical depression, anxiety/fear of future health effects, neurasthenia (fatigue, weakness, dizziness, headaches), and post-traumatic stress symptoms. The symptoms were found to persist for many years after these events. Risk factors included exposure to death and illness, physical and social stigma, community disruption, and personal vulnerability factors such as gender and a history of psychopathology. The need for systematic, psychiatric epidemiologic research among Hiroshima/Nagasaki survivors and long-term follow-up studies of all three cataclysms is emphasized.

INTRODUCTION

From a public health perspective, the psychological effects of nuclear catastrophes may be equally, if not more, prevalent than their physical health consequences. The bombing of Hiroshima and Nagasaki was an extraordinary trauma for the survivors. Yet, epidemiologic research into its mental health consequences is virtually nonexistent. By contrast, during the last few decades, there have been a growing number of epidemiologic studies of the psychiatric, psychological, and social effects of natural and technological disasters and war-related traumas (Bromet and

283

Schulberg, 1987; Solomon, 1989; Kulka et al., 1990; Bromet and Dew, 1995). A recent meta-analysis of this body of disaster research concluded that the average excess psychological morbidity was 17% (Rubonis and Bickman, 1991). However, some of the most horrific events in modern history, such as the bombing of Hiroshima and Nagasaki and other disasters occurring in third world and eastern European countries, had not been studied and hence were not included in the review. It seems reasonable to surmise that the average excess morbidity rate of 17% might have been higher had such events been considered.

Psychiatric epidemiology is concerned with understanding the distribution of psychiatric disorders in the population and the risk factors associated with their onset and course (Bromet and Parkinson, 1992). Mental health disaster research is one important component of this field. The impetus and basic tools needed to advance research in psychiatric epidemiology stemmed from several significant experiences that occurred during World War II. First, psychologists in the armed forces developed a psychological symptom questionnaire, the Neuropsychiatric Screening Adjunct, to screen recruits. More recruits were rejected because of psychological impairment than because of physical impairment. It was therefore obvious that treated cases represented the tip of the iceberg. Thus, following World War II, a number of prevalence studies were carried out to determine the true prevalence of mental disorder in the population, starting with symptom-based research characterizing the 1950s to 1970s (Dohrenwend and Dohrenwend, 1982) and culminating in the recent disorder-based research beginning around 1980 (Robins and Regier, 1991; Kessler et al., 1994; Tsuang et al., 1995). Another event that shaped the focus of much psychiatric epidemiologic work was the observation that many soldiers suffered from "shell shock," or post-traumatic stress syndromes, from the extraordinary stresses to which they were exposed—even though they had screened "negative" for psychological disorder upon recruitment. Disentangling the links between stress and mental disorder became an important area of research and was for the most part the conceptual impetus for the large-scale post-war studies, such as the Midtown Manhattan and the Stirling County studies (Leighton et al., 1963; Robins and Regier, 1991). Unfortunately, the expansion of research in psychiatric epidemiology, including disaster research, did not extend to the group of people who experienced one of the most horrific stressors in history, the survivors of Hiroshima and Nagasaki.

Radiation catastrophes share an important underlying dimension, namely, the perception of risk to health (Slovic, 1991), a perception that for many survivors can turn into an unresolvable fear. This fear may be reinforced when cancers and other illnesses occur among survivors and are attributed, rightly or wrongly, to the radiation exposure. This chapter reviews findings from three unique radiation catastrophes, Hiroshima, Three Mile Island, and Chernobyl. Each differed in form, historical period, sociocultural context, and immediate consequences. Each also differed in the research perspectives and methods that were brought to bear to understand their psychologic effects. Yet the parallels in the findings on psychological reactions to these events are striking. The findings from these three events clearly demonstrate that the psychological legacy of collective radiation exposure—unremitting fear of radiation-related illness—is independent of the extent of exposure; or to put this in statistical language, the psychological effects are poorly correlated with the exposure dose. The following section presents a review of the existing studies of these events, as summarized in Table 18.1. Next, three issues are explored: an integration of the findings on the types of mental health problems arising after radiation exposure; the risk factors associated with negative psychological effects; and the identification of factors that could modify the impact of this exposure.

RESEARCH EVIDENCE

Atomic Bombings of Hiroshima and Nagasaki

Although there are some Japanese-language publications addressing the psychological sequelae of the bombings of Hiroshima and Nagasaki in 1945, only three English-language publications were found, and two were quasi-epidemiological. The first, by Misao and colleagues (1961), described the subjective complaints of 356 survivors (177 females, 179 males), as well as symptoms reported on the Cornell Medical Index completed by 345 survivors (149 female, 196 male). The actual source of the samples was not described in the paper. The findings from the Cornell Medical Index were compared with data from a normal sample of clerks and medical outpatients, but no specific information about the origins of these samples and their data was given. Overall, the authors suggested that the rates of subjective complaints (>50%) and anxiety about possible "A-bomb" diseases (>75%) were elevated. The Cornell Medical Index scores were also noted to be higher in survivors than in controls, although specific data on the controls were not presented.

The second quasi-epidemiologic paper presented findings on the suicide rate from 1950 to 1966 of members of the Life Span Study (n=109,000 exposed and non-exposed individuals) and on symptom responses to a questionnaire modeled after the Cornell Medical Index, which was administered to 10,522 members of the Adult Health Study (AHS) from 1962 to 1965 (Yamada et al., 1991). While no significant difference in the suicide rate was found, symptoms of fatigue, anxiety, and depression were significantly higher in the exposed compared to non-exposed group.

The third study was the detailed report by Lifton (1967) based on in-depth, guided interviews with survivors that he personally conducted in 1962. The generalizability of the findings is open to question because of both the unstructured format of the interview and the nature of the sample, that is, survivors from an unspecified population pool willing to talk at length about highly personal matters with a psychoanalytically oriented American psychiatrist. Nevertheless, this work

TABLE 18.1 English-language publications on the psychological effects of Hiroshima/Nagasaki, Three Mile Island, and Chernobyl.

Authors/Year	Sample and timing	Measures
Hiroshima/Nagasaki	August, 1945	
Misao et al., 1961	356 survivors evaluated in 1955–59 345 survivors evaluated in 1955–59	Subjective complaints Cornell Medical Index[a]
Lifton, 1967	33 randomly selected adults and 42 "articulate" survivors interviewed in 1962	Guided interviews
Yamada et al., 1991	10,522 members of Adult Health Study examined in 1962–65	Questionnaire patterned after the Cornell Medical Index;[a] suicide rate
Three Mile Island (TMI)	March 28, 1979[b]	
Bromet et al., 1990 Dew and Bromet, 1993	>1,000 mothers, workers, and controls, studied from December 1979– April 1989 (mothers)	Schedule for Affective Disorders and Schizophrenia;[c] 90-item Symptom Checklist[d]
Baum et al., 1993	38+ random sample from TMI area and controls evaluated from 1980–89	Symptom questionnaires; physiological and neuropsychological tests
Dohrenwend et al., 1981	Random samples of adults, women, students, workers April–October 1979	Psychiatric Epidemiology Research Instrument[e]
Houts et al., 1988	Cross-sectional random samples and controls telephoned from December 1979–April 1985	Langner Symptom Questionnaire[f]

continued

TABLE 18.1 (cont'd.)   English-language publications on the psychological effects of Hiroshima/Nagasaki, Three Mile Island, and Chernobyl.

Authors/Year	Sample and timing	Measures
Chernobyl	April 26, 1986	
IAEA, 1991	Cross-sectional study of five age cohorts	Unstandardized stress items
Viinamaki et al., 1995	Cross-sectional study of 325 exposed villagers and 278 controls	12-item version of General Health Questionnaire[g]
Havenaar et al., 1996	Cross-sectional study of exposed sample from Gomel	12-item version of General Health Questionnaire (phase 1); Munich Diagnostic Checklist for DSM-III-R (phase 2)

[a] Brodman et al., 1952.
[b] Four major studies of TMI are included in this review. Several other smaller studies have also been conducted (e.g., Prince-Embury and Rooney, 1988, 1995; and Hartsough and Savitsky, 1984).
[c] Endicott and Spitzer, 1978.
[d] Derogatis, 1977.
[e] Dohrenwend et al., 1980.
[f] Langner, 1962.
[g] Goldberg, 1972.

provided a rich resource regarding the psychological experiences of survivors and helped shape the development of current psychiatric nosology for post-traumatic stress disorder. In particular, Lifton's descriptions of psychic numbing (originally referred to as "psychological closure")—that is, the loss of feeling from the extreme agony that resulted from witnessing mass death and dying and being unable to respond to calls for help—provided important insights into this poorly understood phenomenon. This phenomenon, as well as other cardinal symptoms Lifton described, ultimately became criterion symptoms for the classification of post-traumatic stress disorder in the American Psychiatric Association's Diagnostic and Statistical Manual starting with the 1980 edition. The concepts of intrusive and avoidant thinking also were operationalized in the Impact of Events Scale (Horowitz, 1990), which has been widely used in disaster research and studies of war-related traumas.

Three Mile Island (TMI)

Although the accident at the TMI nuclear power plant in central Pennsylvania was regarded at the time as the "worst . . . in the history of commercial nuclear power generation" (Kemeny, 1979, p. 35), radiation health experts concluded that the emission was too small to increase the risk of cancer in the population living near the plant (Upton, 1981; Upton et al., 1992). Recent epidemiologic findings on cancer seem to support that conclusion (Hatch et al., 1990). At the same time, these experts asserted that there was a significant health impact on "anxiety and stress." The Task Force on Behavioral Effects of the President's Commission on the Accident at Three Mile Island concluded, however, that the mental health effects were "short-lived," diminishing sharply by the fall of 1979 (Dohrenwend et al., 1979; Kemeny, 1979; Dohrenwend et al., 1981). This conclusion was subsequently contradicted by several longitudinal studies, including a three-year follow-up of 254 residents who had participated in research subsumed within the Task Force report (Goldsteen et al., 1989). Similarly, high levels of distress were also reported by Houts et al. (1988) among residents participating in telephone interviews by the Pennsylvania Department of Health in January 1980, and by Baum et al. (1993) in a six-year, longitudinal study of a panel sample from the five-mile radius of TMI. The comparison groups in the latter study included residents living near another nuclear reactor, a coal-fired plant, and more than five miles from any power plant. The Baum et al. study found higher rates of a very wide range of symptom areas up to six years after the accident, including somatic complaints, anxiety, and depression; PTSD symptoms of hyperarousal, frequent and bothersome intrusive thoughts about the accident, and avoidance of reminders of it; and physiological symptoms including elevations in blood pressure and in levels of urinary norepinephrine, epinephrine, and cortisol.

Our findings also contradicted the early conclusions of the Task Force report (e.g., Bromet, 1989; Bromet et al., 1990; Bromet, 1991; Dew and Bromet, 1993). This study focused on clinical and subclinical depression and anxiety syndromes in mothers of young children and psychiatric outpatients (studied for one year) who lived within 10 miles of TMI, as well as workers employed there when the accident occurred (Bromet, 1989). Comparison groups were sampled from around a similar nuclear power plant and a coal-fired generating plant. The group who proved most vulnerable to the stresses associated with the accident were the mothers of young children. In four waves of face-to-face interviews conducted from 1979 to 1982, we found that (1) their psychological symptoms were persistently elevated; (2) perceptions of danger became a stronger predictor of distress as time elapsed; (3) pre-TMI accident depression and/or anxiety disorder was the most important prognostic indicator; (4) social support did not moderate the mental health effects of the TMI accident; and (5) only half of the mothers with clinical depression sought help for their episodes (e.g., Bromet et al., 1982; Dew et al., 1987a; Dew et al., 1988). We also found no evidence that the mental health of the mothers' preschool (Cornely and Bromet, 1986) or school-aged children (Bromet et al., 1984) differed significantly from that of comparison site children, but a positive family milieu was found to buffer the stress-mental health relationships in the TMI school-aged sample. After collecting two additional mail-back questionnaires from the TMI mothers in 1985 and 1989, we further observed that the restart of the undamaged reactor had an adverse psychological effect (Dew et al., 1987b) and that by 1989, approximately 35% of the mothers had consistently high levels of distress over the ten-year period of observation (Dew and Bromet, 1993). The distress was associated with continued concerns about TMI and with living within five miles of the plant when the accident occurred (Bromet et al., 1990). In fact, the evacuation experience and perception that TMI was dangerous in 1979 "remained among the strongest contributors to long-term distress, even after other important background characteristics were statistically controlled" (p. 54).

Thus, while TMI proved to be among the least dangerous of the nuclear catastrophes since World War II, it is perhaps the best studied and hence best understood.

Chernobyl

Like Hiroshima and Nagasaki, the effects of the explosion at Chernobyl are complicated to evaluate. Thousands of families living near the Chernobyl plant were exposed to radiation. The evacuation of the 30-kilometer zone was chaotic and at times brutal, and the settings to which evacuees were relocated were openly unreceptive and even hostile. Anecdotes about persistent fears of radiation effects (given the label of "radiophobia" by Russian professionals) and chronic stress associated with feelings of stigma (similar to Lifton's descriptions of social stigma) have frequently been reported. While there have been presentations at international conferences (see Ginzburg and Reis, 1991 for a synopsis of one such meeting) and reports in Russian-language research journals, the only published English-language studies to date of the psychological well-being of community residents exposed to radiation from Chernobyl were a small section of the International Chernobyl Project technical report from the International Atomic Energy Agency (International Atomic Energy Agency, 1991), a study of a small exposed village (Viinamaki et al., 1995), and a general population study in Gomel, Belarus (Havenaar et al., 1996). The first study was the largest (perhaps only) epidemiologic study to date that evaluated the health status of five groups who remained in rural contaminated communities (2-year-olds born in 1988; 5-year-olds born in 1986; 10-year-olds born in 1980; 40-year-olds born in 1950; and 60-year-olds born in 1930). The team concluded that these populations were not significantly different from controls on hematological, thyroid, and general health problems (except abdominal problems), but that 3–5% of all children had health problems that required treatment. However, the study excluded evacuees and cleanup workers ("liquidators") whose exposure after the accident was much greater. Although psychological stress was reported to be more prevalent in the settlements contaminated by radiation, the study did not include standard psychological scales to measure distress. The second and third studies found more impairment (based on responses to the 12-item

General Health Questionnaire) in exposed groups than in controls (Viinamaki et al., 1995; Havenaar et al., 1996).

There are three ongoing, western-based studies of populations exposed to Chernobyl which will ultimately yield important new data on the psychological consequences of this event. The first is the neurological and neuropsychological screening of exposed children who were in utero at the time of the accident (World Health Organization, 1995). A preliminary report in Russian appears to suggest that compared to controls, the mothers of these children have higher psychological (stress) symptom scores, and the children show poorer performance on some of the neuropsychological measures. The second ongoing study is a clinical assessment of liquidators (cleanup workers) living near Moscow (Lasko et al., 1993). The third study addresses the mental health effects of the stress of the Chernobyl experience on evacuee children living in Kiev (Bromet et al., 1994). The latter study was funded in September 1995 by the National Institute of Mental Health.

Two other reports about the psychological sequelae of Chernobyl should be noted. One was based on a study of a small sample of exposed adolescent girls, subsequently brought to Israel, who showed high levels of somatization but no detectable underlying medical conditions at the time of evaluation (personal communication, Lerner and Zilber, 1992). The second study was conducted by two Ukrainian investigators in spring 1992 and focused on the self-reported health status of 249 9th to 11th grade students evacuated to Kiev, along with several comparison groups (personal communication, Panina and Golovakha, 1992). Compared to controls drawn from Kiev, evacuated adolescents were more likely to report frequent headaches, chronic illness, poorer overall health, and lower satisfaction with personal health. On the other hand, they did not report a greater numbers of colds, more days confined to bed because of illness, greater fatigue at the end of the school day, or higher anxiety symptoms [assessed with the Spielberger (1973) scale]. Compared to controls living in Slavuta, a town built to house workers from the Khmelnitsk nuclear power station, the evacuees manifested significantly greater anxiety but were otherwise similar in their psychological profiles.

OVERVIEW OF THE TYPES OF MENTAL HEALTH PROBLEMS

The most common symptoms reported across the studies described above were somatization, depression, post-traumatic stress symptoms, anger/hostility, anxiety (particularly about future health), and, similar to somatization, a combination of symptoms that was once labeled neurasthenia—namely, fatigue, irritability, weakness, and headaches. New and unfamiliar terms have been used to describe the psychological and physical symptoms of radiation survivors, including A-bomb neurosis in Japan and vascular dystony (Stiehm, 1992) and radiophobia (Ginzburg and Reis, 1991) in the former USSR. Table 18.2 summarizes the symptoms and conditions reported in the described studies.

TABLE 18.2 Types of symptoms and conditions reported.

Symptoms/ conditions	Description
Symptoms	Somatization
	Depression/hopelessness
	Anxiety/trauma-specific fears, especially fear of future health problems, such as cancer
	"Neurasthenia"—fatigue, irritability, weakness, headaches
	Post-traumatic stress symptoms or disorder (psychic numbing; flashbacks; sleep disorders)
	Anger/hostility
Conditions	"A-bomb disease"—early effects, cancer, and non-specific complaints of fatigue, weight loss during summer, colds, gastrointestinal problems
	"A-bomb neurosis"—extreme anxiety over physical symptoms resulting from exposure, fear of cancer, and "A-bomb disease"
	"Vascular dystony"—-diagnosis given to Chernobyl evacuees reporting weakness, irritability, headache, angina, and fatigue[*]
	"Radiophobia"—diagnosis given to Chernobyl evacuees with "inappropriate[!]" fear of radioactive material[**]

[*]Stiehm, 1992.
[**]Ginzburg and Reis, 1991.

While the specific rates of disorder vary because of methodological differences in study design, measures, samples, and timing of data collection, the most remarkable aspect about the findings is the intractable nature, or persistence, of these symptoms. In our TMI research, as noted above, 35% of mothers of young children continued to show elevated rates of distress over a ten-year period (Dew and Bromet, 1993). Similarly, the temporal persistence of fearful perceptions was highly significant (Dew et al., 1987c). The Misao et al. (1961) study of A-bomb survivors found that more than 50% reported fatigue, headache, anxiety, and somatic complaints 14 years later. When Yamada et al. (1991) assessed AHS members 17 years later, they found elevated rates of anxiety, fear of disease, fatigue, and depression. Finally, the IAEA finding of elevated rates of distress was based on data collected five years after the catastrophe took place (International Atomic Energy Agency, 1991).

Thus a strong case can be made for extended, long-term follow-up research focused on the psychological effects of these catastrophic events. As *Newsweek* magazine (July 27, 1995) stated about the survivors of Hiroshima and Nagasaki, "Nothing will ever remove the physical and psychic pain of the hibakusha [A-bomb survivors]."

RISK FACTORS

A number of personal and environmental risk factors have been found to elevate the rate of psychological disorder from both technological and natural disasters (Table 18.3). These include greater involvement and/or injury, exposure to death and dying—particularly in family members, exposure to cancer, physical and social stigma, community disruption, economic decline, unresolved fear of the disaster's future effects, and personal vulnerability factors, particularly female gender and having a personal history of psychiatric disorder (Baum, 1987; Bromet and Schulberg, 1987; Solomon, 1992; Dew and Bromet, 1993). It can be hypothesized that the less severe the disaster, the stronger will be the contribution of personal vulnerability factors (Davidson and Foa, 1993). Thus, at TMI, our research showed that personal psychiatric history was the strongest predictor of post-TMI psychopathology. Conceivably, disaster-related stressors may override the moderating impact of these personal vulnerability factors in the A-bomb survivors. Although there are few studies of the specificity of these risk factors for different outcomes (Solomon et al., 1993), it would be important to understand which risk factors predict different types of outcomes so that more specific prevention efforts can be designed.

MODIFYING FACTORS

A number of variables have been suggested that can be conceptualized as modifying factors. That is, their presence may either influence the magnitude of or moderate the effects of the stress from these events. One such factor is media coverage. As early as 1961, Misao et al. claimed that A-bomb survivors learned about symptom effects from the mass media. Specifically, they asserted that the media's lack of caution in describing the sequelae was in part responsible for the level of anxiety in the survivors. At a 1990 symposium on psychological effects of radiation accidents, the President of Radiation Management Consultants introduced a session by urging scientists to "determine how, and how much, stress is related to the media's misrepresentation of the accident and the long-term effects of radiation" (Linnemann, 1991, p. 58). While it is easy to attribute the appearance

TABLE 18.3 Risk factors for psychological effects.

Involvement/injury

Exposure to death and dying
Exposure to cancer
Physical and social stigma
Community disruption
Economic decline
Unresolved fear of future effects
Personal vulnerability factors (female gender; personal
history of psychopathology)

or persistence of symptoms to miscommunication on the part of the mass media, it should be recognized that scientists often made little effort to communicate their findings directly to the affected populations. If we assumed this responsibility more assiduously, the impact of inaccurate communications from the mass media might be lessened.

Social psychologists and psychiatric epidemiologists have also enumerated a number of other variables that could attenuate the effect of the stress. These include cognitive control over intrusive thoughts which can be gained from professional treatment (Baum, 1990; Baum et al., 1993), practical support from government agencies (in the form of medical benefits, social security, and housing; Solomon et al., 1987), and social support from family, social network members, health care and religious professionals (Fleming et al., 1982; Baum and Fleming, 1993). The issue of social support is particularly important. After the A-bomb, the hibakusha tended to marry each other and to maintain lower socioeconomic status because they were perceived as undesirable for marriage, child-bearing, and employment. Similarly, evacuees from the 30-kilometer zone around Chernobyl were stigmatized and received with hostility by the communities where they were relocated because they were feared, and looked upon as sickly. To some extent, the persistence of psychological symptoms might be attributed to these secondary adversities occurring in the aftermath of these catastrophes.

CONCLUSIONS

Better data are needed about the prevalence and risk factors for different types of psychiatric disturbance following the radiation catastrophes of the twentieth century. While it appears that certain types of symptoms may be widespread, their true prevalence is unknown. Because these events have been shown to have long-lasting effects, it is still timely to initiate a long-term outcome study of A-bomb survivors as well as individuals affected by Chernobyl and TMI. Data on risk factors, as well as factors that protect against deleterious mental health outcomes, would help us design targeted and useful intervention strategies. As nuclear power plants continue to age and plans for radioactive waste disposal and nuclear weapons disposal are enacted, these populations become increasingly important to evaluate. The long-term threat to health is a reality that must be understood. By comparing the long-term effects of several radiation-related events, we may be able to elucidate the common effects of exposure, or threatened exposure, to ionizing radiation (Collins and de Carvalho, 1993). Thus, it is crucial that research with both practical and theoretical underpinnings be designed to address the natural history of psychological response to this form of extreme stress. These studies need to address both the survivors and their offspring, because the fear extends beyond the survivor and into the next generation.

ACKNOWLEDGMENTS

I wish to thank Julia Bromet Pilkonis for her exhaustive search and thoughtful summary of the literature on the effects of the bombings of Hiroshima and Nagasaki. Her assistance in this area was invaluable.

19

Atomic Bomb Survivors Relief Law

ITSUZO SHIGEMATSU

SUMMARY

The first official measures for the relief of atomic bomb survivors were taken by the Japanese government in 1957, 12 years after the atomic bombings in Hiroshima and Nagasaki. The Atomic Bomb Survivors Medical Care Law, enacted in this year, provided for health examination of survivors and medical care for health disturbances due to atomic bomb radiation. Thereafter, the need to provide not only medical care but also support to improve their living standards was recognized. The Atomic Bomb Survivors Special Measures Law was effected in 1968 to stabilize survivors' lives and improve their welfare. Since 1995 marked the fiftieth anniversary of the atomic bombings, the survivors and the Japanese people strongly requested implementation of comprehensive relief measures in the areas of health care, treatment, and welfare of the aging survivors. Thus, in July 1995, the Japanese government passed the new Atomic Bomb Survivors Relief Law, which integrates the two previous laws into one and improves and strengthens the measures in force.

INTRODUCTION

The first official measures for the relief of atomic bomb survivors were taken by the Japanese government in 1957, 12 years after the atomic bombings in Hiroshima and Nagasaki. In this year, the Atomic Bomb Survivors Medical Care Law was implemented, providing for medical examinations and care for atomic bomb survivors. Thereafter, the need to provide not only medical care but also support for improving living standards was recognized, and the Atomic Bomb Survivors Special Measures Law was passed in 1968 in order to stabilize survivors' lives and improve their welfare.

Measures for atomic bomb survivors in Japan centered around these two laws, but since 1995 marked the fiftieth anniversary of the atomic bombings, the survivors and the Japanese people strongly requested implementation of comprehensive relief measures in the areas of health care, treatment, and welfare of the aging survivors. Thus, in July 1995, the Japanese government effected the new Atomic Bomb Survivors Relief Law, which integrates the two laws into one and improves and strengthens the measures in force.

CIRCUMSTANCES LEADING TO THE
ESTABLISHMENT OF THE NEW LAW

In August 1945, when the atomic bombs were dropped on Hiroshima and Nagasaki, the populations of those two cities were 330,000 and 250,000, respectively, and totalled 580,000. It is estimated that about one-third of these people died instantaneously from the blast or thermal rays or were acute deaths due to atomic radiation; about another one-third were injured; and the remaining one-third were unaffected. However, the number of atomic bomb survivors was not known until 1950 when a national census was first conducted in Japan after the Second World War. In this national census, 284,000 people claimed to have experienced the atomic bombings.

Although these atomic bomb survivors had various anxieties and problems concerning their health and social life, no official relief measures were taken for them because Japan was then occupied by the allied forces and government measures were subjected to many restrictions, with an especially strict control placed on information concerning the atomic bombs.

However, after the peace treaty was concluded in 1952 and the occupation of Japan was lifted, the government began to take positive measures for atomic bomb survivors, and studies on the methods of treating health disturbances due to the atomic bombs were begun with the establishment of the Council to Study Atomic Bomb Diseases in 1953. In March 1954 the crew of the "No. 5 Lucky Dragon," a Japanese fishing boat, were exposed to radioactive fallout from a hydrogen bomb test conducted by the US in the Bikini Atoll. This episode increased the interest of the people in the issue of radiation exposure. The Atomic Bomb Survivors Medical Care Law was established in March 1957 and health examinations as well as medical care for those designated as survivors by the law began in April of the same year.

As shown in Figure 19.1, the enforcement of this law resulted in the increase of recognized atomic bomb survivors every year from about 200,000 in 1957 to about 280,000 in 1965. In parallel with this, requests not only for health care measures but also for relief measures for improving living standards increased. Thus, in November 1965, the government conducted a nationwide survey on the actual status of atomic bomb survivors. This survey revealed a difference between the survivors and the general population not only in their health conditions but

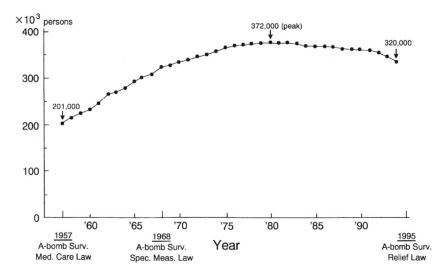

FIGURE 19.1 Enumerated atomic bomb survivors, 1957–1995.

also in their income level and employment status. Therefore, the Atomic Bomb Survivors Special Measures Law was established in 1968.

The number of recognized atomic bomb survivors continued to rise with the improvement of measures for survivors, and peaked at about 372,000 in 1980, but it gradually decreased thereafter and was about 320,000 in 1995. As mentioned before, the two laws for the relief of atomic bomb survivors were integrated into a new law called the Atomic Bomb Survivors Relief Law in 1995.

OUTLINE OF THE NEW LAW

Differences Between the New Law and the Previous Two Laws

Responsibility of the Nation

Because the health disturbances due to atomic bomb radiation are special damages different from other disturbances due to the war, it was stated in the preamble that relief measures for atomic bomb survivors are the responsibility of the nation.

Funeral Allowances

Funeral allowances will be paid to the bereaved families of the atomic bomb victims. (One hundred thousand yen to one member of a bereaved family recognized as an atomic bomb survivor.)

Commemorative Facility

A commemorative facility will be established where people may pay tribute to the memory of the atomic bomb victims and pray for peace.

Income Restriction

The income restrictions provided under the old laws for the payment of the allowances are abolished.

Welfare Activities

Welfare activities such as nursing homes for atomic bomb survivors, dispatch of home helpers, and so forth, are instituted.

Research Support

Support for research studies on the health effects of atomic bomb radiation and medical care is provided.

Major Stipulations Included in the Two Old Laws Which are Also Prescribed in the New Law

Definition of Survivors

Atomic bomb survivors as defined by the law are provided health handbooks. These handbooks are issued by prefectural governors and the mayors of Hiroshima City and Nagasaki City upon application by atomic bomb survivors who conform to any of the following four conditions:

- Directly exposed: a person who at the time of the atomic bomb detonation was directly exposed within Hiroshima City or Nagasaki City or within a specified adjacent area of these cities.

- Early entrant: a person who entered within two weeks after the atomic bomb detonation a zone about two kilometers from the hypocenter of the atomic bomb.

- Relief party: a person who engaged in relief of atomic bomb survivors or other activities physically affected by radiation.

- In utero exposed: a person who at the time of the atomic bomb detonation was in utero of a mother who meets any of the above three conditions.

Budget for Relief Measures

Funds are also provided for research and investigation related to atomic bomb survivors. The total budgetary outlay in the fiscal year of 1995, that is, from April 1995 to March 1996, was 151 billion Japanese yen (US $ 1.5 billion). The Japanese government provides to health handbook holders the following medical care, allowances, and welfare measures:

- The budget for medical care was 32.5 billion yen which was allocated as follows:

    0.1 billion yen for medical treatment of designated diseases.

    29.3 billion yen for medical treatment of diseases in general. For treatment of diseases other than designated diseases, the government bears that portion of the expenses not covered by health insurance.

    3.1 billion yen for health examination of atomic bomb survivors. Health examinations are available four times a year.

- Various allowances as described below were provided to atomic bomb survivors, with the amount totaling 110 billion yen.

    Special Medical Care Allowance (135,400 yen per month for 2,005 persons).

    Special Allowance (50,000 yen per month for 1,581 persons).

    Atomic Bomb Microcephaly Allowance (46,600 yen per month for 25 persons).

    Health Management Allowance (33,000 yen per month for 242,318 persons).

    Health Allowance (14,200 yen per month for 16,184 persons and 28,400 yen per month for 2,458 handicapped).

    Nursing Aid Allowance (up to 103,150 yen per month for 6,036 cases).

    Family Aid Allowances (21,300 yen per month for 30,832 cases).

    Funeral Allowance (149,000 yen for each funeral of an atomic bomb survivors, to be paid for 6,890 cases).

- As welfare measures, a total of 2.7 billion yen is made available as subsidies for the following:

    Atomic bomb hospitals and homes for atomic bomb survivors.

    Home helper and consultation services for atomic bomb survivors.

Atomic bomb memorial ceremonies.

- For research and investigation of atomic bomb survivors, a budget of 2.8 billion yen was provided for the following:

    Research program of the Radiation Effects Research Foundation.

    Research and investigation of atomic bomb diseases.

- A budget of 3.0 billion yen was provided as a subsidy to local governmental institutions for atomic bomb survivors.

COMMENTS

The outline of the Atomic Bomb Survivors Relief Law enacted in 1995 was introduced here. It is true that some of the stipulations in this law are not necessarily consistent with the present scientific knowledge on the health effects of atomic bomb radiation, but it should also be considered that this knowledge was not available when the law for the medical care of atomic bomb survivors was first established in 1957. From the viewpoint of the scientific studies on the health effects of atomic bomb radiation, it is a good news that the financial support for these studies was included in this new law.

PART 5
FUTURE RESEARCH

20

The Ultimate Questions: Future Research at RERF

JOHN D. ZIMBRICK

SUMMARY

As RERF enters its second half-century of research on the human health effects of acute radiation exposure, it stands poised to provide answers to a number of important questions related to the delayed incidence of cancers and other diseases in the atomic bomb survivors and their offspring. One of the most important of these is whether humans irradiated in utero, as infants, or as juveniles are more sensitive to the delayed effects of radiation than humans irradiated as adults. Since over 90% of the atomic bomb survivors who were irradiated at or below nine years of age are still alive in 1995, it will take more years of follow-up and study before sufficient incidence and mortality data are available to answer this question. Some other important questions center on whether acute radiation exposure will lead eventually to delayed health effects other than cancers, such as cardiovascular disease. New genetic and molecular biologic methodologies will allow questions about variations in radiation sensitivity among individual survivors and genetic changes in first-generation offspring to be addressed. Tissue samples from living survivors as well as stored autopsy materials from deceased survivors will be used to develop and test new "biomarkers" of exposure and disease. RERF scientists are uniquely qualified to help conduct similar follow-up studies on other radiation-exposed populations, such as those found in the countries of the former Soviet Union. RERF will be able to accomplish this extraordinary and unique set of tasks only if it continues to enjoy adequate funding and a positive management climate which fosters collaborations with extramural scientists, encourages some non-Japanese staff scientists and directors to spend extended periods at RERF in order to develop and care for long-term research projects, and emphasizes maintenance of high-quality laboratories and state-of-the-art data management systems.

INTRODUCTION

The little I have hitherto learned is almost nothing in comparison with that of which I am ignorant.

—Descartes

Descartes aptly stated what I believe many of us feel about the RERF program. It is at once the longest-running human radiobiology research program in history as well as the source of the greatest amount of data on the health consequences of acute external radiation in humans. Nevertheless, in the context of contemporary knowledge of molecular etiology of diseases such as cancer, it can be argued that the RERF program has yielded only the first glimpses of the enormously complicated processes involved in the induction of disease by ionizing radiation. This chapter will present and discuss some of the major questions that RERF is uniquely positioned to address in order to shed more light on these processes. But first we must finish the discussion of events leading to the present difficult situation in which RERF finds itself, because the resolution of this situation may have a significant influence on RERF's future research program.

THE PRESENT SITUATION, CONTINUED

At this syposium, Frank Putnam provided a vivid and accurate history of ABCC, RERF, and the constant involvement of the National Academy of Sciences (NAS) in the development and management of both institutions. The history is rich with periods of uncertainty and events, some of which brought ABCC and RERF to a crossroads very near the brink of disaster. We now find ourselves at another crossroads, brought to this point by a combination of management issues at the Department of Energy (DoE) and the NAS, and by a crisis in funding due to the sharp weakening of the dollar against the Japanese yen and the limitations in funding provided for RERF by DoE.

   Dr. Putnam summarized the events leading up to the historical NAS business meeting of April 1995, during which a resolution expressing concern over a DoE plan to abruptly terminate calling for continued NAS involvement with RERF was introduced and discussed. The resolution was signed immediately after the meeting by 191 NAS members and transmitted with a supporting letter by NAS President Bruce Alberts to Secretary of Energy Hazel O'Leary. Two months later, Secretary O'Leary responded to President Alberts, stating in her letter that she continued to support the termination of the NAS management of RERF. The letter contained a series of additional statements, one of which could be interpreted as alleging that the NAS conducted secret research in the early days of ABCC to support the nuclear weapons program. The Secretary copied her letter to numerous parties so that it quickly was circulated throughout the scientific community, Capitol Hill, and the press. Since no written evidence was presented that ABCC ever conducted secret research, and since the pioneers of ABCC/RERF, notably Jim Neel who was intimately involved from the very beginning, provided verbal assurance that ABCC conducted its programs openly and published its results regularly, the O'Leary letter prompted a major outcry from numerous quarters.

Meanwhile during this period, RERF held its annual Scientific Council meeting in April and its Board of Directors meeting in June 1995. The Science Councillors, worried about the many negative management and budgetary issues which had been raised over the past many months, recommended that a blue ribbon panel of international experts be established by the US and Japanese governments to provide an independent review of the RERF scientific programs. The Scientific Council also recommended that DoE defer any decision to remove the NAS from its management of RERF until after the blue ribbon panel had submitted its final report to both governments. These recommendations were presented to the RERF Board of Directors at its June 1995 meeting. After considerable discussion, the Board voted to accept them.

Within hours after the RERF Board's decision to accept the Scientific Council's recommendations, NAS President Bruce Alberts received a call from DoE's Assistant Secretary for Environment, Health and Safety, Tara O'Toole, informing him that DoE had decided to accept the recommendations made by the RERF Scientific Council and approved by the RERF Board. NAS would receive a two-year extension of its RERF management activities (October 1, 1995, through September 30, 1997) and DoE would defer its decision on whether to replace the NAS to a later time. DoE would support the formation of the blue ribbon panel, an action that the NAS President had recommended to the Secretary during a meeting in November, 1994.

After the decision by DoE discussed above, the Japanese government also endorsed the RERF Scientific Council recommendations. The two governments have now formed an independent blue ribbon panel composed of nine members with Dr. Roger Clarke as Chair. Dr. Clarke is President of the International Commission on Radiological Protection and Chairman of the National Radiation Protection Board of the UK.

The NAS negotiated the scope of its management activities for the two-year extension period with DoE. While it appears that many of the activities that the NAS has always conducted (e.g., recruiting of non-Japanese staff and directors, dissemination of RERF reports, and arranging for audits of financial matters at RERF) will be continued, there is one important activity which will not be. The US funding provided to RERF under the 50:50 sharing agreement with Japan passed through the NAS as it always was until April 1995; after that, it was sent by DoE through the US Embassy in Tokyo to RERF. This pass-through arrangement was always considered to be a major facet of the "buffer" between the US government and RERF, and it provided the consideration for a subcontract between RERF and the NAS. Without the financial consideration, there is no longer a basis for a formal subcontract. Nevertheless, a "memorandum of understanding" was negotiated in good faith between the two parties that formalized all of the NAS activities related to RERF except the pass-through of funds from the US government.

As it always has in the past, RERF will move away from its present position at a crossroads. During this new journey, RERF's strengths and its overall ability to address and answer the questions outlined below will depend on a number of issues whose outcomes are not clear at present. Most important of these are: (1) the financial crisis, which must be resolved so that RERF will have sufficient staff in critical positions to continue its major scientific programs at a high enough level of quality that the data and analyses it produces are accepted by the international radiation science community; (2) the results and recommendations of the blue ribbon panel review and how these are accepted and implemented by the Japanese and US governments; (3) the stabilizing of the US management and oversight of RERF such that: (a) the independence and objectivity of the RERF data and analyses are ensured; and (b) a positive environment is maintained which encourages outside collaborations, extended terms for a core of non-Japanese scientists who can develop and maintain long-term research projects, and the maintenance of high-quality laboratories and state-of-the-art equipment and data processing facilities.

FUTURE RESEARCH AT RERF

Radiosensitivity of Juveniles Relative to Adults

A crucial question yet unanswered by the RERF studies is whether humans exposed to ionizing radiation in utero, as infants, or as young children or teenagers, are more sensitive to the induction of delayed health effects than are humans exposed as adults. The answer to this question is important because of its possible implications for the setting of radiation exposure standards for the public and for the use of radiation to treat diseases in juveniles.

Excess Mortality Risk from Cancers

The Life Span Study (LSS) at RERF has shown that excess cancer mortality risks from exposure to acute external ionizing radiation depend upon certain modifying factors such as sex, attained age, and age at exposure. Further, it appears that for solid cancers the excess risk persists after radiation exposure and mortality rates increase throughout life, whereas for leukemias the risks persist throughout life for those exposed as adults but decrease for those exposed as children.

Although the LSS data are complete for those survivors who were exposed at or above the age of 50, they are far from complete for cohorts exposed at ages below 50. In fact, about 51% of all of the survivors (who have assigned radiation

TABLE 20.1 Observed and projected surviving fraction of A-bomb survivors as a function of calendar year.

	Calendar year							
Age ATB*	1950	1990	1995	2000	2005	2010	2015	2020
0-9	100	94.0	92.3	89.7	85.8	80.1	71.3	58.3
10-19	100	86.3	82.6	77.1	68.6	55.8	38.6	20.6

* ATB denotes age at time of bombing.

doses) were alive in 1995. Table 20.1 shows cohort sizes for the youngest survivors projected over time until the year 2020.

It can be seen in Table 20.1 that over 90% of the survivors exposed between the ages of 0 to 9 years were still alive in 1995. The cancer mortality in this cohort will increase rapidly in the coming years, and must be studied in order to gain more understanding about the age variations in risks for radiation-induced solid cancers and leukemias in survivors exposed as juveniles versus survivors exposed as adults.

With regard to survivors who were in utero at the time of the bombings, this relatively small cohort thus far exhibits a cancer mortality rate similar to that of the survivors exposed as children. More follow-up will be necessary to gain more quantitative understanding of the cancer mortality rates in this group.

Noncancer Mortality Risks

Another important question which will be addressed in future studies at RERF relates to whether acute radiation exposure induces excess mortality risks from diseases other than cancer.

Recent studies on the data obtained from the LSS have revealed a statistically significant increase in noncancer mortality with increasing radiation dose. Such an increase could be due to confounding factors or bias, rather than to radiation, and further studies are necessary to produce more definitive results. Thus far, only misclassification (classifying a cancer death as a noncancer death) has been ruled out as the cause of the radiation-induced increase in noncancer mortality.

The Adult Health Study (AHS) is likely to contribute key information to help understand the possible induction of noncancer diseases by radiation. This study, one of the longest-running and largest clinical studies in history, initially suggested that acute radiation induces a variety of cardiovascular diseases, such as coronary heart disease. However, when confounding risk factors such as smoking, age, and sex, were factored into the statistical analysis, the radiation effect on cardiovascular diseases was reduced to non-significance, except for atherosclerosis. Hyperparathyroidism has also been demonstrated to be a radiation-induced disease, and its effects may be related to the observed atherosclerosis incidence. Further, chronic liver disease and myoma uteri (benign tumors) have been shown to be induced by radiation, and it is possible that radiation exposure induces earlier onset of female menopause, senile dementia, and late onset of cataracts. As the AHS continues, its careful clinical studies on the young survivors (those shown in the table above) are expected to provide valuable new data to help increase our understanding of the possible relationships between radiation dose and these noncancer diseases, including the involvement of confounding variables such as alcohol consumption, smoking, and infections.

Genetic Studies on First-Generation Offspring of A-Bomb Survivors

Are heritable excess cancer or noncancer health risks found in the first-generation offspring (F_1) of A-bomb survivors? Answers to this question have been sought since the very early days of ABCC. To date, no excess risks of any kind have been found in a very large cohort of over 80,000 F_1. However, their average age is 39 years, so follow-up must be continued for a number of decades in order to determine whether any delayed effects, thus far undetected, will occur. Thus far, follow-up has been done like that in the LSS, that is, through death certificate data from family registries. No clinical examinations like those of the AHS have been done on the F_1 since their first year after birth. An important goal will be to conduct a well-designed clinical study of a subset of F_1 to search for possible inherited health effects. This is because inherited diseases, which might be caused by radiation-induced genetic defects in survivor-parents, are frequently not manifest until after the first 10 years of life, and sometimes not until well into adulthood.

Another major question to be addressed in the F_1 studies is whether germ-line mutations can be detected by state-of-the-art molecular biologic techniques (e.g., deletions or insertions of DNA fragments) in the cells of irradiated parent-child families (called trios). RERF has a cell bank which already contains immortalized B lymphocyte cells from 800 families (composed of 1,600 parents and 1,200 children), and their goal is to obtain cells from 1,000 families, half of them from families with radiation-exposed parents.

Low-Dose Radiation Effects and Mechanisms of Radiation Injury

The RERF survivor-volunteers and bank of stored biologic materials provide an invaluable and unique resource which can be used to address some very important mechanistic questions, such as:

- What is the lowest assigned radiation dose for which subcellular changes, such as chromosome aberrations and DNA changes, can be reliably detected in the biologic samples from the survivors? What is the shape of the dose-response relationship for these changes?

• Can state-of-the-art molecular biologic techniques be used to identify and quantify systematic, radiation-induced changes in genes identified as being involved in the process of carcinogenesis?

• Will biomarkers developed in model cellular and animal systems prove useful in humans as determined by testing them on the AHS cohort and/or the stored biologic samples at RERF?

• Will it be possible to determine the susceptibility of an RERF survivor to radiation-induced injury by, for example, the use of biomarkers to measure his/her capacity to repair radiation damage in DNA? Does the measured susceptibility relate to the observed health status of the survivor?

• What are the shapes of the dose-response curves for various radiation injury endpoints such as site-specific solid cancers, and how can they contribute to the development of new predictive models as well as to the increased under-standing of mechanisms underlying the observations of radiation-induced injury?

The AHS cohort will be extremely valuable for use in obtaining answers to the above questions. So will the bank of stored biologic specimens, which contains 110,000 serum samples, 11,000 plasma samples, and 13,000 lymphocyte specimens from the AHS, along with 7,500 autopsy specimens, 13,000 surgical specimens from the LSS, and transformed lymphocytes from 800 family trios in the F_1 studies.

Collaborative Studies on Other Exposed Populations

The collective expertise gained by ABCC/RERF scientists during the nearly 50 years of study on the A-bomb survivors at Hiroshima and Nagasaki has given them unique qualifications to advise and help conduct similar studies on other exposed populations, such as those found in the countries of the former Soviet Union. Thus it seems inevitable, and in fact desirable, that RERF will evolve into a unique international center for human radiobiology research. RERF directors and staff were among the very first foreign scientists invited to Russia to survey and advise Russian scientists after the Chernobyl accident in 1986. Since then, RERF scientists have worked with other internationally based groups, such as the World Health Organization and the International Atomic Energy Agency, to provide assistance to Russian scientists in studies on exposed workers and members of the general population living in the southern Ural mountains of the Russian Federation.

Epidemiologic studies on the exposed populations in the Russian Federation will be especially valuable in furthering our understanding of radiation effects on humans when the radiation doses are delivered over an extended period of time, and from sources of radiation both external to the body, as well as from internally deposited radioactive materials. Comparisons of the dose-response data obtained from the Russian victims with the analogous data obtained from the RERF studies on the acutely irradiated A-bomb survivors should provide crucial information on possible variations in health effect risks due to changes in rate of delivery of radiation dose.

ACKNOWLEDGMENTS

The views expressed herein are those of the author and are not necessarily those of the NRC or any of its constituent units. The author is grateful for helpful discussions with Jim Neel, Frank Putnam, Jack Schull, Gil Beebe, Seymour Jablon, Stu Finch, Charlie Edington, Mort Mendelsohn, Seymour Abrahamson, Don Harkness, Cathie Berkley, and the staff at RERF.

Bibliography

Abramowitz, M., Stegun, I.A. Handbook of Mathematical Functions. Washington, (DC), National Bureau of Standards (1964).

Akiba, S., Lubin, J., Ezaki, H., Ron, E., Ishimaru, M., Asano, Y., Simizu, Y., Kato, H. Thyroid cancer incidence among atomic bomb survivors in Hioroshima and Nagasaki, 1958–1979. Radiation Effects Research Foundation Technical Report 5-91 [RERF TR 5-91]. Hiroshima, RERF (1991).

Akiyama, M., Umeki, S., Kusunoki, Y.; Kyoizumi, S., Nakamura, N.; Mori, T., Ishikawa, Y., Yamakido, M., Ohama, K., Kodama, T., Endo, K., Cologne, J.B. Somatic-cell mutations as a possible predictor of cancer risk. Health Physics. 68:643–649 (1995).

Albertini, R.J., Allen, E.F., Quinn, A.S., Albertini, M.R. Human somatic cell mutation: in vivo variant lymphocyte frequencies determined by 6-thioguanine-resistant lymphocytes, pp. 235–263. In: Population and Biological Aspects of Human Mutation, Hook, E.B., Potter, J.H. (eds.). New York, Academic Press (1981).

Albertini, R.J., Castle, K.L., Borcherding, W.R. T-cell cloning to detect the mutant 6-thioguanine-resistant lymphocytes present in human peripheral blood. Proceedings of the National Academy of Sciences. 79:6617–6621 (1982).

Albertini, R.J., O'Neill, J.P., Nicklas, J.A., Heintz, N.H., Kelleher, P.C. Alterations of the *hprt* gene in human in vivo derived 6-thioguanine-resistant T lymphocytes. Nature. 316:369–371 (1985).

Albertini, R.J., Sullivan, L.S., Berman, J.K., Greene, C.J., Stewart, J.A., Silveira, J.M., O'Neill, J.P. Mutagenicity monitoring in humans by autoradiographic assay for mutant T lymphocytes. Mutation Research. 204:481–492 (1988).

Albertini, R.J., Nicklas, J.A., O'Neill, J.P., Robison, S.H. In vivo somatic mutations in humans: Measurement and analysis. Annual Review of Genetics. 24:305–326 (1990).

Allen, L.R. A General Report on the ABCC-JNIH Joint Research Program, 1947–1975. Hiroshima, ABCC (1978).

Ammenheuser, M.M., Ward, J.B., Jr., Whorton, E.B., Jr., Killian, J.M., Legator, M.S. Elevated frequencies of 6-thioguanine-resistant lymphocytes in multiple sclerosis patients treated with cyclophosphamide: A prospective study. Mutation Research. 204:509–520 (1988).

Ammenheuser, M.M., Au, W.W., Whorton, E.B., Jr., Belli, J.A., Ward, J.B., Jr. Comparison of *hprt* variant frequencies and chromosome aberration frequencies in lymphocytes from radiotherapy and chemotherapy patients: A prospective study. Environmental and Molecular Mutagenesis. 18:126–135 (1991).

Ammenheuser, M.M., Berenson, A.B., Stiglich, N.J., Whorton, E.B., Jr., Ward, J.B., Jr. Elevated frequencies of *hprt* mutant lymphocytes in cigarette-smoking mothers and their newborns. Mutation Research. 304:285–294 (1994).

Angel, P., Karin, M. The role of *Jun, Fos,* and the *AP-1* complex in cell proliferation and transformation. Biochimica et Biophysica Acta. 1072:129–157 (1991).

Anstee, D.J., Parsons, S.F., Mallinson, G., Spring, F.A., Judson, P.A., Smythe, J. Characterization of monoclonal antibodies against erythrocyte sialoglycoproteins by serological analysis, immunoblotting, and flow cytometry. Review Francaise de Transfusion. 31:317–332 (1988).

Antoku, S., Russell, W.J. Dose to the active bone marrow, gonads, and skin from roentgenography and fluoroscopy. Radiology. 101:669–678 (1971). [ABCC TR 20-70].

Antoku, S., Yoshinaga, H., Russell, W.J., Ihno, Y. Dosimetry, diagnostic medical x-rays. Exposure of ABCC subjects in community hospitals and clinics. Atomic Bomb Casualty Commission Technical Report 6-65 [ABCC TR 6-65]. Hiroshima, ABCC (1965).

Antoku, S., Russell, W.J., Milton, R.C., Yoshinaga, H., Takeshita, K., Sawada, S. Dose to patients from roentgenography. Health Physics. 23:291–299 (1972). [ABCC TRs 4-67, 5-68, 21-70].

Antoku, S., Sawada, S., Russell, W.J., Wakabayashi, T., Mizuno, M., Suga, Y. Radiation quality and anode effect in diagnostic roentgenography in Hiroshima and Nagasaki. Health Physics. 24:611–617 (1973). [ABCC TRs 26-67,23-67].

Antoku, S., Sawada, S., Russell, W.J. Dose from Hiroshima mass radiologic gastric surveys. Health Physics. 38:735–742 (1980). [RERF TR 12-78].

Antoku, S., Hoshi, M., Sawada, S., Russell, W.J. Hospital and Clinic Survey Estimates of Medical X-ray Exposures in Hiroshima and Nagasaki. Part 2. Technical exposure factors. Radiation Effects Research Foundation Technical Report 6-86 [RERF TR 6-86]. Hiroshima, RERF (1986).

Aplan, P.D., Lombardi, D.P., Ginsberg, A.M., Cossman, J., Bertness, V.L., Kirsch, I.R. Disruption of the human SCL locus by "illegitimate" V-(D)-J recombinase activity. Science. 250:1426–1429 (1990).

Aplan, P.D., Lombardi, D.P., Kirsch, I.R. Structural characterization of *sil,* a gene frequently disrupted in T-cell acute lymphoblastic leukemia. Molecular and Cellular Biology. 11:5462–5469 (1991).

Aplan, P.D., Lombardi, D.P., Reaman, G.H., Sather, H.N., Hammond, G.D., Kirsch, I.R. Involvement of the putative hematopoietic transcription factor SCL in T-cell acute lymphoblastic leukemia. Blood. 79:1327–1333 (1992).

Arakawa, E.T. Radiation dosimetry in Hiroshima and Nagasaki atomic bomb survivors. New England Journal of Medicine. 263:488–493 (1960).

Argyris, T.S. Regeneration and the mechanism of epidermal tumor promotion. CRC Critical Reviews in Toxicology. 14:211–258 (1985).

Armitage, P., Doll, R. The age distribution of cancer and a multi-stage theory of carcinogenesis. British Journal of Cancer. 8:1–12 (1954).

Armitage, P., Doll, R. A two-stage theory of carcinogenesis in relation to the age distribution of human cancer. British Journal of Cancer. 9:161–169 (1957).

Asakawa, J., Kuick, R., Neel, J.V., Kodaira, M., Satoh, C., Hanash, S.M. Genetic variation detected by quantitative analysis of end-labeled genomic DNA fragments. Proceedings of the National Academy of Sciences. 91:9052–9056 (1994).

Asakawa, J., Kuick, R., Neel, J.V., Kodaira, M., Satoh, C., Hanash, S.M. Quantitative and qualitative genetic variation in two-dimensional DNA gels of human lymphocytoid cell lines. Electrophoresis. 16:241–252 (1995).

Ashley, C.T., Warren, S.T. Trinucleotide repeat expansion and human disease. Annual Review of Genetics. 29:703-728 (1995).

Auxier, J.A. Ichiban: Radiation Dosimetry for the Survivors of the Bombings of Hiroshima and Nagasaki. TID-27080, U.S. Energy Research and Development Administration, Technical Information Center. Oak Ridge (TN), DoE (1977).

Auxier, J.A., Cheka, J.S., Haywood, F.F., Jones, T.D., Thorngate, J.H. Free-field radiation dose distributions for the Hiroshima and Nagasaki bombs. Health Physics. 12:425–429 (1966).

Avila, M., Otero, G., Cansado, J., Dritschilo, A., Velasco, J.A., Notario, V. Activation of phospholipase D participation in signal transduction pathways responsive to gamma radiation. Cancer Research. 53:4474–4476 (1993).

Awa, A.A. Cytogenetic and oncogenic effects of the ionizing radiations of the atomic bombs, pp. 637–674. In: Chromosomes and Cancer. German, J.L. (ed.). New York, John Wiley and Sons (1974).

Awa, A.A. Chromosome aberration in somatic cells. Journal of Radiation Research, S. 122–131 (1975).

Awa, A.A. Persistent chromosome aberrations in the somatic cells of A-bomb survivors, Hiroshima and Nagasaki. Journal of Radiation Research (Tokyo). 32S:265–274 (1991).

Awa, A.A., Sawada, S. Study on the effects of exposure to low dose radiation using chromosome aberration as an end point–contamination effects of medical radiation, pp. 23–25. In: FY-1989 Report of A-bomb Disease Research Teams, by Joint Field Research Team (the Shigematsu Team). Tokyo, Nippon Koushueisei Kyokai (Japan Public Health Association) (1990).

Baeuerle, P.A., Henkel, T. Function and activation of NF-kB in the immune system. Annual Review of Immunology. 12:141–179 (1994).

Barhoumi, R., Bowen, J.A., Stein, L.S., Echolls, J., Burghardt, R.C. Concurrent analysis of intracellular glutathione content and gap junctional communication. Cytometry. 14:747–756 (1993).

Barhoumi, R., Bailey, R.H., Hutchinson, R.W., Bowen, J.A., Burghardt, R.C. Enhancement of melphalan toxicity by octanol in ovarian adenocarcinoma cell lines: Effects of altered cell-cell communication, glutathione levels, and plasma membrane fluidity. Fundamental and Applied Toxicology. 25:70–79 (1995).

Baudier, J., Delhin, C., Grunwald, D., Khochbin, S., Lawrence, J.J. Characterization of the tumor suppressor protein *p53* as a protein kinase C substrate and a S100b-binding protein. Fundamental and Applied Toxicology. 89:11627–11631 (1992).

Baum, A. Toxins, technology, and natural disasters, pp. 9–53. In: Cataclysms, Crises, and Catastrophes: Psychology In Action. Vanden Bos, G.R., Bryant, B.K. (eds.). Washington DC, American Psychological Association (1987).

Baum, A. Stress, intrusive imagery, and chronic distress. Health Psychology. 9:653–675 (1990).

Baum, A., Fleming, I. Implications of psychological research on stress and technological accidents. American Psychologist. 48:665–672 (1993).

Baum, A., Cohen, L., Hall, M. Control and intrusive memories as possible determinants of chronic stress. Psychosomatic Medicine. 55:274–286 (1993).

Beebe, G.W. Reflections on the work of the Atomic Bomb Casualty Commission in Japan. Epidemiologic Reviews 1:184–210 (1979).

Beebe, G.W., Ishida, M., Jablon, S. Life Span Study. Report Number 1: Description of Study. Mortality in the Medical Subsample, October 1950–June 1958. Atomic Bomb Casualty Commission Technical Report 5-61 [ABCC TR 5-61]. Hiroshima, ABCC (1961).

Beebe, G.W., Usagawa, M. The Major ABCC Samples. Atomic Bomb Casualty Commission Technical Report 12-68 [ABCC TR 12-68]. Hiroshima, ABCC (1968).

Beebe, G.W., Kato, H., Land, C.E. Studies of the mortality of A-bomb survivors. 6. Mortality and radiation dose, 1950–1974. Radiation Research. 75:138–201 (1978). [RERF TR 1-77].

Belsky, J.L., Ishimaru, T., Ichimaru, M., Steer, A., Uchino, H. Operations Manual for the Detection of Leukemia and Related Disorders, Hiroshima and Nagasaki. Atomic Bomb Casualty Commission, Manual 1-72. Hiroshima, ABCC (1972).

Bennett, M.V.L., Burrio, L.C., Borgiello, T.A., Spray, D.C., Hertzberg, E., Saez, J.C. Gap junctions: New tools, new answers, new questions. Neuron. 6:305–320 (1991).

Beral, V., Fraser, P., Carpenter, L., Booth, M., Brown, A., Rose, G. Mortality of employees of the Atomic Weapons Establishment, 1951–1982. British Medical Journal. 297:757–770 (1988).

Bergelson, S., Pinkus, R., Daniel, V. Intracellular glutathione levels regulate *Fos/Jun* induction and activation of glutathione S-transferase gene expression. Cancer Research. 54:36–40 (1994).

Bernard, O., Lecointe, N., Jonveaux, P., Souyri, M., Mauchauffe, M., Berger, R., Larsen, C.J., Mathieu-Mahul, D. Two site-specific deletions and t(1:14) translocation restricted to human T-cell acute leukemias disrupt the 5′ part of the *tal-1* gene. Oncogene. 6:1477–1488 (1991).

Bishop, J.M. Cancer: The rise of the genetic paradigm. Genes and Development. 9:1309–1315 (1995).

Blot, W.J., Akiba, S., Kato, H. Ionizing radiation and lung cancer: A review including preliminary results from a case-control study among A-bomb survivors, pp. 235–248. In: Atomic Bomb Survivor Data: Utilization and Analysis. Prentice, R.L., Thompson, D.J. (eds.). Philadelphia, Society for Industrial and Applied Mathematics (1984).

Boehm, T., Rabbitts, T.H. The human T-cell receptor genes are targets for chromosomal abnormalities in T-cell tumors. FASEB Journal. 3:2344–2359 (1989).

Boice, J.D., Rosenstein, M., Trout, E.D. Estimation of breast doses and breast cancer risk associated with reported fluoroscopic chest examinations of women with tuberculosis. Radiation Research. 73: 373–390 (1978).

Boice, J.D., Day, N.E., Andersen, A., Brinton, L.A., Brown, R., Choi, N.W., Clarke, E.A., Coleman, M.P., Curtis, R.E., Flannery, J.T., Hakama, M., Hakulinen, T., Howe, G.R., Jensen, O.M., Kleinerman, R.A., Magnin, D., Magnus, K., Makela, K., Malker, B., Miller, A.B., Nelson, N., Patterson, C.C., Pettersson, F., Pompe-Kirn, V., Primic-Žakelj, M., Prior, P., Ravnihar, B., Skeet, R.G., Skjerven, J.E., Smith, P.G., Sok, M., Spengler, R.F., Storm, H.H., Stovall, M., Tomkins, G.W.O., Wall, C. Second cancers following radiation treatment for cervical cancer. An international collaboration among cancer registries. Journal of the National Cancer Institute. 74:955–975 (1985).

Boice, J.D., Preston, D.L., Davis, F.G., Monson, R.R. Frequent chest X-ray fluoroscopy and breast cancer incidence among tuberculosis patients in Massachusetts. Radiation Research. 125:214–222 (1991).

Bond, V.P., Cronkite, E.P., Lippencott, S.W. Shellebarger, C.J. Studies on radiation-induced mammary gland neoplasia in the rat. III. Relation of the neoplastic response to dose of total-body radiation. Radiation Research. 12:276–285 (1960).

Bond, V.P., Thiessen, J.W. (eds.). Reevaluations of Dosimetric Factors: Hiroshima and Nagasaki. DE81026279, US Department of Energy, Technical Information Center. Oak Ridge (TN), DoE (1982).

Bond, V.P., Thiessen, J.W., (eds.). Reevaluations of Dosimetric Factors, Hiroshima and Nagasaki, Conf-810928, US Department of Energy. Brookhaven, DoE (1982).

Bond, V.P., Varma, M.N., Sondhaus, C.A., Feinendegen, L.E. An alternative to absorbed dose, quality, and RBE at low exposures. Radiation Research. 104(Suppl.8): S553–S557 (1985).

Bond, V.P., Benary, V., Sondhaus, C.A. A different perception of the linear, nonthreshold hypothesis for low-dose irradiation. Proceedings of the National Academy of Sciences. 88:8666–8670 (1991).

Bond, V.P., Benary, V., Sondhaus, C.A., Feinendegen, L.E. The meaning of linear dose-response relations, made evident by use of absorbed dose to the cell. Health Physics. 68:786–792 (1995).

Bond, V.P., Wielopolski, L., Shani, G. Current misinterpretations of the linear, nothreshold hypothesis. Health Physics. 70: 877–882 (1996).

Boothman, D.A., Lee, S. Regulation of gene expression in mammalial cells following ionizing radiation. Yokohama Medical Bulletin. 42:137–149 (1991).

Branda, R.F., O'Neill, J.P., Sullivan, L.M., Albertini, R.J. Factors influencing mutation at the *hprt* locus in T lymphocytes: Women treated for breast cancer. Cancer Research. 51:6603–6607 (1991).

Brash, D.E., Rudolph, J.A., Simon, J.A., Lin, A., McKenna, G.J., Baden, H.P., Halperin, A.J., Ponten, J. A role for sunlight in skin cancer: UV-induced *p53* mutations in squamous cell carcinoma. Fundamental and Applied Toxicology. 88:10124–10128 (1991).

Breit, T.M., Mol, E.J., Wolvers-Tettero, I.L.M., Ludwig, W.-D., van Wering, E.R., Van Dongen, J.J.M. Site-specified deletions involving the *tal-1* and *sil* genes are restricted to cells of the T-cell α receptor lineage: T-cell receptor δ gene deletion mechanism affect multiple genes. Journal of Experimental Medicine. 177:965–977 (1993).

Bridges, B., Cole, J., Arlett, C.F., Green, M.H.L., Waugh, A.P.W., Beare, D., Henshaw, D.L., Last, R.D. Possible association between mutant frequency in peripheral lymphocytes and domestic radon concentrations. Lancet. 337:1187–1189 (1991).

Brill, A.B., Tomonaga, M., Heyssel, R.M. Leukemia in humans following exposure to ionizing radiation: Summary of findings in Hiroshima-Nagasaki and comparison with other human experience. Annals of Internal Medicine. 56:590–609 (1962). [RERF TR 15-59].

Brissett, J.L., Kumar, N.M., Gilula, N.B., Hall, J.E., Dotto, G.P. Switch in gap junctions' protein expression is associated with selective changes in junctional permeability during keratinocyte differentiation. Fundamental and Applied Toxicology. 91:6453–6457 (1994).

Brodman, K., Erdmann, A.J., Lorge, I., Gershenson, C.P., Wolff, H.G. The Cornell Medical Index-Health Questionnaire. III. The evaluation of emotional disturbances. Journal of Clinical Psychology. 8:119–124 (1952).

Bromet, E.J. The nature and effects of technological failures, pp. 120–139. In: Psychosocial Aspects of Disaster. Gist, R., Lubin, B. (eds.). New York, Wiley (1989).

Bromet, E.J. Psychological effects of the radiation accident at TMI, pp. 61–70. In: The Medical Basis for Radiation Preparedness III: The Psychological Perspective. Ricks, R., Berger, M.E., O'Hara, F.M. (eds.). New York, Elsevier (1991).

Bromet, E.J., Cornely, P.J. Correlates of depression in mothers of young children. Journal of the American Acadamy of Child Psychiatry. 23:335–342 (1984).

Bromet, E.J., Schulberg, H.C. Epidemiologic findings from disaster research, pp. 676–689. In: American Psychiatric Association Annual Review, volume 6. Hales, R., Frances, A. (eds.). Washington, DC, American Psychiatric Press (1987).

Bromet, E., Parkinson, D. Psychiatric disorders, Chap. 57, pp. 947–961. In: Maxcy–Rosenau Public Health and Preventive Medicine, 13th edition. Last, J., Wallace, R. (eds.). East Norwalk (CT), Appleton and Lange (1992).

Bromet, E.J., Dew, M.A. A review of psychiatric epidemiologic research on disasters. Epidemiologic Reviews. 17:113–119 (1995).

Bromet, E.J., Parkinson, D., Schulberg, H., Dunn, L., Gondek, P. Mental health of residents near the Three Mile Island reactor: A comparative study of selected groups. Journal of Preventive Psychiatry. 1:225–276 (1982).

Bromet, E., Hough, L., Connell, M. Mental health of children near the Three Mile Island reactor. Journal of Preventive Psychiatry. 2:275–301 (1984).

Bromet, E.J., Parkinson, D.K., Dunn, L.O. Long-term mental health consequences of the accident at Three Mile Island. International Journal of Mental Health. 19:48–60 (1990).

Bromet, E.J., Carlson, G., Goldgaber, D.G. Mental health of children in the aftermath of the Chernobyl catastrophe. Unpublished grant application, 1994.

Bronner, C.E., Baker, S.M., Morrison, P.T., Warren, G., Smith, L.G., Lescoe, M.K., Kane, M., Earbino, C., Lipford, J., Lindblom, A., Tannergard, P., Bollag, R.J., Godwin, A.R., Ward, D.C., Nordenskjold, M., Fishel, R., Kolodner, R., Lasker R.M. Mutation in the DNA mismatch repair gene homologue hMLH1 is associated with hereditary nonpolyposis colon cancer. Nature. 368:258–261 (1994).

Brown, L., Cheng, J.-T., Chen, Q., Siciliano, M.J., Crist, W., Buchanan, G., Baer, R. Site-specific recombination of the *tal-1* gene is a common occurrence in human T cell leukemia. EMBO Journal. 9:3343–3351 (1990).

Buckton, K.E., Hamilton, G.E., Paton, L., Langlands, A. Chromosome aberrations in irradiated ankylosing spondylitis patients, pp. 142–150. In: Mutagen-Induced Chromosome Damage in Man. Evans, H.J., Lloyd, D.C. (eds.). Edinburgh, Edingburgh University Press (1978).

Burch, W., Oberhammer, F., Schulte-Hermann, R.S. Cell death by apoptosis and its protective role against disease. Trends in Pharmacological Sciences. 13:245–251 (1992).

Burkhardt-Schultz, K., Thomas, C.B., Thompson, C.L., Strout, C.L., Brinson, E., Jones, I.M. Characterization of in vivo somatic mutations at the hypoxanthine phosphoribosyltransferase gene of a human control population. Environmental Health Perspectives. 101:68–74 (1993).

Buttke, T.M., Sandstrom, P.A. Redox regulation of programmed cell death in lymphocytes. Free Radical Research. 22:389–397 (1995).

Cameron, H.M., McGoogan, E. A prospective study of 1,152 hospital autopsies: Inaccuracies in death certification. Journal of Pathology. 133:273–300 (1981).

Cardis, E., Esteve, J. Uncertainties in recorded doses in the nuclear industry: Identification, quantification, and implications for epidemiologic studies. Journal of Radiation Protection Dosimetry. 36:279–285 (1991).

Cardis, E., Esteve, J. International Collaborative Study of Cancer Risk Among Nuclear Industry Workers. I. Report of the Feasibility Study. Internal Report No. 92/001, International Agency for Research on Cancer (IARC), World Health Organization. Lyon, IARC (1992).

Cardis, E., Gilbert, E.S., Carpenter, L., Howe, G., Kato, I., Fix, J., Salmon, L., Cowper, G., Armstrong, B.K., Beral, V., Douglas, A., Fry, S.A, Kaldor, J., Lavé, C., Smith, P.G., Voelz, G., Wiggs, L. Combined Analyses of Cancer Mortality Among Nuclear Industry Workers in Canada, the United Kingdom, and the United States of America. Technical Report No. 25, International Agency for Research on Cancer, World Health Organization. Lyon, IARC (1995a).

Cardis, E., Gilbert, E., Carpenter, L., Howe, G., Kato, I., Armstrong, B., Beral, B., Cowper, G., Douglas, A., Fix, J., Fry, S., Kaldor, J., Lavé, C., Salmon, L., Smith, P.,

Voelz, G., Wiggs. L. Effects of low doses and low dose rates of external ionizing radiation: Cancer mortality among nuclear industry workers in three countries. Radiation Research. 142:117–132 (1995b).

Cariello, N.F. Software for the analysis of mutations at the human *hprt* gene. Mutation Research. 312:173–185 (1994).

Cariello, N.F., Skopek, T.R. Analysis of mutation occurring at the human *hprt* locus. Journal of Molecular Biology. 231:41–57 (1993).

Cariello, N.F., Craft, T.R., Vrieling, H., van Zeeland, A.A., Adams, T., Skopek, T.R. Human *hprt* mutant database: Software for data entry and retrieval. Environmental and Molecular Mutagenesis. 20:81–83 (1992).

Carpenter, L., Higgins, C., Douglas, A., Fraser, P., Beral, V., Smith, P. Combined analysis of mortality in three United Kingdom nuclear industry workforces, 1946–1988. Radiation Research. 138:224–238 (1994).

Cerutti, P., Ghosh, R., Oya, Y., Amstad, P. The role of the cellular antioxidant defense in oxidant carcinogenesis. Environmental Health Perspectives. 102:123–129 (1994).

Chakraborty, R., Neel, J.V. Description and validation of a method for simultaneous estimation of effective population size and mutation rate from human population data. Proceedings of the National Academy of Sciences. 86:9407–9411 (1989).

Chang, C.C., Trosko, J.E., El-Fouly, M.H., Gibson-D'Ambrosio, R.E., D'Ambrosio, S.M. Contact insensitivity of a subpopulation of normal human fetal kidney epithelial cells and of human carcinoma cell lines. Cancer Research. 47:1634–1645 (1987).

Checkoway, H., Pearce, N., Crawford-Brown, D.J., Cragle, D.L. Radiation doses and cause-specific mortality among workers at a nuclear materials fabrication plant. American Journal of Epidemiology. 127:255–266 (1988).

Cheka, J.S., Sanders, F.W., Jones, T.D., Shinpaugh, W.H. Distribution of Weapons Radiation in Japanese Residential Structures. CEX-62.11, U.S. Atomic Energy Commission (AEC), Washington, DC, AEC (1965).

Chen, Q., Cheng, J.-T., Tsai, L.-H., Schneider, N., Buchanan, G., Carroll, A., Crist, W., Ozanne, B., Siciliano, M.J., Baer, R. The *tal* gene undergoes chromosome translocation in T-cell leukemia and potentially encodes a helix-loop-helix protein. EMBO Journal. 9:415–424 (1990).

Chen, R.-H., Maher, V.M., McCormick, J.J. Effect of excision repair by diploid human fibroblasts on the kinds and locations of mutations induced by (7,8)α-dihydroxy-9α,10α-epoxy-7,8,9,10-tetrahydrobenzo[a]pyrene in the coding region of the *hprt* gene. Proceedings of the National Academy of Sciences. 87:8680–8684 (1990).

Clark, L.S., Nicklas, J.A. TCR PCR from crude preparations for restriction digest or sequencing. Environmental and Molecular Mutagenesis. 27:34–38 (1996).

Clarke, A.R., Gledhill, S., Hooper, M.L., Bird, C.C., Wyllie, A.H. *p53* dependence of early apoptotic and proliferative response within the mouse intestinal epithelial following gamma irradiation. Oncogene. 9:1767–1773 (1994).

Clifton, K.H., Crowley, J. Effects of radiation type and role of glucocorticoids, gonadectomy and thyroidectomy in mammary tumor induction in MtT-grafted rats. Cancer Research. 38:1507–1513 (1978).

Clifton, K.H., Sridharan, B.N., Douple, E.B. Mammary carcinogenesis-enhancing effect in irradiated rats with pituitary tumor MtT-F4. Journal of the National Cancer Institute. 55:485–487 (1975).

Coffey, D. Chaos theory for toxicologists. Center for Alternative Animal Testing. 12:7–8 (1995).

Cole, J., Skopek, T.R. ICPEMC Committee on spontaneous mutation: Working paper 3: Somatic mutant frequency, mutation rates, and mutational spectra in the human population in vivo. Mutation Research. 304:33–106 (1994).

Collins, D.L., de Carvalho, A.B. Chronic stress from the Goiania ^{137}Cs radiation accident. Behavioral Medicine. 18:149–157 (1993).

Collins, C., Kuo, W.L., Segraves, R., Fuscoe, J., Pinkel, D., Gray, J.W. Construction and characterization of plasmid libraries enriched in sequences from single human chromosomes. Genomics. 11:997–1006 (1991).

Confort, A., Sinibaldi, P., Lazzaro, D. Ultrastructural and biological observations on neoplastic growth control obtained by contact inhibition. International Journal of Tissue Reactions. 8:213–217 (1986).

Cornely, P., Bromet, E.J. Prevalence of behavior problems in three-year-old children living near Three Mile Island: A comparative analysis. Journal of Child Psychology and Psychiatry and Allied Disciplines. 27:489–498 (1986).

Cox, D., Hinckley, D. Theoretical Statistics. London, Chapman and Hall (1974).

Cox, L.S., Lane, D.B. Tumour suppressors, kinases, and clamps: How p53 regulates the cell cycle in response to DNA damage. Bioassay. 17:501–508 (1995).

Cox, R. Molecular mechanisms of radiation oncogenesis. International Journal of Radiation Biology. 65:57–64 (1994).

Cragle, D.L., McLain, R.W., Qualters, J.R., Hickey, J.L.S., Wilkinson, G.S., Tankersley, W.G., Lushbaugh. C.C. Mortality among workers at a nuclear fuels production facility. American Journal of Industrial Medicine. 14:379–401 (1988).

Crow, J.F., Kaplan, H.S., Marks, P.A., Miller, R.W., Storer, J.B., Upton, A.C., Jablon, S. Report of the Committee for Scientific Review of ABCC, February, 1975. Atomic Bomb Casualty Commission Technical Report 21-75 [ABCC TR 21-75]. Hiroshima, ABCC (1975).

Davidson, J.R.T., Foa, E.B. (eds.). Post-Traumatic Stress Disorder: DSM IV and beyond. Washington, (DC), American Psychiatric Press (1993).

Den Otter, W., Koten, J.W., van der Vegt, B.J.H., Beemer, F.A., Boxma, O.J., Derkinderen, D.J., de Graaf, P.W., Huber, J., Lips, C.J.M., Roholl, P.J.M., Sluijter, F.J.H., Tan, K.E.W.P., van der Heyden, K., van der Ven, L., van Unnik, J.A.M. Oncogenesis by mutations in anti-oncogenes: A view. Anticancer Research. 10:475–488 (1990).

Derkinderen, D.J., Boxma, O.J., Koten, J.W., Den Otter, W. Stochastic theory of oncogenesis. Anticancer Research. 10:497–504 (1990).

Dermietzel, R., Hwang, T.K., Spray, D.S. The gap junction family: Structure, function, and chemistry. Anatomy and Embryology. 182:517–528 (1990).

Derogatis, L.R. The SCL-90 Manual I: Scoring, Administration, and Procedures for the SCL-90. Baltimore (MD), Clinical Psychometric Research (1977).

Devary, Y., Gottlieb, R.A., Smeal, T., Karin, M. The mammalian ultraviolet response is triggered by activation of src tyrosine kinase. Cell. 71:1081–1091 (1992).

Dew, M.A., Bromet, E.J. Predictors of temporal patterns of psychiatric distress during 10 years following the nuclear accident at Three Mile Island. Social Psychiatry and Psychiatric Epidemiology. 28:49–55 (1993).

Dew, M.A., Bromet, E.J., Schulberg, H.C. A comparative analysis of two community stressors' long-term mental health effects. American Journal of Community Psychology. 15:167–1884 (1987a).

Dew, M.A., Bromet, E.J., Schulberg, H.C., Dunn, L.O., Parkinson, D.K. Mental health effects of the Three Mile Island nuclear reactor restart. American Journal of Psychiatry. 144:1074–1077 (1987b).

Dew, M.A., Bromet, E.J., Schulberg, H.C. Application of a temporal persistence model to community residents' long-term beliefs about the Three Mile Island nuclear accident. Journal of Applied Social Psychology. 17:1071–1091 (1987c).

Dew, M.A., Dunn, L., Bromet, E.J., Schulberg, H.C. Factors affecting help-seeking during depression in a community sample. Journal of Affective Disorders. 14:223–234 (1988).

Dohrenwend, B.P., Dohrenwend, B.S. Perspectives on the past and future of psychiatric epidemiology. American Journal of Public Health. 72:1271–1279 (1982).

Dohrenwend, B.P., Dohrenwend, B.S., Kasl, S. Technical staff analysis report on behavioral effects. Presented to the President's Commission on the Accident at Three Mile Island. Unpublished manuscript, 1979.

Dohrenwend, B.P., Shrout, P.E., Egri, G., Mendelsohn, F.S. Measures of nonspecific psychological distress and other dimensions of psychopathology. Archives of General Psychiatry. 37:1220–1236 (1980).

Dohrenwend, B.P., Dohrenwend, B.S., Warheit, G., Bartlett, G., Goldsteen, R., Goldsteen, K., Martin, J. Stress in the community: A report to the President's Commission on the Accident at Three Mile Island. In: The Three Mile Island Nuclear Accident: Lessons and Implications. Moss, T.H., Sills, D.L. (eds.). Annals of the New York Academy of Sciences. 365:159–174 (1981).

Doll, R. The age distribution of cancer: Implications for models of carcinogenesis. Journal of the Royal Statistical Society, Series A. 132:133–166 (1971).

Doll, R., Peto, R. The Causes of Cancer. New York, Oxford University Press, Oxford Medical Publications (1981).

Douglas, A.J., Omar, R.Z., Smith, P.G. Cancer mortality and morbidity among workers at the Sellafield plant of British Nuclear Fuels. British Journal of Cancer. 70:1232–1243 (1994).

Draper, N., Smith, H. Applied Regression Analysis. New York, John Wiley and Sons (1981).

Dumaz, N., Drougard, C., Sarasin, A., Daya-Grosjean, L. Specific UV-induced mutation spectrum in the *p53* gene of skin tumors from DNA-repair-deficient xeroderma pigmentosum patients. Fundamental and Applied Toxicolology. 90:10529–10533 (1993).

Dupree, E.A., Cragle, D.L., McLain, R.W., Crawford-Brown, D.J., Teta, M.J. Mortality among workers at a uranium processing facility, the Linde Air Products Company Ceramics Plant, 1943–1949. Scandinavian Journal of Work, Environment, and Health. 13:100–107 (1987).

Eghbali, B., Kessler, J.H., Reid, I.M., Roy, C., Spray, D.C., Involvement of gap junctions in tumorigenesis: Transfection of tumor cells with connexin43 cDNA retards growth in vivo. Fundamental and Applied Toxicology. 88:10701–10705 (1991).

Endicott, J., Spitzer, R. A diagnostic interview: The schedule for affective disorders and schizophrenia. Archives of General Psychiatry. 33:766–771 (1978).

Epp, E.R., Weiss, H., Laughlin, J.S. Measurement of bone marrow and gonadal dose from x-ray examinations of the pelvis, hip, and spine as a function of field size, field alignment, tube kilovoltage, and filtration. British Journal of Radiology. 34:85–100 (1961).

Epp, E.R., Heslin, J.M., Weiss, H., Laughlin, J.S., Sherman, R.S. Measurement of bone marrow and gonadal dose from x-ray examinations of the pelvis, hip, and spine as a function of field size, tube kilovoltage, and added filtration. British Journal of Radiology. 36:247–265 (1963).

Evans, W.H. Gap junctions: Towards a molecular structure. Bioessays. 8:3–10 (1988).

Fakharzadeh, S.S.S., Trusko, S.P., George, D.L. Tumorigenic potential associated with enhanced expression of a gene that is amplified in a mouse tumor cell line. EMBO Journal. 10:1565–1569 (1991).

Fauser, A.A., Messner, H.A. Granuloerythropoietic colonies in human bone marrow, peripheral blood, and cord blood. Blood. 52:1243–1248 (1978).

Fearon, E.R., Vogelstein, B. A genetic model for colorectal tumorigenesis. Cell. 61:759–767 (1990).

Feinendegen, L.E., Loken, M.K., Booz, J., Muhlensiepen, H., Sondhaus, C.A., Bond, V.P. Cellular mechanisms of protection and repair induced by radiation exposure and their consequence for cell system responses. Stem Cells. S13:7–20 (1995).

Felber, M., Burns, F., Garte, S.J. Amplification of the *c-myc* oncogene in radiation-induced rat skin tumors as a function of linear energy transfer and dose. Radiation Research. 131:297–301 (1992).

Fialkow, P.J. Clonal origin of human tumors. American Reviews in Medicine. 30:135–176 (1979).

Finch, S.C. Review of "Suffering Made Real" by Lindes, M.S. New England Journal of Medicine. 333:198–199 (1995).

Finch, S.C., Hoshino, T., Hrubec, Z., Itoga, T., Nefzger, M.D. Operations Manual for the Detection of Leukemia and Related Disorders, Hiroshima and Nagasaki.

Atomic Bomb Casualty Commission Technical Report 1-65 [ABCC TR 1-65]. Hiroshima, ABCC (1965a).

Finch S.C., Hrubec, Z., Nefzger, M.D., Hoshino, T., Itoga, T. Detection of Leukemia and Related Disorders, Hiroshima-Nagasaki; Research Plan. Atomic Bomb Casualty Commission Technical Report 5-65 [ABCC TR 5-65]. Hiroshima, ABCC (1965b).

Finette, B.A., Poseno, T., Albertini, R.J. V(D)J recombinase mediated *hprt* mutations in peripheral blood lymphocytes of normal children. Cancer Research. 56:1405-1412 (1996).

Finette, B.A., Sullivan, L.M., O'Neill, J.P., Nicklas, J.A., Vacek, P.M., Albertini, R.J. Determination of *hprt* mutant frequencies in T lymphocytes from a healthy pediatric population: Statistical comparison between newborn, children and adult mutant frequencies, cloning efficiency, and age. Mutation Research. 308:223-231 (1994).

Finger, L.R., Harvey, R.C., Moore, R.C.A., Showe, L.C., Croce, C.M. A common mechanism of chromosomal translocation in T- and B-cell neoplasia. Science. 234:982-985 (1986).

Fischer, S.G., Lerman, L.S. Length-independent separation of DNA restriction fragments in two-dimensional gel electrophoresis. Cell. 16:191-200 (1979).

Fleming, R., Baum, A., Gisriel, M.M., Gatchel, R.J. Mediating influences of social support on stress at Three Mile Island. Journal of Human Stress. 8:14-22 (1982).

Fliedner, T.M., Cronkite, E.P., Bond, V.P. (eds.) Assessment of Radiation Effects by Molecular and Cellular Approaches. Dayton (OH), Alpha Medical Press (1995).

Folley, J.H., Borges, W., Yamawaki, T. Incidence of leukemia in atomic bomb survivors, Hiroshima-Nagasaki. American Journal of Medicine. 13:311-321 (1952). [RERF TR 30-A-59].

Ford, D., Easton, D.F., Bishop, D.T., Narod, S.A., Goldgar, D.E. Risks of cancer in BRCA1-mutation carriers. Lancet. 343:692-695 (1994).

Francis, T., Jr., Jablon, S., Moore, F.E. Report of Ad Hoc Committee for Appraisal of ABCC Program, November 6, 1955. Atomic Bomb Casualty Commission Technical Report 33-59 [ABCC TR 33-59]. Hiroshima, ABCC (1959).

Frankfurt, O.S., Seekinger, D., Sugarbaker, E.V. Intercellular transfer of drug resistance. Cancer Research. 51:1190-1195 (1991).

Fraser, P., Carpenter, L., Maconochie, N., Higgins, C., Booth, M., Beral, V. Cancer mortality and morbidity in employees of the United Kingdom Atomic Energy Authority, 1946-1986. British Journal of Cancer. 67:615-624 (1993).

Fritch, P., Richard-LeNaour, H., Denis, S., Menetrier, F. Kinetics of radiation-induced apoptosis in the cerebellum of 14-day-old rats after acute or during continuous exposure. International Journal of Radiation Biology. 66:111-117 (1994).

Fujita, S. Versions of DS86. Radiation Effects Research Foundation (RERF) Update 1:3 (1989). Hiroshima, RERF (1989).

Fuscoe, J.C., Zimmerman, L.J., Lippert, M.L., Nicklas, J.A., O'Neill, J.P., Albertini, R.J. V(D)J recombinase-like activity mediates *hprt* gene deletion in human fetal T lymphocytes. Cancer Research. 51:6001–6005 (1991).

Garber, D., Dunford, C., Pearlstein, S. Data formats and procedures for the evaluated nuclear data file, ENDF, ENDF-102. Brookhaven National Laboratory. Brookhaven (NY), DoE (1975).

Gardner, M.J., Hall, A.J., Downes, S., Terrell, J.D. Follow-up study of children born to mothers resident in Seascale, West Cumbria (birth cohort). British Medical Journal. 295:822–827 (1987).

Genetics Conference, Committee on Atomic Casualties, National Research Council. Genetic effects of the atomic bombs in Hiroshima and Nagasaki. Science. 106:331–333 (1947).

Gerber, G.B., Taylor, D.M., Cardis, E., Theissen, J.W. (eds.). The Future of Human Radiation Research. London, British Institute of Radiology (1991).

Gibbs, R.A., Nguyen, P.N., Edwards, A., Civitello, A.B., Caskey, C.T. Multiplex DNA deletion detection and exon sequencing of the hypoxanthine phosphoribosyltransferase gene in Lesch-Nyhan families. Genomics. 7:235–244 (1990).

Gilbert, E. S. What can be learned from epidemiologic studies of persons exposed to low doses of radiation?, pp. 155–170. In: Biological Effects of Low Level Exposures. Calabrese, E.J. (ed.). Boca Raton (FL), Chemical Rubber Company (1994).

Gilbert, E.S. Radiation worker epidemiology and risk, pp. 553–580. In: Reactor Health Physics. Hershey (PA), Health Physics Society (1995).

Gilbert, E.S., Fix, J.J. Accounting for bias in dose estimates in analyses of data from nuclear worker studies. Health Physics. 68(5):650–660 (1995).

Gilbert, E.S., Cragle, D.L., Wiggs, L.D. Updated analyses of combined mortality data on workers at the Hanford site, Oak Ridge National Laboratory, and Rocky Flats Weapons Plant. Radiation Research. 136:408–421 (1993a).

Gilbert, E.S., Fry, S.A., Wiggs, L.D., Voelz, G.L., Cragle, D.L., Petersen, G.R. Analyses of combined mortality data on workers at the Hanford Site, Oak Ridge National Laboratory, and Rocky Flats Nuclear Weapons Plant. Radiation Research. 120:19–35 (1989).

Gilbert, E.S., Omohundro, E., Buchanan, J.A., Holter, N.A. Mortality of workers at the Hanford site: 1945–1986. Health Physics. 64(6):577–590 (1993b).

Gilbert, E.S., Fix, J.J., Baumgartner, W.V. An approach to evaluating bias and uncertainty in estimates of external dose obtained from personal dosimeters. Health Physics. 70:336–345 (1996).

Gimlich, R.L., Kumar, N.M., Gilula, N.B. Differential regulation of the levels of three gap junction mRNA's in Xenopus embryo. Journal of Cell Biology. 110:597–605 (1990).

Ginzburg, H.M., Reis, E. Consequences of the nuclear power plant accident at Chernobyl. Public Health Reports. 106:32–40 (1991).

Glasser, J.H. The quality and utility of death certificate data. American Journal of Public Health. 71:231–233 (1991).

Goldberg, D.P. The Detection of Psychiatric Illness by Questionnaire. London, Oxford University Press (1972).

Goldberg, G.S., Martyn, K.D., Lau, A.F. A connexin43 antisense vector reduces the ability of normal cells to inhibit the foci formation of transformed cells. Molecular Carcinogenesis. 11:106–114 (1994).

Goldsteen, R., Schorr, J.K., Goldsteen, K.S. Longitudinal study of appraisal at Three Mile Island: Implications for life event research. Social Science and Medicine. 28:389–398 (1989).

Goodhead, D.T. Initial events in the cellular effects of ionizing radiations: Clustered damage in DNA. International Journal of Radiation Biology. 65:7–17 (1994).

Gotz, C., Montenarh, M. p53 and its implications in apoptosis. International Journal of Oncology. 6:1129–1135 (1995).

Grant, S.G., Langlois, R.G., Jensen, R.H., Bigbee, W.L. Somatic mutations and segregation in normal individuals as determined using the in vivo GPA assay: Implications for oncogenesis and aging. American Journal of Human Genetics. 49, 448 (1991).

Green, E.L., Schlager, G., Dickie, M.M. Natural mutation rates in the house mouse: Plan of study and preliminary estimates. Mutation Research. 2:457–465 (1965).

Gribbin, M.A., Weeks, J.L., Howe, G.R. Cancer mortality (1956–1985) amongst male employees of Atomic Energy of Canada Limited with respect to occupational exposure to external low linear energy transfer ionizing radiation. Radiation Research. 133:375–380 (1993).

Grimwood, P., Charles, M.W. The possible repercussions of further revisions in the Japanese atomic bomb dosimetry, pp. 213–216. In: Proceedings of the 17th IRPA Regional Congress, Portsmouth, June 6–10, 1994. Ashford (UK), Nuclear Technology Publishing (1994).

Hadjimichael, O.C., Ostfeld, A.M., D'Atri, D.A. Brubaker, R.E. Mortality and cancer incidence experience of employees in a nuclear fuels fabrication plant. Journal of Occupational Medicine. 25(1):48–61 (1983).

Hainaut, P., Milner, J. Redox modulation of p53 confirmation and sequence-specific DNA binding in vitro. Cancer Research. 53:4469–4473 (1993).

Hakoda, M., Akiyama, M., Hirai, Y., Kyoizumi, S., Awa, A.A. In vivo mutant T-cell frequency in atomic bomb survivors carrying outlying values of chromosome aberration frequencies. Mutation Research. 202:203–208 (1988a).

Hakoda, M., Akiyama, M., Kyoizumi, S., Awa, A.A., Yamakido, M., Otake, M. Increased somatic cell frequency in atomic bomb survivors. Mutation Research. 201:39–48 (1988b).

Hakoda, M., Hirai, Y., Kyoizumi, S., Akiyama, M. Molecular analyses of in vivo hprt mutant T cells from atomic bomb survivors. Environmental and Molecular Mutagenesis. 13:25–33 (1989).

Hanash, S.M., Boehnke, M., Chu, E.H.Y., Neel, J.V., Kuick, R.D. Nonrandom distribution of structural mutants in ethylnitrosourea-treated cultured human lymphoblastoid cells. Proceedings of the National Academy of Sciences. 85:165–169 (1988).

Harada, T., Ishida, M. Neoplasms among atomic bomb survivors in Hiroshima City: First report of the Research Committee on Tumor Statistics, Hiroshima City Medical Association, Hiroshima, Japan. Journal of the National Cancer Institute. 25:1253–1264 (1960). [ABCC TR 10-59].

Harada, T., Ide, M., Ishida, M., Troup, G.M. Tumor Registry Data, Hiroshima and Nagasaki 1957–1959: Malignant Neoplasms. Report of the Research Committee on Tumor Statistics, Hiroshima and Nagasaki City Medical Associations. Atomic Bomb Casualty Commission Technical Report 23-63 [ABCC TR 23-63]. Hiroshima, ABCC (1963).

Hartsough, D.M., Savitsky, J.C. Three Mile Island: Psychological and environmental policy at a crossroads. American Psychologist. 10:1113–1122 (1984).

Hashizume, T., Maruyama, T., Shirogi, A., Tanaka, E., Izawa, M., Kawamura, S., Nagaoka, S. Estimation of the air dose from the atomic bombs in Hiroshima and Nagasaki. Health Physics. 13:335–339 (1967).

Hatada, I., Hayashizaki, Y., Hirotsune, S., Komatsubara, H., Mukai, T. A genomic scanning method for higher organisms using restriction sites as landmarks. Proceedings of the National Academy of Sciences. 88:9523–9527 (1991).

Hatch, M.C., Beyea, J., Nieves, J.W., Susser, M. Cancer near the Three Mile Island nuclear plant: radiation emissions. American Journal of Epidemiology. 132:397–412 (1990).

Havenaar, J.M., Van Den Brink, W., Van Den Bout, J., Kasyanenko, A.P., Poelijoe, N.W., Wohlfarth, T., Meijler-Iljina, L.I. Mental health problems in the Gomel region (Belarus): An analysis of risk factors in an area affected by the Chernobyl disaster. Psychol. Med. 26:845–855 (1996).

Herrlick, P., Rahmsdorf, H.J. Transcriptional and post-translational responses to DNA-damaging agents. Current Opinion in Cell Biology. 6:425–431 (1994).

Heyssel, R.M., Brill, A.B., Woodbury, L.A., Nishimura, E.T., Ghose, T., Hoshino, T., Yamasaki, M. Leukemia in atomic bomb survivors, Hiroshima. Blood. 15:313–331 (1960). [RERF TR 2-59].

Hill, C.S., Treisman, R. Transcriptional regulation by extracellular signals: Mechanisms and specificity. Cell. 80:199–211 (1995).

Hirai, Y., Kusunoki, Y., Kyoizumi, S., Awa, A.A., Pawel, D.J., Nakamura, N., Akiyama, M. Mutant frequency at the $hprt$ locus in peripheral blood T lymphocytes of atomic bomb survivors. Mutation Research. 329:183–196 (1996).

Hiyama, K., Kodaira, M., Satoh, C. Detection of deletions, insertions, and single nucleotide substitutions in cloned β-globin genes and new polymorphic nucleotide substitutions in β-globin genes in a Japanese population using ribonuclease cleavage at mismatches in RNA:DNA duplexes. Mutation Research. 231:219–231 (1990).

Hoel, D.G. Questions about radiation and risk estimates that remain after BEIR V. Radiation Research. 136:137–145 (1993).

Holder, J.W., Elmore, E., Barrett, J.C. Gap junction function and cancer. Cancer Research. 53:3475–3485 (1993).

Hollstein, M., Sidransky, D., Vogelstein, B., Harris, C.C. *p53* mutations in human cancers. Science. 253:49–53 (1991).

Hollstein, M., Rice, K., Greenblatt, M.S., Soussi, T., Fuchs, R., Sørlie, T., Hovig, E., Smith-Sørensen, B., Montesano, R., Harris, C.C. Database of *p53* gene somatic mutations in human tumors and cell lines. Nucleic Acids Research. 17:3551–3555 (1994).

Horowitz, M. Post-traumatic stress disorders: Psychosocial aspects of the diagnosis. International Journal of Mental Health. 19:21–36 (1990).

Hoshi, M., Ichikawa, Y., Nagatomo, T. Thermoluminescence measurement of gamma rays at about 2,000 m from the hypocenter, pp. 149–152. In: U.S.-Japan Joint Reassessment of Atomic Bomb Radiation Dosimetry in Hiroshima and Nagasaki, Final Report, Vol. 2. Roesch, W.C. (ed.). Hiroshima, RERF (1987).

Hossain, M.Z., Wilkens, L.R., Metha, P.P., Loewenstein, W.R., Bertram, J.S. Enhancement of gap junctional communication by retinoids correlates with their ability to inhibit neoplastic transformation. Carcinogenesis. 10:1743–1748 (1989).

Houts, P.S., Cleary, P.D., Hu, T-W. The Three Mile Island crisis: Psychological, Social, and Economic Impacts on the Surrounding Population. University Park, The Pennsylvania State University Press (1988).

Hu, J., Cotgreave, I.A. Glutathione depletion potentiates 12-O-tetradecanoylphorbol-13-Acetate (TPA)-induced inhibition of gap junctional intercellular communication in WB-F344 rat liver epithelial cells: Relationship to intracellular oxidation stress. Chemico-Biological Interactions. 93:291–307 (1995).

Hurst, G.S. "R.H. Ritchie: the grand attractor." Proceedings of the Symposium in Honor of R.H. Ritchie on His Seventieth Birthday, Oak Ridge, TN, October 23–25, 1994. Nuclear Instruments and Methods. B96:xix–xxiii (1995).

Ichikawa, Y., Higashimura, T., Sidei, T. Thermoluminescence dosimetry of gamma rays from atomic bombs in Hiroshima and Nagasaki. Health Physics. 12:395–405 (1966).

Ichikawa, Y., Nagatomo, T., Hoshi, M., Kondo, S. Thermoluminescence measurement of gamma rays at 1,270 to 1,460 m from the hypocenter, pp. 125–136. In: U.S.-Japan Joint Reassessment of Atomic Bomb Radiation Dosimetry in Hiroshima and Nagasaki: Final Report, Vol. 2. Roesch, W.C. (ed.). Hiroshima, RERF (1987).

Ichimaru, M., Ishimaru, T., Belsky, J.L., Tomiyasu, T., Sadamori, N., Hoshino, T., Tomonaga, M., Shimizu, N., Okada, H. Incidence of leukemia in atomic bomb survivors, Hiroshima and Nagasaki, 1959–71, by radiation dose, years after exposure, age, and type of leukemia. Journal of Radiation Research (Tokyo). 19:262–282 (1978). [RERF TR 10-76].

Ichimaru, M., Ishimaru, T., Mikami, M., Yamada, Y., Ohkita, T. Incidence of Leukemia in a Fixed Cohort of Atomic Bomb Survivors and controls, Hiroshima and Nagasaki, October 1950–December 1978. Radiation Effects Research Foundation Technical Report 13-81 [RERF TR-13-81]. Hiroshima, RERF (1981).

Ide, M., Shimizu, K., Sheldon, W.F., Ishida, M. Tumor Registry Data, Hiroshima and Nagasaki 1957–61. Malignant Neoplasms. Report 3. Report of the Research Committee on Tumor Statistics, Hiroshima and Nagasaki City Medical Associations. Atomic Bomb Casualty Commission Technical Report 3-65 [ABCC TR 3-65]. Hiroshima, ABCC (1965).

Ihno, Y., Russell, W.J. Dose to the Gonads and Bone Marrow in Radiographic Examinations at ABCC. Atomic Bomb Casualty Commission Technical Report 24-63 [ABCC TR 24-63]. Hiroshima, ABCC (1963).

Ihno, Y., Russell, W.J., Ishimaru, T. ABCC-JNIH Adult Health Study Hiroshima and Nagasaki 1962–63. Exposure to Medical X-ray. Community Hospital and Clinic Survey. Atomic Bomb Casualty Commission Technical Report 11-63 [ABCC TR 11-63]. Hiroshima, ABCC (1963).

Ikeya, M. New Application of Electron Spin Resonance: Dating, Dosimetry, and Microscopy. London, World Scientific (1993).

International Agency for Research on Cancer (IARC). Study group on cancer risk among nuclear industry workers. Direct estimates of cancer mortality due to low doses of ionizing radiation: An international study. Lancet. 344:1039–1043 (1994).

International Atomic Energy Agency (IAEA). The International Chernobyl Project: An Overview (Assessment of Radiological Consequences and Evaluation of Protective Measures). Vienna, IAEA (1991).

International Commission on Radiological Protection (ICRP). Annals of the ICRP 21 (1-3). Oxford, Pergamon Press (1991).

International Commission on Radiation Units and Measurements (ICRU). ICRU Report 11, Radiation Quantities and Units. Bethesda (MD), ICRU (1968).

International Commission on Radiation Units and Measurements (ICRU). ICRU Report 51, Quantities and Units in Radiation Protection Dosimetry. Bethesda (MD), ICRU (1993).

Ishida, M., Beebe, G.W. Research Plan for Joint NIH-ABCC Study of Lifespan of A-Bomb Survivors. Radiation Effects Research Foundation Technical Report 4-59 [RERF TR 4-59]. Hiroshima, RERF (1959).

Ishida, M., Zeldis, L.J., Jablon, S. Tumor Registry Study in Hiroshima and Nagasaki. Research Plan. Atomic Bomb Casualty Commission Technical Report 2-61 [ABCC TR 2-61]. Hiroshima, ABCC (1961).

Ishimaru, T., Russell, W.J. ABCC-JNIH Adult Health Study, Hiroshima and Nagasaki, 1961. Exposure to Medical X-ray. Preliminary Survey. Atomic Bomb Casualty Commission Technical Report 7-62 [ABCC TR 7-62]. Hiroshima, ABCC (1962).

Ishimaru, T., Hoshino, T., Ichimaru, M., Okada, H., Tomiyasu, T., Tsuchimoto, T., Yamamoto, T. Leukemia in atomic bomb survivors, Hiroshima and Nagasaki, 1 October 1950–30 September 1966. Radiation Research. 45:216–233 (1971). [RERF TR 25-69].

Ishimaru, T., Ichimaru, M., Mikami, M., Yamada, Y., Tomonaga, Y. Distribution of Onset of Leukemia Among Atomic Bomb Survivors in the Leukemia Registry by Dose, Hiroshima and Nagasaki, 1946–75. Radiation Effects Research Foundation Technical Report 12-81 [RERF TR 12-81]. Hiroshima, RERF (1981a).

Ishimaru, T., Otake, M., Ichimaru, M., Mikami, M. Dose-Response Relationship of Leukemia Incidence Among Atomic Bomb Survivors and Their Controls by Absorbed Marrow Dose and Two Types of Leukemia, Hiroshima and Nagasaki, October 1950-December 1978. Radiation Effects Research Foundation Technical Report 10-81 [RERF TR 10-81]. Hiroshima, RERF (1981b).

Ito, T., Seyama, T., Iwamoto, K.S. In vitro irradiation is able to cause *ret* oncogene rearrangement. Cancer Research. 53:2940–2943 (1993).

Iwasaki, M., Miyazawa, C., Niwa, K. Differences in the radiation sensitivity of human tooth enamel in individual and among individuals. Radioisotopes. 44:785–788 (1995).

Jablon, S., Ishida, M., Beebe, G.W. JNIH-ABCC Life Span Study. Report 2. Mortality in Selections I and II. October 1950–September 1959. Atomic Bomb Casualty Commission Technical Report 1-63 [ABCC TR 1-63]. Hiroshima, ABCC (1963).

Jensen, R.H., Bigbee, W.L. Direct immunofluorescence labeling provides an improved method for the glycophorin A somatic cell mutation assay. Cytometry. 23:337–343 (1996).

Jensen, O.M., Storm, H.H. Purposes and uses of cancer registration, pp. 7–21. In: Cancer Registration Principles and Methods. Jensen, O.M., Parkin, D.M., Maclennan, R., Muir, C.S., Skeet, R.G. (eds.). International Agency for Research on Cancer (IARC), World Health Organization. Lyon, IARC (1991).

Jensen, R.H., Langlois, R.G., Bigbee, W.L., Grant, S.G., Moore, D. II, Pilinskaya, M., Vorobtsova, I. and Pleshanov, P. Elevated frequency of glycophorin A mutations in erythrocytes from Chernobyl accident victims. Radiation Research. 141:129–135 (1995).

Jones, P.A., Buckley, J.D., Henderson, B.E., Ross, R.K., Pike, M.C. From gene to carcinogen: A rapidly evolving field in molecular epidemiology. Cancer Research. 51:3617–3620 (1991).

Jones, T.D., Auxier, J.A., Cheka, J.S., Kerr, G.D. In vivo dose estimates for A-bomb survivors shielded by typical Japanese houses. Health Physics. 28:367–381 (1975).

Jou, Y.S., Matesic, D.F., Dupont, E., Lu, S.C., Rupp, H.L., Madhukar, B.V., Oh, S.Y., Trosko, J.E., Chang, C.C. Restoration of gap junctional intercellular communication in a communication-deficient rat liver cell mutant by transfection with connexin43 cDNA. Molecular Carcinogenenesis. 8:234–245 (1993).

Jung, M., Notario, V., Dritschilo, A. Mutations in the *p53* gene in radiation-sensitive and -resistant human squamous carcinoma cells. Cancer Research. 52:6390–6393 (1992).

Jung, M., Zhang, Y., Lee, S., Dritschilo, A. Correction of radiation sensitivity in ataxia telangiectasia cells by a tunicated 1kB-a. Science. 268:1619–1621 (1995).

Kadhim, M.A., Lorimore, S.A., Hepburn, M.D., Goodhead, D.T., Buckle, V.J., Wright, E.G. α-particle-induced chromosomal instability in human bone marrow cells. Lancet. 344:987–988 (1994).

Kadhim, M.A., MacDonald, D.A., Goodhead, D.T., Lorimore, S.A., Marsden, S.J., Wright, E.G. Transmission of chromosomal instability after plutonium α-particle irradiation. Nature. 355:738–740 (1992).

Kanno, Y. Modulation of cell communication and carcinogenesis. Japan Journal of Physiology. 35:693–707 (1985).

Kao, C.Y., Nomata, K., Oakley, C.C., Welsch, C.W., Chang, C.C. Two types of normal breast epithelial cells derived from reduction mammoplasty: Phenotypic characterization and response to SV40 transformation. Carcinogenesis. 16:531–535 (1995).

Kato, H., Schull, W.J. Studies of the mortality of A-bomb survivors, Life Span Study Report 7. Mortality 1950–1978: Part 1. Cancer mortality. Radiation Research. 90:395–432 (1982). [RERF TR 12-80].

Kato, H., Schull, W.J., Awa, A.A., Akiyama, M., Otake, M. Dose-response analyses among atomic bomb survivors exposed to low-level radiation. Health Physics. 52:645–652 (1987).

Kato, K., Antoku, S., Sawada, S., Russell, W.J. Organ doses received by atomic bomb survivors during radiological examinations at the Radiation Effects Research Foundation. British Journal of Radiology. 64:720–727 (1991b). [RERF TR 19-89].

Kato, K., Antoku, S., Sawada, S., Russell, W.J. Calibration of Mg_2SiO_4(Tb) thermoluminescent dosimeters for use in determining diagnostic X-ray doses to Adult Health Study participants. Medical Physics. 18:928–933 (1991a). [RERF TR 11-89].

Kato, K., Antoku, S., Sawada, S., Wada, T., Russell, W.J. Organ doses to atomic bomb survivors during photofluorography, fluoroscopy and computed tomography. British Journal of Radiology. 64:728–733 (1991c). [RERF TR 2-90].

Kaul, D.C., Egbert, S.D. DS86 Uncertainty and Bias Analysis, Draft Report. Science Applications International Corporation (SAIC). San Diego, SAIC (1989).

Kaul, D.C., Jarka, R. Radiation Dose Deposition in the Active Marrow of Reference Man. Washington, DC, Defense Nuclear Agency (DNA-4442F) (1977).

Kavanagh, T.J., Martin, G.M., Livesey, J.C., Rubinovitch, P.S. Direct evidence of intercellular sharing of glutathione via metabolic cooperation. Journal of Cellular Physiology. 137:353–359 (1988).

Kellerer, A.M., Barclay, D. Age dependencies in the modelling of radiation carcinogenesis. Radiation Protection Dosimetry. 41:273–281 (1992).

Kelsey, J.L., Gammon, M.D., John, E.M. Reproductive factors and breast cancer. Epidemiologic Reviews. 15:36–47 (1993).

Kemeny J.G. Report of the President's Commission on the Accident at Three Mile Island. Washington, (DC), US Government Printing Office (1979).

Kendall, G.M., Muirhead, C.R., MacGibbon, B.H., O'Hagan, J.A., Conquest, A.J., Goodill, A.A., Butland, B.K., Fell, T.P., Jackson, D.A., Webb, M.A., Haylock, R.G.E., Thomas, J.M., Silk, T.J. Mortality and occupational exposure to radiation: First analysis of the National Registry for Radiation Workers. British Medical Journal. 304:220–225 (1992).

Kerr, G.D. Organ dose estimates for the Japanese atomic-bomb survivors. Health Physics. 37:487–508 (1979).

Kerr, G.D. The new radiation dosimetry for the A-bombs in Hiroshima and Nagasaki, pp. 5–18. In: Low Dose Radiation: Biological Bases of Risk Assessment. Baverstock, K.F., Stather, J.W. (eds.). London, Taylor and Francis (1989).

Kerr, G.D., Pace, J.V. III, Mendelsohn, E., Loewe, W.E., Kaul, D.C., Farhad, D., Egbert, S.D., Gritzner, M., Scott, W.H., Jr., Marcum, J., Kosako, T., Kanda, K. Transport of initial radiations in air over ground, pp. 66–142. In: US-Japan Joint Reassessment of Atomic Bomb Radiation Dosimetry in Hiroshima and Nagasaki: Final Report, Vol. 1. Roesch, W.C. (ed.). Hiroshima, RERF (1987).

Kerr, G.D., Dyer, F.F., Pace, J.V. III, Brodzinski, R.L., Marcum, J. Activation of Cobalt by Neutrons from the Hiroshima Bomb. ORNL-6590, Oak Ridge National Laboratory. Oak Ridge (TN), DoE (1990).

Kerr, G.D., Ritchie, R.H, Kaye, S.V., Wassom, J.S. A Brief History of the Health and Safety Research Division at Oak Ridge National Laboratory - 50th Anniversary Oak Ridge National Laboratory. ORNL/M-2108, Oak Ridge National Laboratory. Oak Ridge (TN), DoE (1992).

Kerr, J.F.R., Searle, J. Apoptosis: Its nature and kinetic role, pp. 367–384. In: Radiation Biology in Cancer Research. Meyn, R.E. Withers, H.R. (eds.). New York, Raven Press (1979).

Kessler, R., McGonagle, K.A., Zhao, S., Nelson, C.B., Hughes, M., Eshleman, S., Wittchen, H-U., Kendler, K.S. Lifetime and 12-month prevalence of DSM-III-R psychiatric disorders in the United States. Archives of General Psychiatry. 51:8–19 (1994).

Kimura, T., Hamada, T. Determination of specific activity of ^{60}Co in steel samples exposed to the atomic bomb in Hiroshima. Radioisotopes. 42:19–22 (1993).

Knudson Jr., A.G. Mutation and cancer: Statistical study of retinoblastoma. Proceedings of the National Academy of Sciences. 68:820–823 (1971).

Knudson Jr., A.G. Hereditary cancer, oncogenes, and antioncogenes. Cancer Research. 45:1437–1443 (1995).

Kodaira, M., Satoh, C., Hiyama, K. Toyama, K. Lack of effects of atomic bomb radiation on genetic instability of tandem-repetitive elements in human germ cells. American Journal of Human Genetics. 57:1275–1283 (1995).

Kodama, Y., Nakano, M., Ohtaki, K., Delongchamp, R., Nakamura, N. Estimation of minimal size of translocated chromosome segments detected by fluorescence in situ hybridization. International Journal of Radiation Biology. 71:35–39 (1997).

Kondo, S. Carcinogenesis in relation to the stem-cell mutation hypothesis. Differentiation. 24:1–8 (1983).

Kondo, S. (ed). Health Effects of Low-Level Radiation. Osaka, Kinki University Press (1993).

Kopecky, K.L., Nakashima, E., Yamamoto, T., Kato, H. Lung Cancer, Radiation and Smoking Among Atomic Bomb Survivors, Hiroshima and Nagasaki. Radiation Effects Research Foundation Technical Report 13-86 [RERF TR 13-86]. Hiroshima, RERF (1986).

Koshurnikova, N.A., Buldakov, L.A., Bysogolov, G.D., Bolotnikova, M.G., Komleva, N.S., Peternikova, V.S. Mortality from malignancies of the hematopoietic and lymphatic tissues among personnel of the first nuclear plant in the USSR. The Science of the Total Environment. 142:19–23 (1994).

Kossenko, M.M., Degteva, M.O. Cancer mortality and radiation risk evaluation for the Techa river population. The Science of the Total Environment. 142:73–89 (1994).

Kuick, R.D., Skolnick, M.M., Hanash, S.M., Neel, J.V. A two-dimensional electrophoresis-related laboratory information processing system: Spot matching. Electrophoresis. 12:736–746 (1991).

Kuick, R., Asakawa, J., Neel, J.V., Satoh, C., Hanash, S.M. High yield of restriction fragment length polymorphisms in two-dimensional separations of human genomic DNA. Genomics. 25:345–353 (1995).

Kulka, R.A., Schlenger, W.E., Fairbank, J.A., Hough, R.L., Jordan, K.B., Marmar, C.R., Weiss, D.S. Trauma and the Vietnam War Generation. New York, Brunner/Mazel (1990).

Kusunoki, Y., Kodama, Y., Hirai, Y., Kyoizumi, S., Nakamura, N., Akiyama, M. Cytogenetic and immunologic identification of clonal expansion of stem cells into T and B lymphocytes in one atomic-bomb survivor. Blood. 86:2106–2112 (1995).

Kyoizumi, S., Nakamura, N., Hakoda, M., Awa, A.A., Bean, M.A., Jensen, R.H., Akiyama, M. Detection of somatic mutations at the glycophorin A locus in erythrocytes of atomic bomb survivors using a single-beam flow sorter. Cancer Research. 49:581–588 (1989).

Land, C.E. A nested case-control approach to interactions between radiation dose and other factors as causes of cancer. Radiation Effects Research Foundation Commentary and Review Series 1-90 [CR 1-90]. Hiroshima, RERF (1990).

Land, C.E., Hayakawa, N., Machado, S., Yamada, Y., Pike, M.C., Akiba, S., Tokunaga, M. A case-control interview study of breast cancer among Japanese A-bomb survivors: I. Main effects. Cancer Causes and Control. 5:157–165 (1994a).

Land, C.E., Hayakawa, N., Machado, S., Yamada, Y., Pike, M.C., Akiba, S., Tokunaga, M. A case-control interview study of breast cancer among Japanese A-bomb survivors: II. Interactions between epidemiological factors and radiation dose. Cancer Causes and Control. 5:167–176 (1994b).

Land, C.E., Saku, T., Hayashi, Y., Takahara, O., Matsuura, H., Tokuoka, S., Tokunaga, M., Mabuchi K. Salivary gland tumor incidence among atomic bomb survivors, 1950–87. II. Evaluation of radiation-related risk. Radiation Research. 146:28–36 (1996).

Lane, D.P. *p53*, guardian of the genome. Nature 358:15–16 (1992).

Lange, R.D., Moloney, W.C., Yamawaki, T. Leukemia in atomic bomb survivors: 1. General observations. Blood. 9:574–585 (1954). [RERF TR 25-A-59].

Langlois, R.G., Bigbee, W.L., Jensen, R.H. Measurements of the frequency of human erythrocytes with gene expression loss phenotypes at the glycophorin A locus. Human Genetics. 74:353–362 (1986).

Langlois, R.G., Bigbee, W.L., Kyoizumi, S., Nakamura, N., Bean, M.A., Akiyama, M., Jensen, R.H. Evidence for increased somatic cell mutations at the glycophorin A locus in atomic bomb survivors. Science. 236:445–448 (1987).

Langlois, R.G., Nisbet, B.A., Bigbee, W.L., Ridinger, D.N., Jensen, R.H. An improved flow-cytometric assay for somatic mutations at the glycophorin A locus in humans. Cytometry. 11:513–21 (1990).

Langlois, R.G., Akiyama, M., Kusunoki, Y., DuPont, B.R., Moore, D.H., Bigbee, W.L., Grant, S.G., Jensen, R.H. Analysis of somatic cell mutations at the glycophorin A locus in atomic bomb survivors: A comparative study of assay methods. Radiation Research. 136:111–117 (1993).

Langner, T.S. A 22-item screening scale of psychiatric symptoms indicating impairment. Journal of Health and Human Behavior. 3:269–276 (1962).

Lasko, N., Orr, S.P., Pitman, R.K., Tarabrina, J.V., Lazebnaya, E.O., Zelenova, M.E., Grafinina, N.A. Psychophysiologic responses of Chernobyl liquidation workers during personal script-driven imagery. Presented at the 9th Annual Meeting of the International Society for Traumatic Stress Studies, San Antonio, Texas (1993).

LeBeau, M.M., Espinosa, R., Neuman, W.L., Stock, W., Roulston, D., Larson, R.A., Keinanen, M., Westbrook, C.A. Cytogenetic and molecular delineation of the smallest commonly deleted region of chromosome 5 in malignant myeloid diseases. Proceedings of the National Academy of Sciences. 90:5484–5488 (1993).

Leenhouts, H.P., Chadwick, K.H. A two-mutation model of radiation carcinogenesis: Application to lung tumours in rodents and implications for risk evaluation. Journal of Radiological Protection. 14:115–130 (1994).

Lehnert, S. Gene induction and adaptive responses in irradiated cells: Mechanisms and clinical implications. Radiation Research. 141:108–123 (1995).

Leighton, D.C., Harding, J.S., Macklin, D.B., Hughes, C.C., Leighton, A.H. The Character of Danger: The Stirling County Study (Vol. 3). New York, Basic Books (1963).

Lerman, L.S., Silverstein, K., Grinfeld, E. Searching for gene defects by denaturing gradient gel electrophoresis. Cold Spring Harbor Symposium on Quantitative Biology. 51:285–297 (1986).

Lewis, E.B. Leukemia and ionizing radiation. Science. 125:965–972 (1957).

Lifton, R.J. Death in Life: Survivors of Hiroshima. New York, Random House (1967).

Lin, C.S., Goldthwait, D.A., Samols, D. Induction of transcription from the long terminal repeat of Maloney murine sarcoma provirus by UV-irradiation, X-irradiation, and phorbol ester. Fundamental and Applied Toxicology. 87:36–40 (1990).

Lindee, M.S. Suffering Made Real: American Science and the Survivors of Hiroshima. Chicago (IL), The University of Chicago Press (1994).

Linnemann, R. Introductory remarks, p. 59. In: The Medical Basis of Radiation Preparedness III: The Psychological Perspective. Ricks, R.C., Berger M.E., O'Hara, F.M. (eds.). New York, Elsevier (1991).

Lippert, M.J., Albertini, R.J., Nicklas, J.A. Physical mapping of the *hprt* chromosomal region (Xq26). Mutation Research. 326:39–49 (1995a).

Lippert, M.J., Nicklas, J.A., Hunter, T.C., Albertini, R.J. Pulsed field analysis of *hprt* T-cell large deletions: Telomeric region breakpoint spectrum. Mutation Research. 326:51–64 (1995b).

Lippert, M.J., Rainville, I.R., Nicklas, J.A., Albertini, R.J. Large deletions partially external to the human *hprt* gene result in chimeric transcripts. Mutagenesis. 12:185-190 (1997).

Little, J.B. Changing views of cellular radiosensitivity. Radiation Research. 140:299–311 (1994).

Little, M.P. Risks of radiation-induced cancer at high doses and dose rates. Journal of Radiological Protection. 13:3–25 (1993).

Little, M.P. Are two mutations sufficient to cause cancer? Some generalizations of the two-mutation model of carcinogenesis of Moolgavkar, Venzon, and Knudson, and of the multistage model of Armitage and Doll. Biometrics. 51:1278–1291 (1995).

Little, M.P. Generalisations of the two-mutation and classical multistage models of carcinogenesis fitted to the Japanese atomic bomb survivor data. Journal of Radiological Protection. 16:7–24 (1996).

Little, M.P., Charles, M.W. Time variations in radiation-induced relative risk and implications for population cancer risks. Journal of Radiological Protection. 11:91–110 (1991).

Little, M.P., Hawkins, M.M., Shore, R.E., Charles, M.W., Hildreth, N.G. Time variations in the risk of cancer following irradiation in childhood. Radiation Research. 126:304–316 (1991).

Little, M.P., Hawkins, M.M., Charles, M.W., Hildreth, N.G. Fitting the Armitage-Doll model to radiation-exposed cohorts and implications for population cancer risks. Radiation Research. 132:207–221 (1992).

Little, M.P., Hawkins, M.M., Charles, M.W., Hildreth, N.G. Corrections to the paper "Fitting the Armitage-Doll model to radiation-exposed cohorts and implications for population cancer risks" (letter). Radiation Research. 137:124–128 (1994).

Little, M.P., Muirhead, C.R., Boice, J.D., Kleinerman, R.A. Using multistage models to describe radiation-induced leukaemia. Journal of Radiological Protection. 15:315–334 (1995).

Lo, C.W., Gilula, N.B. Gap junctional communication in the preimplanation mouse embryo. Cell. 18:411–422 (1979).

Loewe, W.E. Hiroshima and Nagasaki: Delayed neutron contributions and comparison of calculated and measured cobalt activations. Nuclear Technology. 68:311–318 (1985).

Loewe, W.E., Mendelsohn, E. Revised Estimates of Dose at Hiroshima and Nagasaki, and Possible Consequences for Radiation-Induced Leukemia (Preliminary). Technical Report D-80-14, Lawrence Livermore Laboratory. Livermore (CA), DoE (1980).

Loewenstein, W.R. Permeability of membrane junctions. Annals of the New York Academy of Sciences. 137:441–472 (1966).

Loewenstein, W.R. Junctional intercellular communication and the control of growth. Biochimica et Biophysica Acta. 560:1–65 (1979).

Lubin, J.H., Steindorf, K. Cigarette use and the estimation of lung cancer attributable to radon in the United States. Radiation Research. 141:79–85 (1995).

Lubin, J.H., Boice Jr., J.D., Edling, C., Hornung, R.W., Howe, G.R., Kunz, E., Kusiak, R.A., Morrison, H.I., Radford, E.P., Samet, J.M., Tirmarche, M., Woodward, A., Yao, S.X., Pierce, D.A. Lung cancer in radon-exposed miners and estimation of risk from indoor exposure. Journal of the National Cancer Institute. 87:817–827 (1995).

Lucas, J.N., Tenjin, T., Straume, T., Pinkel, D., Moore, D.Z., Litt, M., Gray, J.W. Rapid human chromosome aberration analysis using fluorescence in situ hybridization [published erratum appears in International Journal of Radiation Biology, 56(2):201 (1989)]. International Journal of Radiation Biology. 56:35–44 (1989).

Lucas, J.N., Awa, A., Straume, T., Poggensee, M., Kodama, Y., Nakano, M., Ohtaki, K., Weier, H.U., Pinkel, D., Gray, J., Littlefield, G. Rapid translocation frequency analysis in humans decades after exposure to ionizing radiation. International Journal of Radiation Biology. 62:53–63 (1992).

Lucke-Huhle, C., Gloss, B., Herrlich, P. Radiation-induced gene amplification in rodent and human cells. Acta Biologica Hungarica. 41:159–171 (1990).

Luebeck, E.G., Curtis, S.B., Cross, F.T., Moolgavkar, S.H. Two-stage model of radon-induced malignant lung tumors in rats: Effects of cell killing. Radiation Research. 145:163-173 (1996).

Mabuchi, K., Soda, M., Ron, E., Tokunaga, M., Ochikubo, S., Sugimoto, S., Ikeda, T., Terasaki, M., Preston, D.L., Thompson, D.E. Cancer incidence in atomic bomb survivors. Part I: Use of the tumor registries in Hiroshima and Nagasaki for incidence studies. Radiation Research. 137S:1–16 (1994). [RERF TR 3-91].

Macilwain, C., Swinbanks, D. US energy department backs down on plan for A-bomb victim studies. Nature. 375:709 (1995).

Manchester, D.K., Nicklas, J.A., O'Neill, J.P., Lippert, M.J., Langlois, R.G., Grant, S.G., Jensen, R.H., Albertini, R.J., Bigbee, W.L. Sensitivity of somatic mutations in umbilical cord blood to maternal environments. Environmental and Molecular Mutagenesis. 26:203–212 (1995).

Maruyama, T., Kumamoto, Y., Noda, Y. Reassessment of gamma-ray doses using thermo- luminescence measurements, pp. 113–124. In: US-Japan Joint Reassessment of Atomic Bomb Radiation Dosimetry in Hiroshima and Nagasaki: Final Report, Vol. 2. Roesch, W.C. (ed.). Hiroshima, RERF (1987).

Matsuo, T., Tomonaga, M., Bennett, J.M., Kuriyama, K., Imanaka, F., Kuramoto, A., Kamada, N., Ichimaru, M., Finch, S.C., Pisciotta, A.V., Ishimaru, T. Reclassification of leukemia among A-bomb survivors by French-American-British (FAB) classification: 1. Concordance of diagnosis in Nagasaki cases by RERF members and a member of FAB cooperative group. Japanese Journal of Clinical Oncology. 18:91–96 (1988). [RERF TR 4-87].

Matsuura, H., Yamamoto, T., Sekine, I., Ochi, Y., Otake, M. Pathological and Epidemiologic Study of Gastric Cancer in Atomic Bomb Survivors, Hiroshima and Nagasaki, 1950–77. Radiation Effects Research Foundation Technical Report 12-83 [RERF TR 12-83]. Hiroshima, RERF (1983).

McGinniss, M.J., Falta, M.T., Sullivan, L.M., Albertini, R.J. In vivo *hprt* mutant frequencies in T cells of normal human newborns. Mutation Research. 240:117–126 (1990).

McGinniss, M.J., Nicklas, J.A., Albertini, R.J. Molecular analyses of in vivo *hprt* mutations in human T lymphocytes: IV. Studies in newborns. Environmental and Molecular Mutagenesis. 14:229–237 (1989).

McMichael, A.J. Standardized mortality ratios and the "healthy worker effect": Scratching beneath the surface. Journal of Occupational Medicine. 18:165–168 (1976).

Meister, A. Glutathione, ascorbate, and cellular protection. Cancer Research. 54:1969–1975 (1994).

Mendelsohn M.L. Some mechanistic insights from RERF cancer epidemiology, pp. 69–88. In: Proceedings of the 30th Anniversary Conference of the Nuclear Safety Research Association. Tokyo (Japan), NSRA (1994).

Mendelsohn, M.L. RERF Interoffice Memorandum, February 10, 1995.

Mendelsohn M.L. A simple reductionist model for cancer risk in atom bomb survivors, pp. 185–192. In: Modeling of Biological Effects and Risks of Radiation Exposure. Inaba, J., Kobayashi, S. (eds.). National Institute of Radiological Sciences. Chiba (Japan), NIRS (1995).

Messing, K., Ferraris, J., Bradley, W.E.C., Swartz, J., Seifert, A.M. Mutant frequency of radiotherapy technicians appears to be associated with recent dose of ionizing radiation. Health Physics. 57:537–544 (1989).

Metha, P.P., Hotz-Wagenblath, A., Rose, B., Shalloway, D., Loewenstein, W.R. Incorporation of the gene for a cell-cell channel protein into transformed cells leads to normalization of growth. Journal of Membrane Biology. 124:207–235 (1991).

Meyn, R.E., Stephens, L.C., Voehringer, D.W., Story, M.D., Mirkovic, N., Milas, L. Biochemical modulation of radiation-induced apoptosis in murine lymphoma cells. Radiation Research. 136:327–334 (1993).

Miki, Y., Swenson, J., Shattuck-Eidens, D., Futreal, P.A., Harshman, K., Tavtigian, S., Liu, Q., Cochran, C., Bennett, L.M., Ding, W., Bell, R., Rosenthal, J., Hussey, C., tran, T., McClure, M., Frye, C., Hattier, T., Phelps, R., Haugen-Strano, A., Katcher, H., Yakumo, K., Gholami, Z., Shaffer, D., Stone, S., Bayer, S., Wray, C., Bogden, R., Dayananth, P., Ward, J., Tonin, P., Narod, S., Bristow, P.K., Norris, F.H., Helvering, L., Morrison, P., Rosteck, P., Lai, M., Barrett, J.C., Lewis, C., Neuhausen, S., Cannon-Albright, L., Goldar, D., Wiseman, R., Kamb, A., Skolnik, M.H. A strong candidate for the breast and ovarian cancer susceptibility gene BRCA1. Science. 266:66–71 (1994).

Miller, A.B., Howe, G.R., Sherman, G.J., Lindsay, J.P., Yaffe, M.J., Dinner, P.J., Risch, H.A., Preston, D.L. Mortality from breast cancer after irradiation during fluoroscopic examinations in patients being treated for tuberculosis. New England Journal of Medicine. 321:1285–1289 (1989).

Milton, R.C., Shohoji, T. Tentative 1965 Radiation Dose Estimation for Atomic Bomb Survivors. Atomic Bomb Casualty Commission Technical Report 1-68 [ABCC TR 1-68]. Hiroshima, ABCC (1968).

Misao, T., Hattori, K., Shirakawa, M., Suga, M., Ogawa, N., Ohara, Y., Ohno, T., Fukuta, J., Hamada, T., Kuwahara, H., Fukamachi, K., Hitsumoto, S., Kamatani, T. Characteristics of abnormalities observed in atom-bombed survivors. Journal of Radiation Research (Tokyo). 2:85–97 (1961).

Miura, M., Sasaki, T. Role of glutathione in the intrinsic radio resistance of cell lines from a mouse squamous cell carcinoma. Radiation Research. 126:229–236 (1991).

Miyachi, E., Nishikawa, C. Blocking effect of L-arginine on retinal gap junctions by activating guanylate cyclase via generation of nitric acid. Biogenic Amines 10:459–464 (1994).

Mollo, F., Bertoldo, E., Grandi, G., Cavallo, F. Reliability of death certification for different types of cancer. An autopsy survey. Pathology, Research and Practice. 181:442–447 (1986).

Moloney, W.C. Leukemia in atomic bomb survivors. New England Journal of Medicine. 253:88–90 (1955). [RERF TR 25-B-59].

Moloney, W.C., Kastenbaum, M.A. Leukemogenic effects of ionizing radiation on atomic bomb survivors, Hiroshima. Science. 121:308–309 (1955). [RERF TR 25-D-59].

Moloney W.C., Lange R.D. Leukemia in atomic bomb survivors. 2. Observations on early phases of leukemia. Blood. 9:663–85 (1954). [RERF TR 30-B-59].

Montero, R., Norppa, H., Autio, K., Lindholm, C., Ostrosky-Wegman, P., Sorsa, M. Determination of 6-thioguanine-resistant lymphocytes in human blood by immunohistochemical antibromodeoxyuridine staining. Mutagenesis. 6:169–170 (1991).

Moolgavkar, S.H. The multistage theory of carcinogenesis and the age distribution of cancer in man. Journal of the National Cancer Institute. 61:49–52 (1978).

Moolgavkar, S.H., Venzon, D.J. Two-event models for carcinogenesis: Incidence curves for childhood and adult tumors. Mathematical Biosciences. 47:55–77 (1979).

Moolgavkar, S.H., Knudson Jr., A.G. Mutation and cancer: A model for human carcinogenesis. Journal of the National Cancer Institute. 66:1037–1052 (1981).

Moolgavkar, S.H., Luebeck, E.G. Multistage carcinogenesis: Population-based model for colon cancer. Journal of the National Cancer Institute. 84:610–618 (1992).

Moolgavkar, S.H., Luebeck, E.G., Krewski, D., Zielinski, J.M. Radon, cigarette smoke, and lung cancer: A re-analysis of the Colorado Plateau uranium miners' data. Epidemiology. 4:204–217 (1993).

Moolgavkar, S.H., Dewanji, A., Venzon, D.J. A stochastic two-stage model for cancer risk assessment. I. The hazard function and the probability of tumor. Risk Analysis. 8:383–392 (1988).

Moolgavkar, S.H., Cross, F.T., Luebeck, G., Dagle, G.E. A two-mutation model for radon-induced lung tumors in rats. Radiation Research. 121:28–37 (1990).

Morley, A.A., Trainor, K.J., Seshadri, R., Ryall, R.G. Measurement of in vivo mutations in human lymphocytes. Nature. 302:155–156 (1983).

Morstin, K., Bond, V.P., Baum, J.W. Probabilistic approach to obtain hit size effectiveness functions which relate microdosimetry and radiobiology. Radiation Research. 120:383–402 (1989).

Muir, C. Waterhouse, J., Mack, T., Powell, J., Whelan, S. Cancer Incidence in Five Continents, Vol. V. International Agency for Research on Cancer (IARC), World Health Organization. Lyon, IARC (1987).

Muirhead, C.R., Darby, S.C. Modelling the relative and absolute risks of radiation-induced cancers. Journal of the Royal Statistical Society, Series A. 150:83–118 (1987).

Myers, R.M., Lumelsky, N., Lerman, L.S., Maniatis, T. Detection of single base substitutions in total genomic DNA. Nature. 313:495–498 (1985).

Nakamura, N., Sposto, R., Kushiro, J., Akiyama, M. Is interindividual variation of cellular radiosensitivity real or artifactual? Radiation Research. 125:326–330 (1991).

Nakamura, N., Sposto, R., Akiyama, M. Dose survival of G0 lymphocytes irradiated in vitro: A test for a possible population bias in the cohort of atomic bomb survivors exposed to high doses. Radiation Research. 134:316–322 (1993).

Nakamura, N., Iwasaki, M., Miyazawa, C., Akiyama, M., Awa, A.A. Assessing Radiation Dose Recorded in Tooth Enamel. Radiation Effects Research Foundation RERF Update 6, Summer Issue, 6–7 (1994). Hiroshima, RERF (1994).

Nakamura, T.Y., Yamamoto, I., Kanno, Y., Shiba, Y., Goshima, K. Intercellular metabolic coupling of glutathione between mouse and quail cardiac myocyte and its protective role against oxidative stress, pp. 167–170. In: Progress in Cell Research, Volume 4. Rotterdam, Elsevier Science (1995).

Nakatsuka, H., Shimizu, Y., Yamamoto, T., Sekine, I., Ezaki, H., Tahara, E., Takahashi, M., Shimoyama, N., Mochinaga, N., Tomita, M., Tsuchiya, R., Land,
C.E. Colorectal cancer incidence and radiation dose among atomic bomb survivors, Hiroshima and Nagasaki, 1950–80. Journal of Radiation Research (Tokyo).
33:342–361 (1992). [RERF TR 15-92].

Natarajan, A.T., Ramalho, A.T., Vyas, R.C., Bernini, L.F., Tates, A.D., Ploem, J.S.,
Nascimento, C.H., Curado, M.P. Goiania radiation accident: Results of initial dose
estimation and follow-up studies, pp. 145–153. In: New Horizons in Biological
Dosimetry. New York, Wiley-Liss (1991).

National Research Council (NRC). Committee on the Biological Effects of Ionizing
Radiations. The Effects on Populations of Exposure to Low-Levels of Ionizing
Radiation (BEIR I). Washington, DC, National Academy Press (1972).

National Research Council (NRC). Committee on the Biological Effects of Ionizing
Radiations. The Effects on Populations of Exposure to Low-Levels of Ionizing
Radiation (BEIR III). Washington, DC, National Academy Press (1980).

National Research Council (NRC). Committee on the Biological Effects of Ionizing Radiations. Health Effects of Radon and Other Internally Deposited Alpha
Emitters (BEIR IV). Washington, DC, National Academy Press (1988).

National Research Council (NRC). Committee on the Biological Effects of Ionizing
Radiations. Health Effects of Exposure to Low Levels of Ionizing Radiation (BEIR
V). Washington, DC, National Academy Press (1990).

Neel, J.V. A study of major congenital defects in Japanese infants. American Journal
of Human Genetics. 10:398–445 (1958).

Neel, J.V. Physician to the Gene Pool. New York, John Wiley and Sons (1994a).

Neel, J.V. Problem of "false positive" conclusions in genetic epidemiology: Lessons
from the leukemia cluster near the Sellafield nuclear installation. Genetic Epidemiology. 11:213–233 (1994b).

Neel, J.V. The genetic effects of human exposures to ionizing radiation, pp. 115–131.
In: Radiation and Public Perception: Benefits and Risks. Young, J.P., Yalow, R.S.
(eds.). Washington, DC, American Chemical Society (1995a).

Neel, J.V. Invited editorial: New approaches to evaluating the genetic effects of the
atomic bombs. American Journal of Human Genetics. 57:1263–1266 (1995b).

Neel, J.V., Lewis, S.E. The comparative radiation genetics of humans and mice.
Annual Review of Genetics. 24:327–362 (1990).

Neel, J.V., Lewis, S.E. Report of a workshop on the application of molecular genetics
to the study of mutations in the children of atomic bomb survivors. Mutation
Research. 291:1–20 (1993).

Neel, J.V., Schull, W.J. Studies on the potential genetic effects of the atomic bomb.
Acta Genetica. 6:183–196 (1956). [RERF TR 21-B-59].

Neel, J.V., Schull, W.J. The Children of Atomic Bomb Survivors: A genetic study.
Washington, DC, National Academy Press (1991).

Neel, J.V., Rosenblum, B.B., Sing, C.F., Skolnick, M.M., Hanash, S.M., Sternberg, S. Adapting two-dimensional gel electrophoresis to the study of human germ-line mutation rates, pp. 259–306. In: Two-Dimensional Gel Electrophoresis of Proteins: Methods and Applications. Celis, J.E., Bravo, R. (eds.). New York, Academic Press (1984).

Neel, J.V., Satoh, C., Goriki, K., Fujita, M., Takahashi, N., Asakawa, J., Hazama, R. The rate with which spontaneous mutation alters the electrophoretic mobility of polypeptides. Proceedings of the National Academy of Sciences. 83:389–393 (1986).

Neel, J.V., Strahler, J.R., Hanash, S.M, Kuick, R., Chu, E.H.Y. 2-D PAGE and genetic monitoring: Progress, prospects, and problems, pp. 79–89. In: Two-Dimensional Electrophoresis. Endler, A.T., Hanash, S.M. (eds.). Weinheim, VCH Publishers (1989).

Neel, J.V., Schull, W.J., Awa, A.A., Satoh, C., Kato, H., Otake, M., Yoshimoto, Y. The children of parents exposed to atomic bombs: Estimates of the genetic doubling dose of radiation for humans. American Journal of Human Genetics. 46:1053–1072 (1990).

Nelson, S.L., Giver, C.R., Grosovsky, A.J. Spectrum of x-ray-induced mutations in the human *hprt* gene. Carcinogenesis. 15:495–502 (1994).

Neriishi, K., Stram, D.O., Vaeth, M., Mizuno, S., Akiba, S. The observed relationship between the occurrence of acute radiation effects and leukemia mortality among A-bomb survivors. Radiation Research. 125:206–213 (1991)

Neriishi, K., Wong, F.L., Nakashima, E., Otake, M., Kodama, K., Choshi, K. Relationship between cataract and epilation in atomic bomb survivors. Radiation Research. 144:107–113 (1995).

Nicklas, J.A., O'Neill, J.P., Albertini, R.J. Use of T-cell receptor gene probes to quantify the in vivo *hprt* mutations in human T lymphocytes. Mutation Research. 173:65–72 (1986).

Nicklas, J.A., Hunter, T.C., Sullivan, L.M., Berman, J.K., O'Neill, J.P., Albertini, R.J. Molecular analyses of in vivo *hprt* mutations in human T lymphocytes I. Studies of low frequency "spontaneous" mutants by Southern blots. Mutagenesis. 2:341–347 (1987).

Nicklas, J.A., Hunter, T.C., O'Neill, J.P., Albertini, R.J. Molecular analyses of in vivo *hprt* mutations in human T lymphocytes: III. Longitudinal study of *hprt* gene structural alterations and T-cell clonal origins. Mutation Research. 215:147–160 (1989).

Nicklas, J.A., Falta, M.T., Hunter, T.C., O'Neill, J.P., Jacobson-Kram, D., Williams, J., Albertini, R.J. Molecular analysis of in vivo *hprt* mutations in human lymphocytes V. Effects of total body irradiation secondary to radioimmunoglobulin therapy (RIT). Mutagenesis. 5:461–468 (1990).

Nicklas, J.A., Hunter, T.C., O'Neill, J.P., Albertini, R.J. Fine structure mapping of the *hprt* region of the human X chromosome (Xq26). American Journal of Human Genetics. 49:267–278 (1991a).

Nicklas, J.A., O'Neill, J.P., Hunter, T.C., Falta, M.T., Lippert, M.J., Jacobson-Kram, D., Williams, J.R., Albertini, R.J. In vivo ionizing irradiations produce deletions in the *hprt* gene of human T lymphocytes. Mutation Research. 250:383–396 (1991b).

Nicotera, T.M., Privalle, C., Wang, T.C., Oshimura, M., Barrett, J.C. Differential proliferative responses of Syrian hamster embryo fibroblasts to paraquat-generated superoxide radicals depending on tumor-suppressor gene function. Cancer Research. 54:3884–3888 (1994).

Noble, K.B. (ed). Shielding Survey and Dosimetry Plan: Hiroshima-Nagasaki. Atomic Bomb Casualty Commission Technical Report 7-67 [ABCC TR- 7-67]. Hiroshima, ABCC (1967).

Nowell, P.C. The clonal evolution of tumor cell population. Science 194:23–28 (1976).

O'Neill, J.P., McGinniss, M.J., Berman, J.K., Sullivan, L.M., Nicklas, J.A., Albertini, R.J. Refinement of a T lymphocyte cloning assay to quantify the in vivo thioguanine-resistant mutant frequency in humans. Mutagenesis. 2:87–94 (1987).

O'Neill, J.P., Sullivan, L.M., Booker, J.K., Pornelos, B.S., Falta, M.T., Greene, C.J., Albertini, R.J. Longitudinal study of the in vivo *hprt* mutant frequency in human T lymphocytes as determined by a cell cloning assay. Environmental and Molecular Mutagenesis. 13:289–293 (1989).

O'Neill, J.P., Nicklas, J.A., Hunter, T.C., Batson, O.B., Allegretta, M., Falta, M.T., Branda, R.F., Albertini, R.J. The effect of T-lymphocyte "clonality" on the calculated *hprt* mutation frequency occurring in vivo in humans. Mutation Research. 313:215–225 (1994).

Oho, G. Statistical observation on deaths due to malignant neoplasm in A-bomb survivors. Nihon Iji Shimpo (Japanese Medical Journal). 1686:8–19 (1956).

Ohtaki, K. G-banding analysis of radiation-induced chromosome damage in lymphocytes of Hiroshima A-bomb survivors. Japanese Journal of Human Genetics. 37:245–262 (1992).

Ohtaki, K., Shimba, H., Awa, A.A., Sofuni, T. Comparison of type and frequency of chromosome aberrations by conventional and G-staining methods in Hiroshima atomic bomb survivors. Journal of Radiation Research. 23:441–449 (1982).

Olive, P.L., Durand, R.E. Drug and radiation resistance in spheroids: Cell contact and kinetics. Cancer Metastasis Reviews. 13:121–138 (1994).

Ostrosky-Wegman, P., Montero, R.M., Cortinas de Nova, C., Tice, R.R., Albertini, R.J. The use of bromodeoxyuridine labeling in the human lymphocyte HGPRT somatic mutation assay. Mutation Research. 191:211–214 (1988).

Ostrosky-Wegman, P., Montero, R., Palao, A., Cortinas de Nava, C., Hurtado, F., Albertini, R.J. 6-thioguanine-resistant T-lymphocyte autoradiographic assay. Determination of variant frequencies in individuals suspected of radiation exposure. Mutation Research. 232:49–56 (1990).

Oughterson, A.S., Warren, S. Medical Effects of the Atomic Bomb in Japan. New York, McGraw-Hill (1956).

Papadopoulos, N., Nicolaides, N.C., Wei, Y-F., Ruben, S.M., Carter, K.C., Rosen, C.A., Haseltine, W.A., Fleischmann, R.D., Fraser, C.M., Adams, M.D., Venter, J.C., Hamilton, S.R., Petersen, G.M., Watson, P., Lynch, H.T., Peltomaki, P., Mecklin, J-P., de la Chapelle, A., Kinzler, K.W., Vogelstein, B. Mutation of a *mutL* homolog in hereditary colon cancer. Science. 263:1625–1629 (1994).

Parkin, D.M., Muir, C., Whelan, S., Gao, Y-T., Ferlay, J., Powell, J. Cancer Incidence in Five Continents, Volume VI. International Agency for Research on Cancer (IARC), World Health Organization. Lyon, IARC (1992).

Perera, F.P., Motzer, R.J., Tang, D., Reed, E., Parker, R., Warburton, D., O'Neill, J.P., Albertini, R., Bigbee, W.L., Jensen, R.H., Santella, R., Tsai, W.-Y., Simon-Cereijido, G., Randall, C., Bosl, G. Multiple biologic markers in germ cell tumor patients treated with platinum-based chemotherapy. Cancer Research. 52:3558–3565 (1992).

Perera, F.P., Tang, D.L., O'Neill, P., Bigbee, W.L., Albertini, R., Santella, R., Ottman, R., Tsai, W.Y., Dickey, C., Mooney, L.A., Savela, K., Hemminki, K. *hprt* and glycophorin A mutations in foundry workers: Relationship to PAH exposure and to PAH-DNA adducts. Carcinogenesis. 14:969–973 (1993).

Perera, F.P., Dickey, C., Santella, R., O'Neill, J.P., Albertini, R.J., Ottman, R., Tsai, W.Y., Mooney, L.A., Savela, K., Hemminki, K. Carcinogen-DNA adducts and gene mutation in foundry workers with low-level exposure to polycyclic aromatic hydrocarbons. Carcinogenesis. 15:2905–2910 (1994).

Peto, R. Epidemiology, multistage models, and short-term mutagenicity tests, pp. 1403–1428. In: Origins of Human Cancer. H.H. Hiatt, J.A. Winsten (eds.). Cold Spring Harbor, Cold Spring Harbor Laboratory (1977).

Pierce, D.A. An Overview of the Cancer Mortality Data on Atomic Bomb Survivors. Radiation Effects Research Foundation Commentary and Review Series 1-89 [RERF CR 1-89]. Hiroshima, RERF (1989).

Pierce, D.A., Vaeth, M. The shape of the cancer mortality dose-response curve for the A-bomb survivors. Radiation Research. 126:36–42 (1991).

Pierce, D.A., Preston, D.L., Ishimaru T. A Method for Analysis of Cancer Incidence in Atomic Bomb Survivors, with Application to Leukemia. Radiation Effects Research Foundation Technical Report 15-83 [RERF TR 15-83]. Hiroshima, RERF (1983).

Pierce, D.A., Stram, D.O., Vaeth, M. Allowing for random errors in radiation exposure estimates for the atomic bomb survivor data. Radiation Research. 123:275–284 (1990). [RERF TR 5-88].

Pierce, D.A., Shimizu, Y., Preston, D.L., Vaeth, M., Mabuchi, K. Studies of the Mortality of Atomic Bomb Survivors. Report 12, Part 1. Cancer: 1950–90. Radiation Research. 146:1–27 (1996).

Pinkston, J.A., Sekine, I. Postirradiation sarcoma (malignant fibrous histiocytoma) following cervix cancer. Cancer. 49:434–438 (1982). [RERF TR 11-80].

Pinkston, J.A., Antoku, S., Russell, W.J. Malignant neoplasms among atomic bomb survivors following radiation therapy. Acta Radiologica: Oncology, Radiation, Physics, Biology. 20:267–271 (1981). [RERF TR 3-80].

Pirollo, K., Tong, Y.A., Villego, S.Z., Chen, Y., Chung, E.H. Oncogene-transformed NIH 3T3 cells display radiation resistance levels indicative of a signal transduction pathway leading to the radiation-resistance phenotype. Radiation Research. 135:234–243 (1993).

Pitot, H.C., Goldsworthy, T.L., Moran, S. The natural history of carcinogenesis: Implications of experimental carcinogenesis in the genesis of human cancer. Journal of Supramolecular Structure and Cellular Biochemistry. 17:133–146 (1981).

Pitts, J.D. Cancer gene therapy: A bystander effect using the gap junctional pathway. Molecular Carcinogenesis. 11:127–130 (1994).

Pollycove, M. Positive effects of low level radiation in human populations, pp. 171–187. In: Biological Effects of Low Level Exposures. Calabrese, E.J. (ed.). Boca Raton (FL), Chemical Rubber Company Press (1994).

Potten, C.S., Merritt, A., Hickman, J., Hall, P., Faranda, A. Characterization of radiation-induced apoptosis in the small intestine and its biological implications. International Journal of Radiation Biology. 65:71–78 (1994).

Potter, V.R. Probabilistic aspects of the human cybernetic machine. Perspectives in Biology and Medicine. 17:164–183 (1974).

Potter, V.R. Phenotypic diversity in experimental hepatomas: The concept of partially blocked ontogeny. British Journal of Cancer. 38:1–23 (1978).

Potter, V.R. Cancer as a problem in intercellular communication: Regulation by growth-inhibiting factors. Progress in Nucleic Acid Research and Molecular Biology. 29:161–173 (1983).

Preston, D. Cancer risks and biomarker studies in the atomic bomb survivors. Stem Cells. 13:S40–S48 (1995).

Preston, D.L., Pierce, D.A. The effect of changes in dosimetry on cancer mortality risk estimates for the atomic bomb survivors. Radiation Research. 114:437–466 (1988). [RERF TR 9-87].

Preston, D.L., Kato, H., Kopecky, K.J., Fujita, S. Life Span Study Report 10. Part 1. Cancer mortality among A-bomb survivors in Hiroshima and Nagasaki, 1950–82. Radiation Research. 111:151–178 (1987). [RERF TR 1-86].

Preston, D.L., McConney, M.E., Awa, A.A., Ohtaki, K., Itoh, M., Honda, T. Comparison of the Dose-Response Relationships for Chromosome Aberration Frequencies Between the T65D and DS86 Dosimetry. Radiation Effects Research Foundation Technical Report 7-88 [RERF TR 7-88]. Hiroshima, RERF (1988).

Preston, D.L., Lubin, J.H., Pierce, D.A. Epicure User's Guide. Seattle, Hirosoft International, Inc. (1993).

Preston, D.L., Kusumi, S., Tomonaga, M., Izumi, S., Ron, E., Kuramoto, A., Kamada, N., Dohy, H., Matsuo, T., Nonaka, H., Thompson, D.E., Soda, M., Mabuchi, K. Cancer incidence in atomic bomb survivors. Part III: Leukemia, lymphoma, and multiple myeloma, 1950-1987. Radiation Research. 137:S68–S97 (1994). [RERF TR 24-92].

Prince-Embury, S., Rooney, J.F. Psychological symptoms of residents in the aftermath of the Three Mile Island nuclear accident and restart. Journal of Social Psychology. 128:779–790 (1988).

Prince-Embury, S., Rooney, J.F. Psychological adaptation among residents following restart of Three Mile Island. Journal of Traumatic Stress. 8:47–59 (1995).

Putnam, F.W. Hiroshima and Nagasaki Revisited: The Atomic Bomb Casualty Commission and the Radiation Effects Research Foundation. Perspectives in Biology and Medicine. 37:515–545 (1994).

Putnam, F.W. Review of "Suffering Made Real" by Lindes, M.S. Journal of the American Medical Association. 274:430–431 (1995).

Radiation Effects Research Foundation (RERF) Newsletter. Memories of ABCC-RERF: Commemoration of the 40th Anniversary of US-Japan Joint Studies of Late A-bomb Effects, pp. 28-30. RERF Newsletter. Vol. 14(4). Hiroshima, RERF (1988).

Radiation Effects Research Foundation (RERF), 1983a. US-Japan Joint Workshop for Reassessment of Atomic Bomb Radiation Dosimetry in Hiroshima and Nagasaki. Kato, H., Fujita, S., Ellett, W., Jablon, S., Maruyama, T., Radford, E.P., Taira, S., Day, G. (eds.). Hiroshima, RERF Printing Office (1983a).

Radiation Effects Research Foundation (RERF), 1983b. Second US-Japan Joint Workshop for Reassessment of Atomic Bomb Radiation Dosimetry in Hiroshima and Nagasaki with Special Reference to Shielding and Organ Dose. Kato, H., Fujita, S., Ellett, W., Jablon, S., Maruyama, T., Radford, E.P., Taira, S., Day, G. (eds.). Hiroshima, RERF Printing Office (1983b).

Radiation Effects Research Foundation (RERF), 1987. US-Japan Joint Reassessment of Atomic Bomb Radiation Dosimetry in Hiroshima and Nagasaki: Final report. Roesch, W.C. (ed.). Hiroshima, RERF (1987).

Rainville, I.R., Albertini, R.J., Nicklas, J.A. Breakpoints and junctional regions of intragenic deletions in the *hprt* gene in human T cells. Somatic Cell and Molecular Genetics. 21:309–326 (1995).

Ramakrishan, N., McClain, D.E., Catravas, G.N. Membranes as sensitive targets in thymus apoptosis. International Journal of Radiation Biology. 63:693–701 (1993).

Ren, P., de Feijter, A.W., Paul, D.L., Ruch, R.J. Enhancement of liver cell gap junction protein expression by glucocorticoids. Carcinogenesis. 15:1807–1813 (1994).

Renan, M.J. Point mutations, deletions, and radiation carcinogenesis. Radiation Research. 131:227–228 (1992).

Revel, J.P. The oldest multicellular animal and its junction, pp. 135–149. In: Gap Junctions. Hertzburg, E.L., Johnston, R.L. (eds.). New York, Alan R. Liss (1988).

Rhoades, W.A. Development of a code system for determining radiation protection of armored vehicles (the VCS Code), ORNL-TM-4664, Oak Ridge National Laboratory. Oak Ridge (TN), Oak Ridge National Laboratory (ORNL) (1974).

Rhoades, W.A., Barnes, J.M., Santoro, R.T. An explanation of the Hiroshima activation dilemma, pp. 238–244. Proceedings of the 8th International Conference of Radiation Shielding. La Grange Park (IL), American Nuclear Society (1994).

Ritchie, R.H., Hurst, G.S. Penetration of weapons radiation: Application to the Hiroshima-Nagasaki studies. Health Physics. 1:390–404 (1959).

Robins, L.N., Regier, D.A. (eds.). Psychiatric Disorders in America: The Epidemiologic Catchment Area Study. New York, The Free Press (1991).

Robinson, D., Goodall, K., Albertini, R.J., O'Neill, J.P., Finette, B., Sala-Trepat, M., Moustacchi, E., Tates, A.D., Beare, D.M., Green, M.H.L., Cole, J. An analysis of in vivo *hprt* mutant frequency in circulating T lymphocytes in the normal human population: A comparison of four databases. Mutation Research. 313:227–247 (1994).

Roderick, T.H. Using inversions to detect and study recessive lethals and detrimentals in mice, pp. 135–167. In: Utilization of Mammalian Specific Locus Studies in Hazard Evaluation and Estimation of Genetic Risk. de Serres, F.J., Sheridan, W. (eds.). New York, Plenum (1983).

Roesch, W.C. (ed.). US-Japan Joint Reassessment of Atomic Bomb Radiation Dosimetry in Hiroshima and Nagasaki, Final Report, Radiation Effects Research Foundation. Hiroshima, RERF (1987).

Rogot, E., Murray, J.L. Smoking and causes of death among U.S. veterans: 16 years of observation. Public Health Reports. 95:213–222 (1980).

Ron, E., Preston, D.L., Mabuchi, K., Thompson, D.E., Soda, M. Cancer incidence in atomic bomb survivors. Part IV: Comparison of cancer incidence and mortality. Radiation Research. 137:S98–S112 (1994a).

Ron, E., Carter, R., Jablon, S., Mabuchi, K. Agreement between death certificate and autopsy diagnoses among atomic bomb survivors. Epidemiology. 5:48–56 (1994b). [RERF CR 6-92].

Rose, P.F., Dunford, C. (eds.). Data Formats and Procedures for the Evaluated Nuclear Data File, ENDF-6, ENDF-102 (1991). Brookhaven (NY), DoE (1991).

Rossi, H.H., Mays, C.W. Leukemia risk from neutrons. Health Physics. 34:353–360 (1978).

Rubonis, A., Bickman, L. Psychological impairment in the wake of disaster: The disaster-psychopathology relationship. Psychological Bulletin. 190:384–399 (1991).

Russell, W.L. Effect of the interval between irradiation and conception on mutation frequency in female mice. Proceedings of the National Academy of Sciences. 54:1552–1557 (1965).

Russell, W.J., Antoku, S. Radiation therapy among A-bomb survivors. American Journal of Public Health. 66:773–777 (1976). [ABCC TR 36-71].

Russell, W.L., Kelly, E.M. Mutation frequencies in male mice and the estimation of genetic hazards of radiation in men. Proceedings of the National Academy of Sciences. 79:542–544 (1982).

Russell, W.L., Russell, L.B., Kelly, E.M. Radiation dose rate and mutation frequency. Science. 128:1546–1550 (1958).

Russell, W.J., Ishimaru, T., Ihno, Y. ABCC-JNIH Adult Health Study, Hiroshima and Nagasaki, 1962. Exposures to Medical X-ray. July–November 1962; Survey of Subjects. Atomic Bomb Casualty Commission Technical Report 9-63 [ABCC TR 9-63]. Hiroshima, ABCC (1963).

Russell, W.J., Yoshinaga, H., Antoku, S., Mizuno, M. Active bone marrow distribution in the adult. British Journal of Radiology. 39:735–739 (1966). [ABCC TR 28-64].

Saez, J.C., Spray, D.S., Hertzberg, E.L. Gap junctions: Biochemical properties and functional regulation under physiological and toxicological conditions. In Vitro Toxicololgy. 3:69–86 (1990).

Sala-Trepat, M., Cole, J., Green, M.H.L., Rigaud, O., Vilcoq, J.R., Moustacchi, E. Genotoxic effects of radiotherapy and chemotherapy on the circulating lymphocytes of breast cancer. III: Measurement of mutant frequency to 6-thioguanine-resistance. Mutagenesis. 5:593–598 (1990).

Sasaki, T., Yamamoto, M., Yamaguchi, T., Sugiyama, S. Development of multicellular spheroids of HeLa cells co-cultured with fibroblasts and their response to x-irradiation. Cancer Research. 44:345–351 (1984).

Satoh, C., Takahashi, N., Asakawa, J., Hiyama, K., Kodaira, M. Variations among Japanese of the factor IX gene (F9) detected by PCR-denaturing gradient gel electrophoresis. American Journal of Human Genetics. 52:167–175 (1993).

Saunders, E.L., Meredith, M.J., Eisert, D.R., Freeman, M.L. Depletion of glutathione after gamma irridation modifies survival. Radiation Research. 125:267–276 (1991).

Savitsky, K., Bar-Shira, A., Gilad, S., Rotman, G., Ziv, Y., Vanagaite, L., Tagle, D.A., Smith, S., Uziel, T., Sfez, S., Ashkenazi, M., Pecker, I., Frydman, M., Harnik, R., Patanjali, S.H., Simmons, A., Clines, G.A., Sartiel, A., Gatti, R.A., Chessa, L., Sanal, O., Lavin, M.F., Jaspers, N.G.J., Taylor, A.M.R., Arlett, C.F., Miki, T., Weissman, S.M., Lovett, M., Collins, F.S., Shiloh, Y. A single ataxia telangiectasia gene with a product similar to pi-3 kinase. Science. 268:1749–1753 (1995).

Sawada, S., Wakabayashi, T., Takeshita, K., Yoshinaga, H., Russell, W.J. Radiological practice since the atomic bombs, Hiroshima and Nagasaki. American Journal of Public Health. 61:2455–2468 (1971a). [ABCC TR 25-67].

Sawada, S., Takeshita, K., Yamamoto, O., Russell, W.J., Land, C., Fujita, S. Fluoroscopy and Radiation Therapy Exposure Reported by ABCC-JNIH Adult Health Study Subjects, Hiroshima: Pilot Studies. Atomic Bomb Casualty Commission Technical Report 1-71 [ABCC TR 1-71]. Hiroshima, ABCC (1971b).

Sawada, S., Wakabayashi, T., Takeshita, K., Russell, W.J., Yoshinaga, H., Ihno, Y. ABCC-JNIH Adult Health Study, Exposure to Medical X-ray in Community Hospitals and Clinics Survey of Subjects, February 1964–January 1965. Atomic Bomb Casualty Commission Technical Report 24-67 [ABCC TR 24-67]. Hiroshima, ABCC (1967).

Sawada, S., Fujita, S., Russell, W.J., Takeshita, K. Radiological practice in Hiroshima and Nagasaki trends from 1964 to 1970. American Journal of Public Health. 65:622–633 (1975). [ABCC TR 41-72].

Sawada, S., Land, C.E., Otake, M ., Russell, W.J., Takeshita, K., Yoshinaga, H., Hombo, Z. Hospital and Clinic Survey Estimates of Medical X-ray Exposures in Hiroshima and Nagasaki. Part 1. RERF Population and the General Population. Radiation Effects Research Foundation Technical Report 16-79 [RERF TR 16-79]. Hiroshima, RERF (1979).

Schlager, G., Dickie, M.M. Spontaneous mutations and mutation rates in the house mouse. Genetics. 57:319–330 (1967).

Schull, W.J. Song Among the Ruins. Cambridge (MA), Harvard University Press (1990).

Schull, W.J. Effects of Atomic Radiation: A Half Century of Studies from Hiroshima and Nagasaki. New York, John Wiley and Sons (1995).

Schull, W.J., Neel, J.V. The Effects of Inbreeding on Japanese Children. New York, Harper and Row (1965).

Searle, A.G. Mutation induction in mice, pp. 131–207. In: Advances in Radiation Biology. Lett, J.T., Adler, H.I., Zelle, M. (eds.). New York, Academic Press (1974).

Seifert, A.M., Bradley, W.C., Messing, K. Exposure of nuclear medicine patients to ionizing radiation is associated with rises in *hprt*-mutant frequency in peripheral T lymphocytes. Mutation Research. 191:57–63 (1988).

Shattuck-Eidens, D., McClure, M., Simard, J., Labrie, F., Narod, S.A., Couch, F., Hoskins, K., Weber, B., Castilla, L., Erdos, M., Brody, L., Friedman, L., Oster-meyer, E., Szabo, C., King., M-C., Jhanwar, S., Offit, K., Norton, L., Gilewski, T., Lubin, M., Osborne, M., Black, D., Boyd, M., Steel, M., Ingles, S., Haile, R., Lindblom, A., Olsson, H., Borg, A., Bishop, T., Solomon, E., Radice, P., Spatti, G., Gayther, S., Ponder, B., Warren, W., Stratton, M., Liu, Q., Fujimura, F., Lewis, C., Skolnick, M.H., Goldgar, D.E. A collaborative survey of 80 muta-tions in the BRCA1 breast and ovarian cancer susceptibility gene: Implications for presymptomatic testing and screening. Journal of the American Medical Association. 273:535–541 (1995).

Sheridan, J.D. Cell communication and growth, pp. 187–232. In: Cell to Cell Communication. New York, Plenum Press (1987).

Shields, P.G., Harris, C.C. Molecular epidemiology and the genetics of environmental cancer. Journal of the American Medical Association. 266:681–687 (1991).

Shigematsu, I., Kagan, A. Radiation Effects Research Foundation: The First 10 Years, 1975–85. Hiroshima, RERF (1985).

Shigematsu, I., Mendelsohn, M.L. The Radiation Effects Research Foundation of Hiroshima and Nagasaki. Past, present, and future. Journal of the American Medical Association. 274:425–426 (1995).

Shimizu, Y., Kato, H., Schull, W.J., Preston, D.L., Fujita, S., Pierce, D.A. Life Span Study Report 11. Part 1. Comparison of Risk Coefficients for Site-Specific

Cancer Mortality Based on the DS86 and T65DR Shielded Kerma and Organ Doses. Radiation Research. 118:502–524 (1989). [RERF TR 12-87].

Shimizu, Y., Kato, H., Schull, W.J. Life Span Study Report 11, Part 2. Cancer Mortality in the Years 1950–85 Based on the Recently Revised Doses (DS86). Radiation Research. 121:120–141 (1990). [RERF TR 5-88].

Shore, R.E., Hildreth, N., Woodard, E., Dvoretsky, P., Hempelmann, L., Pasternack, B. Breast cancer among women given x-ray therapy for acute postpartum mastitis. Journal of the National Cancer Institute. 77:689–696 (1986).

Sierra-Rivera, E., Meredith, M.J., Summer, M.L., Smith, M.D., Voorhees, G.J, Stoffel, C.M., Freeman, M.L. Genes regulation glutathione concentrations in x-ray-transformed rat embryo fibroblasts: Changes in gamma-glutamylcysteine synthetase and gamma-glutamyltranspeptidase expression. Carcinogenesis. 15:1301–1307 (1994).

Sinclair, W.K. Science, Radiation Protection, and the NCRP. Radiation Effects Research Foundation RERF Update 6:3–5 (1994). Hiroshima, RERF (1994).

Sinclair, W.K., Failla, P. Dosimetry of the atomic bomb survivors. Radiation Research. 88:437–447 (1981).

Siskin, E.E., Gray, T., Barrett, J.C. Correlation between sensitivity to tumor promotion and sustained hyperplasia of mice and rats treated with 12-o-tetradecanoylphorbol-13-acetate. Carcinogenesis. 3:403–407 (1982).

Sklar, M.D. The ras oncogenes increase the intrinsic resistance of NIH 3T3 cells to ionizing radiation. Science. 239:645–647 (1988).

Skolnick, M.M., Neel, J.V. An algorithm for comparing two-dimensional electrophoretic gels, with particular reference to the study of mutation, pp. 55–160. In: Advances in Human Genetics. Harris, H., Hirschhorn, K. (eds.). New York, Plenum Press (1986).

Slichenmyer, W.J., Nelson, W.G., Slebos, R.J., Kasten, M.B. Loss of a p53 associated G1 checkpoint does not decrease cell survival following DNA damage. Cancer Research. 53:4164–4168 (1993).

Slovic, P. Perception of risk from radiation, pp. 211–227. In: The Medical Basis for Radiation Preparedness. III: The Psychological Perspective. Ricks, R., Berger, M.E., O'Hara, F.M. (eds.). New York, Elsevier (1991).

Snell, F.M., Neel, J.V., Ishibashi K. Hematologic studies in Hiroshima and a control city two years after the atomic bomb. Archives of Internal Medicine. 84:569–604 (1949). [RERF TR 27-A-59].

Solomon, S.D. Research issues in assessing disaster's effects, pp. 308–340. In: Psychosocial Aspects of Disaster. Gist, R., Lubin, B. (eds.). New York, John Wiley and Sons (1989).

Solomon, S.D. Mental health effects of natural and human-made disasters. PTSD Research Quarterly. 3:1–7 (1992).

Solomon, S.D., Smith, E.M., Robins, L., Fischbach, R. Social involvement as a mediator of disaster-induced stress. Journal of Applied Social Psychology. 17:1092–1112 (1987).

Solomon, Z., Laor, N., Weiler, D. The psychological impact of the Gulf War: A study of acute stress in Israeli evacuees. Archives of General Psychiatry. 50:320–321 (1993).

Somosy, Z., Kovac, J., Siklos, L., Koteles, G.J. Morphological and histological changes in intercellular junctional complexes in epithelial cells of mouse small intestine upon x-irradiation: Changes of ruthenium red permeability and calcium content. Scanning Microscopy 7:961–971 (1993).

Sondhaus, C.A., Bond, V.P., Feinendegen, L.E. Cell-oriented alternatives to quality factor and dose equivalent for low-level radiation. Health Physics. 59:35–48 (1990).

Spadinger, I., Palcic, B. The relative biological effectiveness of ^{60}Co γ-rays, 55 kVp X-rays, 250 kVp X-rays, and 11 MeV electrons at low doses. International Journal of Radiation Biology. 61:345–353 (1992).

Spielberger, C. State-Trait Anxiety Inventory for Children. Palo Alto (CA), Consulting Psychologists Press (1973).

Sposto, R., Preston, D.L. Correction for Catchment Area Nonresidency in Tumor-Registry-Based Cohort Studies. Radiation Effects Research Foundation Technical Report 1-92 [RERF CR 1-92]. Hiroshima, RERF (1992).

Sposto, R., Stram, D.O., Awa, A.A. An estimate of the magnitude of random error in the DS86 dosimetry from data on chromosome aberrations and severe epilation. Radiation Research. 128:157–169 (1991).

Sposto, R., Preston, D.L., Shimizu, Y., Mabuchi, K. The effect of diagnostic misclassification on non-cancer and cancer mortality dose response in A-bomb survivors. Biometrics. 48:605–617 (1992). [RERF TR 4-91].

Srole, L., Langner, T.S., Michael, S.T., Kirkpatrick, P., Opler, M.K., Rennie, T.A. Mental Health in the Metropolis: The Midtown Manhattan Study. New York, Harper and Row (1962).

Stevenson, M.A., Pollock, S.S., Coleman, C.N., Calderwood, S.K. X-radiation, phorbol esters, and H_2O_2 stimulate mitogen-activated protein kinase activity in NIH-3T3 cells through the formation of reactive oxygen intermediates. Cancer Research. 54:12–15 (1994).

Stewart, A.M., Kneale, G.W. A-bomb radiation and evidence of late effects other than cancer. Health Physics. 58:729–735 (1990).

Stiehm, E.R. The psychologic fallout from Chernobyl. American Journal of Diseases of Children. 146:761–762 (1992).

Stovall, M. Organ doses from radiotherapy of cancer of the uterine cervix, pp. 131–136. In: Second Cancer in Relation to Radiation Treatment for Cervical Cancer. International Agency for Research on Cancer (IARC), World Health Organization. Scientific Publication No. 52. Day, N.E., Boice J.D. Jr. (eds.). Lyon, IARC (1983).

Stram, D.O., Sposto, R., Preston, D., Abrahamson, S., Honda, T., Awa, A.A. Stable Chromosome Aberrations Among A-bomb Survivors: An Update. Radiation Research. 136:29–36 (1993).

Straume, T., Egbert, S.D., Woolson, W.A., Finkel, R.C., Kubik, P.W., Gove, H.E., Sharma, P., Hoshi, M., Neutron discrepancies in the DS86 Hiroshima dosimetry system. Health Physics. 63:421–426 (1992).

Straume, T., Harris, L.J., Marchetti, A.A., Egbert, S.D. Neutrons confirmed in Nagasaki and at the Army pulsed radiation facility: Implications for Hiroshima. Radiation Research. 138:193–200 (1994).

Strauss, G.H., Albertini, R.J. 6-thioguanine-resistant lymphocytes in human peripheral blood, p. 327. In: Progress in Genetic Toxicology: Developments in Toxicology and Environmental Sciences, Volume 2. Scott, D., Bridges, B.A., Sobels, F.H. (eds.). Amsterdam, Elsevier/North Holland Biomedical Press (1977).

Strauss, G.H., Albertini, R.J. Enumeration of 6-thioguanine-resistant peripheral blood lymphocytes in man as a potential test for somatic cell mutations arising in vivo. Mutation Research. 61:353–379 (1979).

Struewing, J.P., Brody, L.C., Erdos, M.R., Kase, R.G., Giambarresi, T.R., Smith, S.A., Collins, F.S., Tucker, M.A. Detection of eight BRCA1 mutations in 10 breast/ovarian cancer families, including one family with male breast cancer. American Journal of Human Genetics. 57:1–7 (1995a).

Struewing, J.P., Aeliovich, D., Paretz, T., Avishai, N., Kaback, M.M., Collins, F.S., Brody, L.C. The carrier frequency of the BRCA1 185delAG mutation is approximately 1 percent in Ashkenazi Jewish individuals. Nature Genetics. 11:198–200 (1995b).

Sugahara, T., Sagan, L.A., Aoyama, T. (eds.). Low-Dose Irradiation and Biological Defense Mechanisms. Amsterdam, Excerpta Medica (1992).

Suzuki, H., Takahashi, T., Kuroishi, T., Suyama, M., Ariyoshi, Y., Takahashi, T., Ueda, R. *p53* mutations in non-small-cell lung cancer in Japan: Association between mutations and smoking. Cancer Research. 52:734–736 (1992).

Takahashi, N., Hiyama, K., Kodaira, M., Satoh, C. An improved method for the detection of genetic variations in DNA with denaturing gradient gel electrophoresis. Mutation Research. 234:61–70 (1990).

Takeshita, K., Antoku, S., Sawada, S. Exposure pattern, surface, bone marrow integral, and gonadal dose from fluoroscopy. British Journal of Radiology. 45:53–58 (1972). [ABCC TR 19-69].

Takeshita, K., Kihara, T., Sawada, S. Medical and dental radiological trends in Japan. Nippon Acta Radiologica 38:682–696 (1978). [ABCC TR 18-75].

Tan, W.-Y. Stochastic Models of Carcinogenesis. New York, Marcel Dekker (1991).

Tates, A.D., Grummt, T., Tornqvist, M., Farmer, P.B., van Dam, F.J., van Mossel, H., Schoemaker, H.M., Osterman-Golkar, S., Uebel, C., Tang, Y.S., Zwinderman, A.H., Natarajan, A.T., Ehrenberg, L. Biological and chemical monitoring of occupational exposure to ethylene oxide. Mutation Research. 250:483–497 (1991).

Thacker, J. The nature of mutants induced by ionizing radiation in cultured hamster cells. III. Molecular characterization of *hprt*-deficient mutants induced by gamma-rays or alpha particles showing that the majority have deletions of all or part of the *hprt* gene. Mutation Research. 160:267–275 (1986).

Thomas, D.C. A model for dose rate and duration of exposure effects in radiation carcinogenesis. Environmental Health Perspectives. 87:163–171 (1990).

Thompson, D.J. (ed.). US-Japan Joint Reassessment of Atomic Bomb Radiation Dosimetry in Hiroshima and Nagasaki. Proceedings of a Workshop Held at Nagasaki, Japan, 16–17 February 1983, Radiation Effects Research Foundation. Hiroshima, RERF (1983a).

Thompson, D.J. (ed.). US-Japan Joint Reassessment of Atomic Bomb Radiation Dosimetry in Hiroshima and Nagasaki. Proceedings of a Workshop Held at Hiroshima, Japan, 8–9 November 1983, Radiation Effects Research Foundation. Hiroshima, RERF (1983b).

Thompson, D.E., Mabuchi, K., Ron, E., Soda, M., Tokunaga, M., Ochikubo, S., Sugimoto, S., Ikeda, T., Terasaki, M., Izumi, S., Preston, D.L. Cancer incidence in atomic bomb survivors. Part II: Solid tumors, 1958–87. Radiation Research. 137:S17–S67 (1994). [RERF TR 5-92].

Timblin, C.R., Janssen, Y.W.M., Mossman, B.J. Transcriptional activation of the proto-oncogene c-jun by asbestos and H_2O_2 is directly related to increased proliferation and transformation of tracheal epithelial cells. Cancer Research. 55:2223–2226 (1995).

Tofilon, P.J., Buckley, W., Deen, D.F. Effect of cell-cell interactions on drug sensitivity and growth of drug-sensitive and resistant tumor cells in spheroids. Science. 226:862–864 (1984).

Tokunaga, M., Land, C.E., Yamamoto, T., Asano, M., Tokuoka, S., Ezaki, H., Nishimori, I. Incidence of female breast cancer among atomic bomb survivors, Hiroshima and Nagasaki, 1950–1980. Radiation Research. 112:243–272 (1987).

Tokunaga, M., Mabuchi, K., Kato, H., Shimaoka, K, Land, C.E. Guidelines for the Conduct of Site-Specific Cancer Incidence Studies Among A-bomb Survivors, Hiroshima and Nagasaki. Radiation Effects Research Foundation Technical Report 9-88 [RERF TR 9-88]. Hiroshima, RERF (1988).

Tokunaga, M., Land, C.E., Tokuoka, S., Nishimori, I., Soda, M., Akiba, S. Incidence of female breast cancer among atomic bomb survivors, Hiroshima and Nagasaki, 1950–1985. Radiation Research. 138:209–223 (1994).

Tokuoka, S., Kawai, K., Shimizu, Y., Inai, K., Ohe, K., Fujikura, T., Kato, H. Malignant and benign ovarian neoplasms among atomic bomb survivors, Hiroshima and Nagasaki, 1950-1980. Journal of the National Cancer Institute. 79:47-57 (1987). [RERF TR 8-86]

Tomonaga, M., Brill, A.B., Itoga, T., Heyssel, R.M. Leukemia in Atomic Bomb Survivors, Nagasaki. Atomic Bomb Casualty Commission Technical Report 11-59 [ABCC TR 11-59]. Hiroshima, ABCC (1959).

Tomonaga, M., Matsuo, T., Carter, R.L., Bennett, J.M., Kuriyama, K., Imanaka, F., Kusumi, S., Mabuchi, K., Kuramoto, A., Kamada, N., Ichimaru, M., Pisciotta, A.V., Finch, S.C. Differential Effects of Atomic Bomb Irradiation in Inducing Major Leukemia Types: Analyses of Open-City Cases Including the Life Span Study Cohort Based Upon Updated Diagnostic Systems and the Dosimetry System 1986

(DS86). Radiation Effects Research Foundation Technical Report 9-91 [RERF TR 9-91]. Hiroshima, RERF (1991).

Trosko, J.E. Possible role of intercellular communication in the modulation of the biological response to radiation. Yokohama Medical Bulletin. 42:151–165 (1991).

Trosko, J.E. Does Radiation "Cause" Cancer? Radiation Effects Research Foundation RERF Update 4:3–5 (1992). Hiroshima, RERF (1992).

Trosko, J.E. Radiation-induced carcinogenesis: Paradigm considerations, pp. 205–241. In: Biological Effects of Low-Level Exposures. Calabrese, E.J. (ed.). Boca Raton (Fl), Lewis Publishers (1994).

Trosko, J.E. Biomarkers for low-level exposure causing epigenetic responses in stem cells. Stem Cells. 13:231-239 (1995).

Trosko, J.E. The role of radiation in multistep carcinogenesis. Health Physics. 70:812–822 (1996).

Trosko, J.E., Chang, C.C., Medcalf, A. Mechanism of tumor promotion: Potential role of intercellular communication. Cancer Investigation. 1:511–526 (1983).

Trosko, J.E., Chang, C.C., Madhukar, B.V., Klaunig, J.E. Chemical, oncogene, and growth factor inhibition of gap junctional intercellular communication: An integrative hypothesis of carcinogenesis. Pathobiology. 58:265–278 (1990).

Trosko, J.E., Chang, C.C., Dupont, E., Madhukar, B.V., Kalimi, G. Chemical modulation of gap junctional intercellular communication in vitro: An in vitro biomarker of epigenetic toxicology, pp. 465–478. In: In Vitro Methods in Toxicology. Jolles, G. and Cordier, A. (eds.). London, Academic Press (1992).

Trosko, J.E., Chang, C.C., Madhukar, B.V., Dupont, E. Oncogenes, tumor suppressor genes, and intercellular communication in the "Oncogeny as Partially Blocked Ontogeny Theory," pp. 181–197. In: New Frontiers in Cancer Research. Iversen, O.H. (ed.). Washington, DC, Taylor and Francis (1993).

Tsuang, M.T., Tohen, M., Zahner, G.E.P. Textbook in Psychiatric Epidemiology. New York, John Wiley and Sons (1995).

Tucker, H.G. A stochastic model for a two-stage theory of carcinogenesis, pp. 347–403. In: Fifth Berkeley Symposium on Mathematical Statistics and Probability, Berkeley, University of California Press (1967).

Turner, D.R., Morley, A.A., Haliandros, M., Kutlaca, B., Sanderson, B.J. In vivo somatic mutations in human lymphocytes frequently result from major gene alterations. Nature. 315:343–345 (1985).

Tycko, B., Sklar, J. Chromosomal translocations in lymphoid neoplasia: A reappraisal of the recombinase model. Cancer Cells. 2:1–8 (1990).

United Nations Scientific Committee on the Effects of Atomic Radiation (UNSCEAR). Sources, Effects, and Risks of Ionizing Radiation, 1988 Report to the General Assembly, with Annexes. New York, United Nations (1988).

United Nations Scientific Committee on the Effects of Atomic Radiation (UNSCEAR). Sources, Effects, and Risks of Ionizing Radiation, 1993 Report to the General Assembly, with Annexes. New York, United Nations (1993).

United Nations Scientific Committee on the Effects of Atomic Radiation (UNSCEAR). Sources, Effects, and Risks of Ionizing Radiation, 1994 Report to the General Assembly, with Annexes. New York, United Nations (1994).

Uckum, F., Tuel-Ahlgren, L., Song, C.W., Waddick, K., Meyers, D.E., Kirihara, J., Ledbetter, J.A., Schieveu, G.L. Ionizing radiation stimulates unidentified tyrosine-specific protein kinase in human B-lymphocyte precursors, triggering apoptosis and clonogenic cell death. Fundamental and Applied Toxicology. 89:9005–9009 (1992).

Ueda, H., Ullrich, S.J., Gangemi, J.D., Kappel, C.A., Ngo, L., Feitelson, M.A., Jay, G. Functional inactivation but not structural mutation of *p53* causes liver cancer. Nature Genetics. 9:41–47 (1995).

Uitterlinden, A.G., Slagboom, P.E., Knook, D.L., Vijg, J. Two-dimensional DNA fingerprinting of human individuals. Proceedings of the National Academy of Sciences. 86:2742–2746 (1989).

Upton, A.C. Health impact of the Three Mile Island accident. In: The Three Mile Island Nuclear Accident: Lessons and Implications. Moss, T.H., Sills, D.L. (eds.). Annals of the New York Academy of Sciences. 365:63–70 (1981).

Upton, A.C. The question of thresholds for radiation and chemical carcinogenesis. Cancer Investigation. 7:267–276 (1989).

Upton, A.C., Shore, R.E., Harley, N.H. The health effects of low-level ionizing radiation. Annual Review of Public Health. 13:127–150 (1992).

Vaeth, M., Preston, D.L., Mabuchi, K. The shape of the cancer incidence dose-response curve for the A-bomb survivors, pp. 75–78. In: Low Dose Irradiation and Biological Defense Mechanisms. Sugahara, T., Sagan, L.A., Aoyama, T. (eds.). Proceedings of the International Conference on Low Dose Irradiation and Biological Defense Mechanisms, Kyoto, Japan, 12–16 July, 1992. Amsterdam, Excerpta Medica (1992).

Valentine, W.N. Present Status of the Study of the Incidence of Leukemia Among Individuals Surviving Exposure to the Atomic Bomb in Hiroshima and Nagasaki. Atomic Bomb Casualty Commission, Semi-Annual Report. Hiroshima, ABCC (1951).

Vergouwen, R.P.F.A., Huiskamp, R., Bas, R.V., Roopers-Gajadien, H.L., de Jong, F.H., van Eerdenburg, F.J.C.M., Davids, J.A.G., de Rooij, D.G. Radiosensitivity of testicular cells in the prepubertal mouse. Radiation Research. 139:316–326 (1994).

Viinamaki, H., Kumpusalo, E., Myllykangas, M., Salomaa, S., Kumpusalo, L., Kolmakov, S., Ilchenko, I., Zhukowsky, G., Nissinen, A. The Chernobyl accident and mental wellbeing: A population study. Acta Psychiatrica Scandinavia. 91:396–401 (1995).

Vrieling, H., Thijssen, J.C.P., Rossi, A.M., van Dam, F.J., Natarajan, A.T., Tates, A.D., van Zeeland, A.A. Enhanced *hprt* mutant frequency but no significant difference in mutation spectrum between a smoking and a non-smoking human population. Carcinogenesis. 13:1625–1631 (1992).

Wakabayashi, T., Kato, H., Ikeda, T., Schull, W.J. Life Span Study Report 9. Part 3. Tumor Registry Data, Nagasaki, 1959–78. Radiation Effects Research Foundation Technical Report 6-81 [RERF TR 6-81]. Hiroshima, RERF (1981).

Wald, N. Leukemia in Hiroshima City atomic bomb survivors. Science. 127:699–700 (1958). [RERF TR 38-A-59].

Wald, N., Truax, W.E., Sears, M.E., Suzuki, G., Yamamoto, T. Hematologic findings in atomic bomb survivors, Hiroshima-Nagasaki; a 10-year review, pp. 382–389. In: Proceedings of the 6th International Congress of the Hematolology International Society. New York, Grune and Stratton (1958).

Ward, J.B., Ammenheuser, M.M., Bechtold, W.E., Whorton, E.B., Jr., Legator, M.S. hprt mutant lymphocyte frequencies in workers at a 1,3-butadiene production plant. Environmental Health Perspectives. 102(supp 9):79–85 (1994).

Ward, J.F. Radiation mutagenesis: The initial DNA lesions responsible. Radiation Research. 142:362–368 (1995).

Waterhouse, J., Muir, C, Shanmugaratnam, K., Powell, J. Cancer Incidence in Five Continents, Vol. IV. International Agency for Research on Cancer (IARC), World Health Organization. Lyon, IARC (1982).

Weichselbaum, R.R., Hallahan, D.E., Sukhatme, V., Dritschilo, A., Sherman, M.L., Kufe, D.W. Biological consequences of gene regulation after ionizing radiation exposure. Journal of the National Cancer Institute. 83:480–484 (1991).

Weiss, H.A., Darby, S.C., Doll, R. Cancer mortality following x-ray treatment for ankylosing spondylitis. International Journal of Cancer. 59:327–338 (1994).

Wiggs, L.D., Cox-DeVore, C.A., Wilkinson, G.S., Reyes, M. Mortality among workers exposed to external ionizing radiation at a nuclear facility in Ohio. Journal of Occupational Medicine. 33:632–637 (1991).

Wiggs, L.D., Johnson, E.R., Cox-Devore, C.A., Voelz, G.L. Mortality through 1990 among white male workers at the Los Alamos National Laboratory: Considering exposure to plutonium and external radiation. Health Physics. 67:577–588 (1994).

Wilkinson, G.S., Tietjen, G.L., Wiggs, L.D., Galke, W.A., Acquavella, J.F., Reyes, M., Voelz, G.L., Waxweiler, R.J. Mortality among plutonium and other radiation workers at a plutonium weapons factory. American Journal of Epidemiology. 125:231–50 (1987).

Williams, M., Rainville, I.R., Nicklas, J.A. Use of inverse PCR to amplify and sequence breakpoints of hprt deletion and translocation mutations. Environmental and Molecular Mutagenesis. 27:34 (1996).

Wilson, R.H., Fix, J.J., Baumgartner, W.V., Nichols, L.L. Description and Evaluation of the Hanford Personnel Dosimeter Program from 1944 Through 1989. Pacific Northwest Laboratory, PNL-7447. Richland (WA), PNL (1990).

Wilson, R.R. Nuclear radiation at Hiroshima and Nagasaki. Radiation Research. 4:349–359 (1956).

Wing, S., Shy, C.M., Wood, J.L., Wolf, S., Cragle, D.L., Frome, E.L. Mortality among workers at Oak Ridge National Laboratory: Evidence of radiation effects in follow-up through 1984. Radiation Research. 265:1397–1402 (1991).

Wolff, S. Is radiation all bad? The search for adaptation. Radiation Research. 131:117–123 (1992).

Woloschak, G.E., Chang-Liu, C.M., Shearin-Jones, P. Regulation of protein kinase C by ionizing radiation. Cancer Research. 50:3963–3967 (1990).

Woodbury, L.A., Holmes, R.R., Scott, J.K. Death Certificate Survey: Hiroshima 1950–1954. Preliminary Report of the Atomic Bomb Casualty Commission. Hiroshima, ABCC (1954).

Woolson, W.A., Gritzner, M.L., Egbert, S.D., Roberts, J.A., Otis, M.D. House and Terrain Shielding. pp. 227–305. In: US-Japan Joint Reassessment of Atomic Bomb Radiation Dosimetry in Hiroshima and Nagasaki: Final Report, Vol. 1. Roesch, W.C. (ed.). Hiroshima, RERF (1987).

Wooster, R., Neuhausen, S.L., Mangion, J., Quirk, Y., Ford, D., Collins, N., Nguyen, K., Seal, S., Tran, T., Averill, D., Fields, P., Marshall, G., Narod, S., Lenoir, G.M., Lynch, H., Feunteun, J., Deville, P., Cornelisse, C.J., Menko, F.H., Daly, P.A., Ormiston, W., McManus, R., Pye, C., Lewis, C.M., Cannon-Albright, L.A., Peto, J., Ponder, B.A.J., Skolnick, M.H., Easton, D.F., Goldar, D.E., Stratton, M.R. Localization of a breast cancer susceptibility gene, BRCA2, to chromosome 13q12-13. Science. 265:2088–2090 (1994).

World Health Organization Pilot Project, Brain damage in utero. Presented by project investigators at "Health Consequences of the Chernobyl and Other Radiological Accidents." Geneva, November 20–23, 1995. Geneva, WHO (1995).

Yamada, M., Kodama, K., Wong, F.L. The long-term psychological sequelae of atomic-bomb survivors in Hiroshima and Nagasaki, pp. 155–163. In: The Medical Basis of Radiation Preparedness III: The Psychological Perspective. Ricks, R.C., Berger, M.E., O'Hara, F.M. (eds.). New York, Elsevier (1991).

Yamamoto, O., Fujita, S. Medical and occupational radiation exposure reported by self-administered questionnaire. Nippon Acta Radiologica. 37:1144–1152 (1977). [ABCC TR 1-73].

Yamamoto, T., Kopecky, K.L., Fujikura, T., Tokuoka, S., Monzen, T., Nishimori, I., Nakashima, E., Kato, H. Lung cancer incidence among A-bomb survivors in Hiroshima and Nagasaki, 1950–80. Radiation Research. 128:156–171 (1987). [RERF TR 12-86].

Yamamoto, O., Antoku, S., Russell, W.J., Fujita, S., Sawada, S. Medical x-ray exposure doses as contaminants of atomic bomb doses. Health Physics. 54:257–269 (1988). [RERF TR 16-86].

Yamasaki, H., Enomoto, T., Hamel, E., Kanno, Y. Membrane interaction and modulation of gene expression by tumor promoters, pp. 221–233. In: Cellular Interactions by Environmental Tumor Promoters. Fujiki, H., Hecker, E., Moore, T., Sugimura, T., Weinstein, I.B. (eds.). Tokyo, Japan Scientific Societies Press (1984).

Yamasaki, H., Hollstein, M., Mesnil, M., Martel, N. Selective lack of intercellular communication between transformed and non-transformed cells as a common property of chemical and oncogene transformation of BALB/c3T3 cells. Cancer Research. 47:5658–5664 (1987).

Yamazaki, J.N., Fleming, L.B. Children of the Atomic Bomb. Durham (NC), Duke University Press (1995).

Yancey, S.B., Bismal, S., Revel, J.P. Spatial and temporal patterns of distribution of the gap junction protein connexin during mouse gastrulation and organogenesis. Development. 114:203–212 (1992).

Yang, J.-L., Maher, V.M., McCormick, J.J. Amplification of cDNA from the lysate of a small clone of diploid human cells and direct DNA sequencing. Genetics. 83:347–354 (1989).

Yi, M., Au, L.C., Ichikawa, N., Tso, P.O. Enhanced resolution of DNA restriction fragments: A procedure by two-dimensional electrophoresis and double labeling. Proceedings of the National Academy of Sciences. 87:3919–3923 (1990).

York, E.N. Dose calculations for Hiroshima and Nagasaki. In communication from M. Morgan, Air Force Special Weapons Center, to G.S. Hurst, ORNL (1957) [Cited by Ritchie and Hurst, 1959].

Yoshimoto, Y., Neel, J.V., Schull, W.J., Kato, H., Soda, M., Eto, R., Mabuchi, K. Malignant tumors during the first two decades of life in the offspring of atomic bomb survivors. American Journal of Human Genetics. 46:1041–1052 (1990).

Zhu, D., Caveney, S., Kidder, G.M., Naus, C.C.G. Transfection of C6 glioma cells with connexin43 cDNA: Analysis of expression, intercellular coupling, and cell proliferation. Fundamental and Applied Toxicology. 88:1883–1887 (1991).

Zhu, D., Kidder, G.M., Caveny, S., Naus, C.C.G. Growth retardation in glioma cells cocultured with cells overexpressing a gap junction protein. Fundamental and Applied Toxicology. 89:10218–10221 (1992).

Contributors

SEYMOUR ABRAHAMSON, *Department of Zoology, University of Wisconsin, 250 North Mills Street, Madison, Wisconsin 53706*

RICHARD J. ALBERTINI, *Genetics Laboratory, University of Vermont, 32 North Prospect, Burlington, Vermont 05401*

JUN-ICHI ASAKAWA, *Department of Genetics, Radiation Effects Research Foundation, 5-2 Hijiyama Park, Minami-ku, 732 Hiroshima, Japan*

AKIO A. AWA, *Department of Genetics, Radiation Effects Research Foundation, 5-2 Hijiyama Park, Minami-ku, 732 Hiroshima, Japan*

VICTOR P. BOND, *Medical Department, Building 490, Brookhaven National Laboratory, P.O. Box 5000, Upton, New York 11973*

EVELYN J. BROMET, *Department of Psychiatry, University at Stony Brook, Putnam Hall South Campus, Stony Brook, New York 11794-8790*

ETHEL S. GILBERT, *Health Protection Department, Battelle, Mail Stop K3-54, P.O. Box 999, Richland, Washington 99352*

RONALD H. JENSEN, *Division of Molecular Cytometry, Department of Laboratory Medicine, Room MCB 230, University of California-San Francisco, San Francisco, California 94103*

SAMIR M. HANASH, *Medical Science II M4708, Department of Human Genetics, University of Michigan, 1500 East Medical Center, Ann Arbor, Michigan 48109-0618*

KAZUO KATO, *Radiation Effects Research Foundation, 5-2 Hijiyama Park, Minami-ku, 732 Hiroshima, Japan*

DEAN C. KAUL, *Science Applications International Corporation, 10260 Campus Point Drive, San Diego, California 92121*

GEORGE D. KERR, *Oak Ridge National Laboratory, 1060 Commerce Park, Oak Ridge, Tennessee 37830-6480*

KAZUNORI KODAMA, *Department of Clinical Studies, Radiation Effects Research Foundation, 5-2 Hijiyama Park, Minami-ku, 732 Hiroshima, Japan*

YOSHIAKI KODAMA, *Department of Genetics, Radiation Effects Research Foundation, 5-2 Hijiyama Park, Minami-ku, 732 Hiroshima, Japan*

356

RORK KUICK, *Medical Science II M4708, Department of Human Genetics, University of Michigan, 1500 East Medical Center, Ann Arbor, Michigan 48109-0618*

CHARLES E. LAND, *Radiation Epidemiology Branch, National Cancer Institute, 6130 Executive Boulevard, EPN-408, Bethesda, Maryland 20892*

MARK P. LITTLE, *National Radiological Protection Board, Chilton, Didcot, Oxfordshire, OX11-0RQ, United Kingdom*

KIYOHIKO MABUCHI, *Department of Epidemiology, Radiation Effects Research Foundation, 5-2 Hijiyama Park, Minami-ku, 732 Hiroshima, Japan*

MORTIMER L. MENDELSOHN, *Biology and Biotechnology Research Program, Lawrence Livermore National Laboratory, P.O. Box 808, L-452, Livermore, California 94551-9900*

CHYUZO MIYAZAWA, *Department of Genetics, Radiation Effects Research Foundation, 5-2 Hijiyama Park, Minami-ku, 732 Hiroshima, Japan*

DANIEL H. MOORE II, *Division of Molecular Cytometry, Department of Laboratory Medicine, Room MCB 230, University of California-San Francisco, San Francisco, California 94103*

NORI NAKAMURA, *Department of Genetics, Radiation Effects Research Foundation, 5-2 Hijiyama Park, Minami-ku, 732 Hiroshima, Japan*

MIMAKO NAKANO, *Department of Genetics, Radiation Effects Research Foundation, 5-2 Hijiyama Park, Minami-ku, 732 Hiroshima, Japan*

JAMES V. NEEL, *Medical Science II M4708, Department of Human Genetics, University of Michigan, 1500 East Medical Center, Ann Arbor, Michigan 48109-0618*

JANICE A. NICKLAS, *Genetics Laboratory, University of Vermont, 32 North Prospect, Burlington, Vermont 05401*

KAZUO OHTAKI, *Department of Genetics, Radiation Effects Research Foundation, 5-2 Hijiyama Park, Minami-ku, 732 Hiroshima, Japan*

J. PATRICK O'NEILL, *Genetics Laboratory, University of Vermont, 32 North Prospect, Burlington, Vermont 05401*

DAVID J. PAWEL, *Department of Genetics, Radiation Effects Research Foundation, 5-2 Hijiyama Park, Minami-ku, 732 Hiroshima, Japan*

LEIF E. PETERSON, *Department of Medicine, Baylor College of Medicine, One Baylor Plaza, ST-924, Houston, Texas 77030*

DONALD A. PIERCE, *Department of Statistics, Radiation Effects Research Foundation, 5-2 Hijiyama Park, Minami-ku, 732 Hiroshima, Japan*

DALE L. PRESTON, *Department of Statistics, Radiation Effects Research Foundation, 5-2 Hijiyama Park, Minami-ku, 732 Hiroshima, Japan*

ELAINE RON, *Radiation Epidemiology Branch, National Cancer Institute, 6130 Executive Boulevard, EPN-408, Bethesda, Maryland 20892*

WALTER J. RUSSELL, *Department of Radiology, Radiation Effects Research Foundation, 5-2 Hijiyama Park, Minami-ku, 732 Hiroshima, Japan*

CHIYOKO SATOH, *Department of Genetics, Radiation Effects Research Foundation, 5-2 Hijiyama Park, Minami-ku, 732 Hiroshima, Japan*

WILLIAM J. SCHULL, *Human Genetics Center, University of Texas School of Public Health, P.O. Box 20334, Houston, Texas 77225*

ITSUZO SHIGEMATSU, *Radiation Effects Research Foundation, 5-2 Hijiyama Park, Minami-ku, 732 Hiroshima, Japan*

JAMES E. TROSKO, *Department of Pediatric and Human Development, Michigan State University, B240 Life Sciences Building, East Lansing, Michigan 48824-1317*

LUCIAN WIELOPOLSKI, *Medical Department, Building 490, Brookhaven National Laboratory, P.O. Box 5000, Upton, New York 11973*

JOHN D. ZIMBRICK, *Board on Radiation Effects Research, National Academy of Sciences, 2101 Constitution Avenue, Suite 342, Washington, D.C. 20418*

Index

Mutagens, 164, 181, 248, 253, 257
Mutant
frequency, 253, 256, 257, 259, 261,
268
mutation distinguished from, 257–
258, 259
sibling, 261
Mutation-growth-mutation model, 214
Mutations, 128. *See also hprt* gene mutations; Somatic-cell mutations;
Two-mutation model
age effect, 256
from apoptotic cells, 192
assays for, 254
and carcinogenesis, 179, 181, 213–
219, 248
complex, 265
disease surrogates for, 167
doubling dose, 159, 162–163
of erythrocyte and blood plasma proteins, 160
frequencies, 261, 262
in glycophorin system, 170
half-life, 215–216, 268
in HGPRT system, 170
hot spots, 263
lethal, 163, 182
methods for detecting, 260
in minisatellite loci lengths, 166
mutants distinguished from, 257–
258, 259
nucleotide substitutions, 166
p53 gene, 179, 265
point, 181, 263, 264
primary frequencies, 254
in reporter regions, 254, 265–266
splice, 263, 264
spontaneous, 161, 162, 164, 166,
267
stem cell, 213, 214, 217, 248
T-cell antigen receptor gene defect,
170
tandem, 264
variant frequency, 255
MVK model, 228–229
generalized, 229–236
Myelodysplastic syndrome, 82

Myeloma, 102, 103
Myoma uteri, 308
Myotonic dystrophy, 167

N

Nagasaki. *See also* City differences
burst height, 7–8
hypocenter, 7–8, 40
leukemia mortality risk, 44–45
neutron measurements, 40, 41
shielding considerations, 21
T65DR, 7–8
Nagasaki Prefectural Cancer Registry, 132–
133
Nagasaki Tumor Registry. *See* Hiroshima
and Nagasaki Tumor Registries
Nakamura, Nori, 73–88
Nakano, Mimako, 73–88
Nara University (Japan), 34, 102
National Academy of Sciences, 174, 304–
306
National Cancer Institute (US), 119
National Institute of Radiological Sciences
(Japan), 17, 34
National Registry of Radiation Workers,
150
National Research Council, 174
Neel, James V., 159–175, 305
Nervous system cancers, 123, 141
Neurasthenia, 290, 291
Neuropsychiatric Screening Adjunct, 284
Neutron radiation. *See also* Gamma-ray/neutron
ratios
cadmium-difference gold measurements, 40, 41
chlorine activation data, 23
DS86 dose, 22, 26, 27, 28, 29, 34,
36, 37, 38, 39, 40–49, 76
free-field dose, 36, 40, 42
Hiroshima, 27, 31, 38, 39, 40, 42,
43, 44, 46, 115
and leukemia, 108, 115
organ dose, 26, 162
relative biological effectiveness, 162,
232
source term, 27
sulfur-activation data, 38, 39, 40,
41, 42, 44